C0-ATX-004

# Recent Advances in the Development and Germination of Seeds

# NATO ASI Series

## Advanced Science Institutes Series

*A series presenting the results of activities sponsored by the NATO Science Committee, which aims at the dissemination of advanced scientific and technological knowledge, with a view to strengthening links between scientific communities.*

The series is published by an international board of publishers in conjunction with the NATO Scientific Affairs Division

| | | |
|---|---|---|
| A | **Life Sciences** | Plenum Publishing Corporation |
| B | **Physics** | New York and London |
| C | **Mathematical and Physical Sciences** | Kluwer Academic Publishers Dordrecht, Boston, and London |
| D | **Behavioral and Social Sciences** | |
| E | **Applied Sciences** | |
| F | **Computer and Systems Sciences** | Springer-Verlag |
| G | **Ecological Sciences** | Berlin, Heidelberg, New York, London, |
| H | **Cell Biology** | Paris, and Tokyo |

***Recent Volumes in this Series***

*Volume 181*—Skin Pharmacology and Toxicology: Recent Advances
edited by Corrado L. Galli, Christopher N. Hensby, and Marina Marinovich

*Volume 182*—DNA Repair Mechanisms and their Biological Implications in Mammalian Cells
edited by Muriel W. Lambert and Jacques Laval

*Volume 183*—Protein Structure and Engineering
edited by Oleg Jardetzky

*Volume 184*—Bone Regulatory Factors: Morphology, Biochemistry, Physiology, and Pharmacology
edited by Antonio Pecile and Benedetto de Bernard

*Volume 185*—Modern Concepts in *Penicillium* and *Aspergillus* Classification
edited by Robert A. Samson and John I. Pitt

*Volume 186*—Plant Aging: Basic and Applied Approaches
edited by Roberto Rodríguez, R. Sánchez Tamés, and D. J. Durzan

*Volume 187*—Recent Advances in the Development and Germination of Seeds
edited by Raymond B. Taylorson

*Series A: Life Sciences*

# Recent Advances in the Development and Germination of Seeds

Edited by

Raymond B. Taylorson

United States Department of Agriculture
Beltsville, Maryland

Plenum Press
New York and London
Published in cooperation with NATO Scientific Affairs Division

Proceedings of the Third International Workshop on Seeds,
sponsored by NATO,
held August 6-12, 1989,
in Williamsburg, Virginia

Library of Congress Cataloging-in-Publication Data

International Workshop on Seeds (3rd : 1989 : Williamsburg, Va.)
Recent advances in the development and germination of seeds / edited by Raymond B. Taylorson.
p. cm. -- (NATO ASI series. Series A, Life sciences ; vol. 187)
"Proceedings of the Third International Workshop on Seeds, sponsored by NATO, held August 6-12, 1989, in Williamsburg, Virginia"--T.p. verso.
Includes bibliographical references.
ISBN 0-306-43521-7
1. Seeds--Development--Congresses. 2. Germination--Congresses. I. Taylorson, Raymond B. II. North Atlantic Treaty Organization. III. Title. IV. Series: NATO ASI series. Series A, Life sciences ; v. 187.
QK661.I56 1989
582'.0467--dc20 90-33668
CIP


A Division of Plenum Publishing Corporation
233 Spring Street, New York, N.Y. 10013



Printed in the United States of America

PREFACE

These Proceedings are a product of the International Workshop on Seeds held in Williamsburg, Virginia, USA, at the College of William and Mary, during the week of August 6-11, 1989. Sixty-eight participants attended. The location provided a scenic and historical setting for the excellent work presented. Good facilities and amenities also contributed to the success of the meeting.

The Proceedings present the substance of the main lectures given at this meeting. In addition, there were 29 brief paper presentations and 30 poster presentations which have been summarized in abstract form in a separate publication.

This meeting represents the third such meeting of a diverse group of scientists interested in the behavior of seeds, both in an agricultural sense and as tools for the advancement of more particular subject matter. The first meeting was held in Jerusalem, Israel in 1980 and the second in Wageningen, The Netherlands in 1985. A fourth meeting is being planned.

The Editor and Organizer wishes to thank not only the contributors to this volume for their efforts but also all the other participants whose combined efforts made this meeting a great success.

R. B. Taylorson
Beltsville, Maryland

CONTENTS

# SEED RESEARCH - PAST, PRESENT AND FUTURE

Michael Black

Division of Biosphere Sciences
King's College (University of London)
Campden Hill Road
London W8 7AH, England

## THE PAST LEADING TO THE PRESENT

Research on seeds has a long history. Most botanists know about the writings of the Greek scholar Theophrastus (372-287 BC). He displayed an awareness of seed physiology that would not disgrace the pages of books written two millenia after his death. Dormancy, reserve deposition, the effects of environmental factors on seed development and germination, seed longevity and priming - topics which are in the forefront of modern research - all received his comments (see Evenari, 1984). Relatively little new knowledge was added until the end of the 18th century: and from the late 19th and early 20th century studies on seeds grew apace to meet the demands of agriculture, horticulture, forestry, malting and brewing and as part of the search for an understanding as to how the physical and biological worlds function.

Seeds have been the objects of many different kinds of research. As they are quantitatively the most important parts of the human diet (cereals, pulses, etc.) nutritionists demanded detailed knowledge of their chemical composition. Applied chemistry (phytochemistry, cereal chemistry) provided information about the starch, protein and fat reserves, which later became the immensely valuable starting point for studies by plant biochemists and molecular biologists on the biochemical and molecular mechanisms involved in reserve deposition.

Most of our important economic crop plants (food plants, forest trees) are grown from seeds and so from a purely practical view it has been necessary to learn about their germination physiology. This includes how germination is affected by physical and chemical factors in the seed's environment - temperature, water, light, inhibitors, etc. - and also how it is regulated by factors within the seed itself, i.e. its dormancy. These considerations are of course important to the ecologist as they can tell him about the processes involved in plant establishment. And the success of seeds also rests on their viability, and on their quick and uniform development into healthy, well-growing seedlings, i.e. their vigour. Both of these depend substantially upon conditions experienced by the seeds in storage (or in nature, in the seed bank) and so much research has been carried out on the relationship between factors such as temperature and humidity, seed moisture content, and viability and longevity.

*Recent Advances in the Development and Germination of Seeds*
Edited by R.B. Taylorson
Plenum Press, New York

As seedling establishment involves the mobilisation of the seeds' storage reserves the processes occurring here have attracted the seed researcher's attention. In an economic context one aspect of mobilisation has received particularly intensive study - the modification of the endosperm in the first few days following germination of barley, i.e. malting. And to this time it is the mobilisation in the cereal grain which of all seed types is best understood. Arising out of this is an example of how seed research has contributed to fundamental knowledge about plant processes. The observations made early in the 20th century concerning embryo/endosperm interactions in malting barley eventually became explicable in terms of action of a hormone secreted by the embryo, gibberellin, upon the aleurone tissue surrounding the starchy endosperm. This is still the best understood hormone/target tissue system in plants, one which has contributed enormously to our understanding of the biochemistry and molecular biology of hormone action and the regulation of gene expression (Jacobsen and Chandler, 1987). It is worth mentioning in this context that the plant hormones themselves were discovered as a result of research on seeds. Auxins, gibberellins and cytokinins were first isolated from seed tissues of higher plants; and seeds are also known as rich sources of abscisic acid and ethylene.

Much fundamental research on plant biochemistry has been carried out with seeds, mostly in respect of metabolic processes concerned with reserve synthesis and utilisation. Hence, much of our knowledge about the synthesis and breakdown of fatty acids, triacylglycerols and starch, and about protein processing comes from studies on seeds. Sub-cellular events involved in reserve metabolism - protein packaging, secretion, spherosome, glyoxysome, peroxisome and lysosome functions - have been elucidated using seed and seedling material.

One of the greatest contributions that seed research has made to botanical science is undoubtedly concerned with the discovery of phytochrome, the pigment involved in a myriad of light-controlled developmental processes occurring at all phases in the plant's life history. As every botanist knows, the first steps on the road to the discovery of phytochrome, in lettuce seeds, were taken in Beltsville, not many miles from the site of this meeting, by Flint and McAlister, culminating in the momentous research of Borthwick, Hendricks, the Tooles and their colleagues (Borthwick, 1972). Seed researchers can justly be proud of the contributions that their system continued to make to plant photophysiology, for example in the discovery of the high energy reaction.

As mentioned above, seeds have been a focus of attention of ecologists. Individual plant and community establishment depend on seed dispersal, and all facets of seed physiology play a part thereafter. The relationship between the seed and its environment is an important determinant of the quality and quantity of the seed bank, and the eventual emergence of seedlings depends upon longevity, germination physiology, dormancy and effective reserve utilisation. The conservationist, too, is much concerned with seeds. The urgent need for genetic conservation is now appreciated and, central to the developing technology for maintaining germ plasm collections, are studies on seed viability and longevity (Roberts, 1989).

It can rightly be claimed that a branch of biological science was founded on studies of seeds. We all know that Mendel used the appearance of pea seeds in his investigation of inheritance. Now we understand the biochemical basis of the wrinkled and smooth surface and, more notably, the *r* locus has recently been cloned. Mendel's gene from pea seeds can now be studied with all the sophistication afforded by molecular genetics.

THE PRESENT LEADING TO THE FUTURE

In modern times the area of activity covered by the term seed research has become fairly well circumscribed to include seed development and maturation, dormancy and germination, viability and longevity, and reserve mobilisation. Of course, these are all approachable in appropriately different ways by ecologists, physiologists, crop scientists, horticulturists, biochemists and molecular biologists.

It has to be said that much of seed research still takes the form of descriptive physiology. Cases in point are dormancy and germination ecophysiology. Seeds have a remarkable sensitivity to the environment unrivalled by any other phase in the plant's life history. Seeds can detect light quality, quantity, photoperiod, temperature, alternations of temperature, oxygen levels, different chemicals, water precipitation and water potential (Bewley and Black 1982). All of these can determine whether or not a seed germinates and the time and place for doing so, so that the resulting seedling is formed in the most clement situation and favourable time to support its further development. Such a complex sensitivity understandably occupies the attention of seed ecologists, through whose efforts we have already learned much, and we will continue to do so, about the germination behaviour of very many plant species. Similar approaches have proved valuable in crop physiology too and we can cite several cases, for example various species of vegetable seeds, whose responses to the environment are now well enough understood to enable growers to produce crops at times and in situations which previously were unsuitable. But although we know much, in an empirical way, about what determines germination we have very little understanding of the mechanisms involved. We cannot explain the temperature regulation of germination and dormancy, the perception and action of temperature alternations and the mechanism of action of promotive or inhibitory light. Similarly, although we can make mathematically-derived predictions about changes in viability (Ellis, 1988) we still have only a relatively superficial understanding about the degenerative changes which can occur in a stored seed leading to diminution in vigour or eventually death.

These remarks are not intended to imply that seed physiologists are satisfied simply with making descriptions about germination. In some ways, physiological mechanisms have become clearer. For example, coat effects in dormancy (leaving aside hard coats which are poorly permeable to water) are very likely to be due to interference with oxygen access to the embryo; and indeed, we know that diffusion of this gas across the enclosing tissues is impeded, sometimes as a result of enzymic oxygen consumption. But the next step in the syndrome - why this leads to the inhibition of germination that we call dormancy - is unfortunately still a mystery. Ingenious hypotheses have been propounded to link the oxygen requirement with the particularities of metabolism of the embryo that are required for ultimate axis extension (the culmination of germination) but the final, integrated explanation still eludes us (Ross, 1984).

The responsibility for progress does not, however, fall entirely on the seed biologist alone. To understand how a seed germinates I believe that we need to know more about the regulation of cell elongation, for it is from the occurrence of this event that we recognise that germination has taken place. This process, fundamental to plant growth, is still incompletely understood. Information is slowly being gathered about wall extensibility, cell wall biochemistry, pH effects, hormonal action, proton secretion, water uptake, and so on, but we are still far from having a unified picture of cell extension. When this is available we should then be in a better position to begin to analyse, with more sophistication than hitherto, the critical events which must proceed before axial elongation can start. We can then begin to identify in more precise physiological, biochemical and molecular terms where the internally and externally derived regulation of germination might act.

There are, however, aspects of seed function which are already amenable to an advanced biochemical understanding. We must include in this context the metabolism of the seed storage reserves. Synthesis and deposition of fats, starch and proteins is now well understood in general outline and, in some cases, in fine detail too. The physiology and biochemistry of reserve mobilisation has also been clarified in broad respects and, in cereals, we have appreciable understanding about regulation of mobilisation at the molecular level. This has happened because in cereals, or more precisely in barley, oats and wheat, there exists an experimentally manipulable system in which a characterised factor - the hormone, gibberellin - participates to regulate enzyme production. Since the target cells for this factor, the aleurone tissue, are separate from the storage tissue and the metabolic sink (the embryonic axis), many complications which affect other seeds (eg those with cotyledon-located reserves) are absent. Hence, we have reached the stage when the regulated genes (isozymes of $\alpha$-amylase) have been isolated, the promoter regions tentatively identified, and hormone-induced DNA-binding, trans-acting factors detected (Ou-Lee et al. 1988).

Many seed biologists would in fact contend that significantly new understanding has been achieved in recent seed research mainly where genetical, molecular, and cell biological approaches have been taken. The use of mutant lines, for example, is beginning to clarify the roles of the endogenous hormones, abscisic acid and gibberellins, in the onset of dormancy and in the germination process (Karssen and Groot, 1987).

But it is in respect of seed development that we recognise the power of the application of molecular biology and genetics. Genes for several storage proteins, for example of dwarf bean, soya bean, wheat, barley and maize, have been isolated and cloned. Promoter sequences have been obtained and used in various gene constructs to transform plants so that regulation of storage protein gene expression can be studied. In most cases, it can be shown that the regulatory sequences include temporal and tissue specificity. We are therefore getting closer to understanding at fundamental molecular level the regulation of reserve protein deposition in developing seeds. The application of these approaches to other reserves, for example to the key enzymes in starch and triacylglycerol synthesis is proceeding well in several laboratories so that in these cases too, we can expect a revolutionary expansion of our knowledge.

It could be argued that, exciting though these advances are, they are not strictly telling us about seed development as such but about certain specific events that occur during seed development. We cannot explain, for example, what regulates the ordered sequence that is initiated by the fertilisation of the ovum, followed by the establishment of a zygotic polarity and continues with the patterns of cell division and enlargement that form an embryo inside the seed, perhaps accompanied by a mature endosperm. On *a priori* grounds we can suppose that there is a sequential expression of genes, but regulation is not yet understood. It is clear, however, that two factors participate - the water potential of the developing embryo, imposed by the osmotic environment, and the hormone abscisic acid. Young embryos move quickly into seedling formation when they are isolated from a developing seed but are maintained in an embryogenetic state by exposure to low water potentials or to abscisic acid (e.g. Finkelstein and Crouch, 1986). Several genes whose expression is regulated by these factors have been cloned, and the ABA-sensitive promoter sequences have been isolated in a few cases. Many of the ABA-regulated genes which in different species are expressed late in development code for polypeptides which have substantial homology and shared peculiarities of structure, for example in being glycine rich and hydrophilic (Dure et al, 1989; Mundy and Chua, 1988). It is considered that the appearance of these proteins prior to maturation/drying is part of the mechanism providing protection against the effects of dessication.

Drying during maturation may affect the transformation of seeds from the developmental to the germinative phase (Kermode, 1989). Here too, patterns of gene expression are altered, especially in the case of enzymes for reserve mobilisation. Responsive 'germination genes' have not yet been identified but this is an area where molecular approaches should encourage rapid progress.

These exciting developments point to one path for our future research, a road which could lead to the establishment of a molecular basis in other fields of seed biology. Some aspects of seed physiology seem to be prime candidates for this kind of approach. Dormancy, for example, where a block on radicle emergence is imposed by tissues enclosing the embryo or by factors within the embryo itself, might plausibly be due to the prevention of expression of certain genes. The techniques of molecular biology enable us to detect qualitative and quantitative differences in gene products which would set the stage for the identification of the genes themselves. Several factors which participate in the dormancy syndrome are known in other plant systems to operate at gene level. High and low temperatures, anoxia, hypoxia, light and hormones are cases in point which may operate similarly in the imposition or termination of dormancy.

This is not to advocate the molecular path as the only one which should be taken. We must consolidate the interfaces among physiology, biochemistry, ecophysiology, genetics and cell and molecular biology in our search for an understanding of seed behaviour. The technology exists to enable us to do this in the pursuit of our aims: the constraints are largely those imposed by the availability of resources and funding.

## REFERENCES


Bewley, J.D. and Black, M. 1982, Physiology and biochemistry of seeds. Vol.2. Viability, dormancy and environmental control. Springer-Verlag, Berlin, Heidelberg, New York. p. 375.

Borthwick, H.A., 1972, History of phytochrome, in: "Phytochrome", K. Mitrakos and W. Shropshire, eds., Academic Press, London, New York, p.3.

Dure, L., Crouch, M., Harada, J., Ho, T.-H.D., Mundy, J., Quatrano, R., Thomas, T. and Sung, Z.R. 1989, Common amino acid sequence domains among the LEA proteins of higher plants, Plant Molecular Biology, 12:475-486.

Ellis, R.H., 1988, The viability equation, seed viability and practical advice on seed storage, Seed Science Technology, 16:29.

Evenari, M., 1984, Seed physiology: its history from antiquity to the beginning of the 20th century, Bot. Rev. 50:119.

Finkelstein, R.R. and Crouch, M.L., 1986, Rape seed embryo development in culture on high osmoticum is similar to that in seeds, Plant Physiol., 81:907.

Jacobsen, J.V. and Chandler, P.M., 1987, Gibberellin and abscisic acid in germinating cereals, in: "Plant hormones and their role in plant growth and development", P.J. Davies, ed., Martinus Nijhoff, Dordrecht. p. 164.

Karssen, C.M. and Groot, S.P.C., 1987, The hormone-balance theory of dormancy evaluated, in: "Monograph no.15 British Plant Growth Regulator Group", N.J. Pinfield and M. Black, eds., p.17.

Kermode, A., 1989 (in press), Regulatory mechanisms involved in the transition from seed development to germination, CRC Press, Boca Raton.

Mundy, J. and Chua, N.-H., 1988, Abscisic acid and water-stress induce the expression of a novel rice gene, EMBO Journal, 17:2279.

Ou-Lee, T.-M., Turgeon, R. and Wu, R., 1988, Interaction of a gibberellin-induced factor with the upstream region of an α-amylase gene in rice aleurone tissue, Proc. Natl. Acad. Sci., U.S.A., 85:6366.

Roberts, E.H., 1989, Seed storage for genetic conservation, Plants Today, 2:12.

Ross, J.D., 1984, Metabolic aspects of dormancy, in: "Seed Physiology. Germination and reserve mobilisation", D.R. Murray, ed., Academic Press, Australia, p.45.

USE OF TRANSGENIC PLANTS FOR STUDIES OF

SEED-SPECIFIC GENE EXPRESSION

Philip A. Lessard, Randy D. Allen,
François Bernier, and Roger N. Beachy

Department of Biology
Washington University
St. Louis, MO, USA, 63130

## INTRODUCTION

In studying the regulated patterns of gene expression in higher plants, molecular biologists employ a variety of *in vitro* tools. The accumulation, distribution and decay of various gene products are monitored through the use of immunological and electrophoretic techniques. The interactions of these gene products can sometimes be observed. Cloned copies of interesting and important genes can be used to follow the accumulation and disappearance of specific messenger RNAs. But in many ways, such studies are merely descriptive. In order to achieve full understanding of the mechanisms which regulate gene expression, researchers need a means to determine cause and effect, a means to introduce a change at the molecular level and to observe its effect(s) on gene expression.

Transgenic plants can be very useful for such studies. Upon introducing foreign genes into plant cells and regenerating whole, fertile plants, many researchers found that the inserted genes were faithfully regulated in their new hosts in a manner similar to that of their natural genome. This feature of the genetic transformation of plants has made a number of regulated gene systems amenable to in depth analysis. By introducing wild-type and specifically altered versions of a gene into plants, it has been possible to identify nucleotide sequences which are responsible for the gene's regulation and expression. In this way, for example, the importance of the CAAT and TATA sequence elements as well as the polyadenylation signal, AATAAA, were determined for plant genes (see Schell, 1987, for review). *In vitro* studies of plant genes and their products often give a limited view of the functions and effects of particular genes. By using transgenic plants, the consequences of introducing or mutating certain genes can be assessed in the intact organism, and developmental and pleiotropic effects can be determined (e.g. Medford et al., 1989).

Transgenic plants have been used to identify regulatory elements of photoregulated genes involved in photosynthesis (Timko et al., 1985; Nagy et al., 1987; Castresana et al., 1988; Kuhlemeier et al., 1988; Dean et al., 1989a,b; Ueda et al., 1989; Kuhlemeier et al., 1989). They have also proven useful in studying genes involved in flower and fruit development (Giovannoni, et al., 1989), hormone action (Medford et al., 1989; Broglie, et al., 1989), and response to stress (Kaulen et al., 1986; Logemann et al., 1989). Transgenic plants have been instrumental in the

*Recent Advances in the Development and Germination of Seeds*
Edited by R.B. Taylorson
Plenum Press, New York

characterization of genes involved in the interactions between plants and either pathogenic (Beachy et al., 1987; Linthorst et al., 1989; Meshi et al., 1989; Schmulling et al., 1989; Elliott et al., 1989) or beneficial (Forde et al., 1989) microbes. In at least one case (Strittmatter and Chua, 1987) regulatory elements from two differentially expressed genes have been combined to produce a gene with a unique pattern of expression in transgenic plants. A number of genes which exhibit tissue-specific patterns of expression have also been studied in transgenic plants (e.g. Stockhaus et al., 1987).

In this review, we will demonstrate how transgenic plants have been useful in our own work on seed storage protein genes. The genes encoding seed storage proteins typically display a very tightly regulated pattern of expression. They are expressed at relatively high levels but only during specific stages of seed development and only in specific tissues (e.g. embryo and/or endosperm). In order to examine the mechanisms underlying this type of regulation, our laboratory has undertaken the study of the genes encoding the subunits of a major seed storage protein in soybean, β-conglycinin.

## β-CONGLYCININ IS AN ABUNDANT TISSUE-SPECIFIC AND DEVELOPMENTALLY REGULATED PROTEIN

The globulin storage proteins of soybean account for 70-80% of the total seed protein. Of this, approximately 30% is composed of β-conglycinin, a 7S, vicillin-like, glycosylated seed storage protein (Meinke et al., 1981). In the protein bodies of soybean seeds, β-conglycinin occurs as a trimer of subunits, α, α' and β. The individual subunits may be distinguished by differences in electrophoretic mobility and amino acid composition. These subunits assemble in random combinations to form the trimeric multimers (Thanh and Shibasaki, 1978; Coates et al., 1985).

The α and α' subunits of β-conglycinin begin to accumulate in soybean embryos 18-20 days after pollination, shortly after cell division has ceased in developing cotyledons. The β subunit, however, does not begin to accumulate until about one week later, during the maturation stage of development (Meinke et al., 1981). In addition to the differences in timing of accumulation for each subunit, there are differences in distribution of the subunits within tissues of the embryo. Whereas all three subunits are found within cotyledons, reduced levels of the α subunit and virtually none of the β subunit accumulate within the embryonic axis (Meinke, et al., 1981; Ladin et al., 1987). When compared to the α and α' subunits, the β subunit has been shown to respond differently to environmental signals such as abscisic acid or nutritional stress (Hollowach et al., 1984; Gayler and Sykes, 1985; Bray and Beachy, 1985).

The subunits of β-conglycinin are each encoded by different members of a gene family (Beachy et al., 1981; Bray et al., 1987; Ladin et al., 1987). This gene family contains at least 15 different members, three of which may encode the α and α' subunits, and six of which may encode the β subunit. The exact identity of the other genes have not yet been determined. These genes are clustered in several chromosomal regions and are highly homologous to one another (Harada et al., 1989). The genes encoding β-conglycinin are regulated primarily at the level of transcription (Harada et al., 1989). Messenger RNAs from these genes accumulate in a highly regulated manner, first appearing at 17 days after pollination, shortly before the onset of storage protein synthesis, peaking at mid-maturation stages and then decreasing in abundance as the seed approaches maturity (Meinke et al., 1981; Naito et al., 1988). As was the case with protein accumulation, mRNAs encoding the β subunit do not begin to accummulate to significant levels until three to five days after α and α' mRNAs appear (Naito et al., 1988).

Because β-conglycinin genes are expressed to high levels, their accumulation is relatively easy to monitor in an experimental environment. The strict temporal and tissue-specific expression of these genes also make them very interesting from a developmental viewpoint. Together, these features make the β-conglycinins an ideal system for studying the mechanisms governing highly regulated gene expression in plants.

## TRANSGENIC PLANTS PRESERVE PATTERNS OF β-CONGLYCININ GENE EXPRESSION

In order to manipulate the β-conglycinin genes experimentally, it was necessary to first isolate copies of specific α' and β subunit genes, determine which sequences in these genes might serve important regulatory functions and then, after making specific alterations, reintroducing these genes into transgenic plants to observe whether or not their specific patterns of expression were conserved. Ideally, the genes would be reintroduced into their natural host, soybeans, but two problems make this unfeasible. The first is that soybeans are notoriously difficult to transform and regenerate. The second is that once the genes were reintroduced into soybean plants, it would be difficult to follow the accumulation of the mutant genes' products because of their similarity to the native storage proteins. To overcome these problems, wild-type and altered copies of the genes can be introduced into plants which are more easily transformed and which do not naturally produce proteins resembling β-conglycinins. Two plant species which are used for such studies are petunia (*Petunia hybrida*) and tobacco (*Nicotiana tabacum L.*).

Individual genes encoding the α' and β subunits of β-conglycinin were cloned and introduced into petunia and tobacco plants (Beachy et al., 1985; Tierney et al., 1987; Bray et al., 1987) via *Agrobacterium*-mediated transformation (Horsch et al., 1985). When the accumulation of mRNA was measured by northern analyses and the accumulation of protein was determined by immunoassays, the expression of these genes in transgenic plants was found to closely reflect that in soybeans (Beachy et al., 1985; Bray et al., 1987). That is, each β-conglycinin subunit was produced only in seeds and only during the correct stages of seed development. *In situ* mRNA hybridization experiments in other labs have supported these results (Barker et al., 1988). In addition, the protein produced from this gene in transgenic plants was antigenically similar to that of soybeans and it was also found to assemble into multimers with sedimentation coefficients of approximately 7S (Beachy et al., 1985; Bray et al., 1987). This preservation of genetic regulation of seed storage protein genes in tobacco, petunia and soybean provides a key justification for using heterologous systems in studying gene expression. Other laboratories have also observed preserved patterns of gene expression for a variety of different seed-specific genes from both dicots (Sengupta-Gopalan et al., 1985; Voelker et al., 1987; Shirsat et al., 1989; Voelker et al., 1989) and monocots (Hoffman et al., 1987; Colot et al., 1987; Schernthaner et al., 1988; Robert, et al., 1989). Because of the high degree of similarity in expression between the different plants, it can be inferred that the mechanisms which regulate these genes are highly conserved, and it strongly suggests that results obtained through the use of transgenic plants can be directly applied to the original system, in the case of the β-conglycinins, soybean.

## TRANSGENIC PLANTS CAN BE USED TO DEFINE REGULATORY ELEMENTS WITHIN THESE GENES

To identify important *cis*-acting regulatory elements, a series of terminal deletions were made in the cloned α' subunit gene. In these tests, progressively greater and greater portions of the 5' (upstream) region of the α′ gene were removed producing mutant α′ genes whose 5′ termini were located at nucleotide positions -904, -457, -257, -208, -159, -69, -42 or +14 (negative numbers indicate nucleotide positions upstream of the transcription start site, positive numbers, those downstream from the start site). These constructs were then introduced into petunia plants and resultant transformants which bore single copies of the genes to be tested were assayed for gene expression (Chen et al., 1986). Remarkably, in these experiments when multiple transgenic plants bearing the same deletion construct were examined, each exhibited similar levels of β-conglycinin protein and mRNA synthesis (Chen et al., 1986). When plants harboring different deletions were compared, a number of interesting results emerged. First, deletion of sequences from -904 to -257 had very little effect on the amount of mRNA or protein produced. Secondly, deletion of an additional 98 base-pairs resulted in a decrease in gene expression of 90-95% relative to control plants. Further deletion of sequences reduced expression of the α′ subunit gene to an undetectable level (Fig. 1). These results suggested that there was a powerful cis-acting transcriptional regulatory sequence between positions -257 and -159 in the α′ gene.

It should be noted here that analyses of this type serve to identify only those regulatory sequences which lie closest to the gene's site of transcription initiation. Because only 5′ terminal sequences are deleted, any redundant regulatory element which may lie upstream of the most proximal element will not be identified.

## THE USE OF REPORTER GENES FACILITATES THE STUDY OF GENE EXPRESSION IN TRANSGENIC PLANTS

Until this point, all of the assays of gene expression in transgenic plants had been done using northern analyses and immunoassays to monitor RNA and protein accumulation, respectively. Because the detection of gene products primarily through these methods is both costly and time consuming, many researchers have made use of reporter genes in studies of gene expression. Reporter genes typically encode proteins which can be easily detected by enzymatic activity (e.g. the ability to acetylate a compound such as chloramphenicol) or some other readily assayed property (e.g.

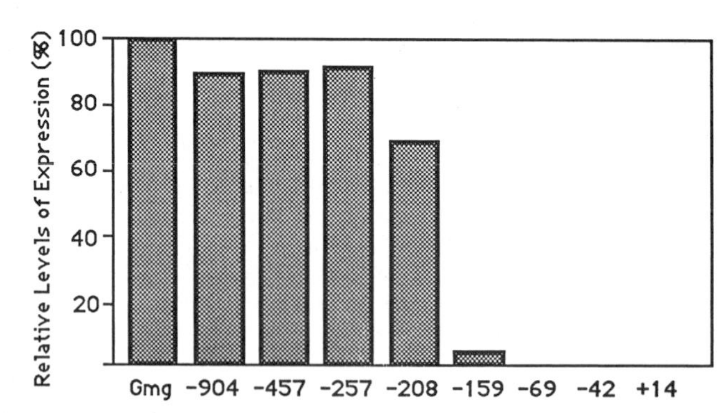


Fig. 1. Schematic diagram representing relative levels of β-conglycinin mRNA in seeds of petunia plants transformed with 5′ terminal deletions of a full-length β-conglycinin α′ subunit genomic clone. Gmg is the full length clone and represents 100% expression. Numbers at the bottom of the diagram indicate end points of the individual deletions. Adapted from Chen et al. (1986).

antibiotic resistance or even taste, as in the case of the thaumatin II reporter gene; Witty, 1989). Instead of detecting the gene product with antibody reactions, a researcher may simply assay the crude plant extracts directly for reporter activity. Among the most common reporters for plant studies are chloramphenicol acetyl transferase (CAT), neomycin phosphotransferase (NPT II), luciferase and β-glucuronidase (GUS). Each of these enzymes is present in very low or negligible amounts in plants. Both CAT and NPT II activities can be detected in transgenic plants by incubating crude tissue extracts with the appropriate radiolabelled substrate (chloramphenicol or a neomycin-type antibiotic such as kanamycin, respectively). The reaction products can be identified by thin-layer chromatography and autoradiography. These assays are easily quantitated, making measurement of the accumulated gene products possible (see Schell, 1987, for review). Reporter proteins such as GUS and luciferase do not require the use of radioactively labelled compounds and are thus sometimes preferred over CAT or NPT II. GUS can be assayed either histochemically to reveal the distribution of the gene product among different cell types or fluorimetrically to give a more quantitative measurement of gene expression (Jefferson et al., 1986). The luciferase reporter can also be used to give quantitative measurements of gene activity (de Wet et al., 1987) or in some cases to determine distribution of gene activity among different cells (Gallie et al., 1989).

## TRANSGENIC PLANTS CAN BE USED TO FUNCTIONALLY DEFINE A TISSUE-SPECIFIC AND TEMPORALLY REGULATED ENHANCER ELEMENT

Once we had established the presence of a strong, _cis_-acting transcriptional regulatory element between positions -257 and -159 of the α′ gene, we wished to better define the function of this sequence. Specifically, we wanted to determine whether the sequence element contained a transcriptional enhancer, whether it determined the tissue-specificity and temporal regulation of the α′ gene as well as its level of transcription, and whether there were negative control elements within the -257 to -159 region which might, for example, decrease expression of that gene in non-seed tissues.

Enhancers are _cis_-acting regulatory sequence elements which can stimulate transcription from a distant promoter. By definition, enhancers function irrespective of orientation and can act upon heterologous promoters (Serfling et al., 1985; Sassone-Corsi and Borelli, 1986; Rosales et al., 1987).

In order to identify whether a transcriptional enhancer lay within the regulatory region of the α′ subunit gene, a chimeric gene was constructed composed of the promoter and upstream regulatory elements from the cauliflower mosaic virus (CaMV) 35S transcription unit fused to the CAT gene coding sequence and ending with the α′ gene's own polyadenylation signal (7S 3′ end; Fig. 2A). The CaMV 35S promoter has been shown to function in most tissues of transgenic plants (Odell et al., 1985; Sanders et al., 1987). A fragment of DNA representing sequences from -257 to -78 was then introduced into the CaMV 35S promoter at position -90 in both orientations. In separate constructs, this DNA fragment was introduced into the chimeric reporter gene in either orientation both between the CAT coding sequence and the 7S 3′ end and downstream of the 7S 3′ end (Fig. 2B; Chen et al., 1988).

These chimeric genes were then introduced into tobacco plants and the resulting transformants assayed for CAT activity. Expression of the CAT gene in control plants, which bore only the 35S promoter-driven construct with no additional sequences from the α′ subunit gene, showed CAT activity to be distributed relatively evenly among the different organs of the transgenic plants (roots, stems, leaves and seeds) and the activity showed very little variation over the course of development. When the α' sequences were introduced at position -90 of the CaMV 35S promoter, however, CAT activity in seeds increased 40- to 50-fold relative to controls, irrespective of orientation of the sequence element. When the α' sequences were introduced between the CAT coding sequence and the polyadenylation signal, a two- to three-fold increase resulted relative to controls. When the sequences were introduced downstream of the polyadenylation signal, virtually no enhancement was observed (Fig. 2C). Furthermore, enhancement associated with inserting the α' sequence into the -90 position of the CaMV 35S promoter was seen only in seeds during mid- to late stages of seed development and was not observed in non-seed tissues. In no case was CAT activity found to decrease in non-seed tissues relative to controls when α' sequences were introduced into the chimeric gene. These results suggested that the -257 to -78 fragment from the the α' subunit gene of β-conglycinin alone contains all the information required to confer tissue-specific and developmentally regulated enhancement of transcription to a heterologous promoter. Because α' sequences did not lower expression of the CAT constructs in non-seed tissues such as roots, stems and leaves, we concluded that the α' fragment bears no detectable negative regulatory activities, i.e. it does not function by decreasing expression of the gene in non-seed tissues; rather it increases gene expression in seeds.

## PROTEINACEOUS FACTORS INTERACT WITH REGULATORY ELEMENTS FROM THE β-CONGLYCININ GENES

Transcriptional enhancers and other upstream activating sequences (UAS) most likely exert their effects by increasing the rate of transcription from the promoters with which they are associated (Sassone-Corsi and Borelli, 1986). It is likely that enhancers act by increasing the binding of RNA polymerase II molecules to the DNA (Treisman and Maniatis, 1985; Weber and Schaffner, 1985). Binding of polymerase may be

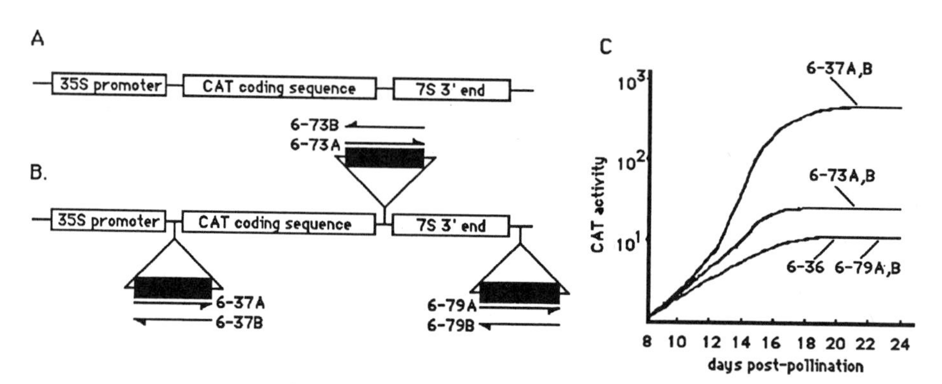


Fig. 2. (A) Schematic diagram of reporter gene construct containing the cauliflower mosaic virus 35S promoter, the coding sequence from the chloramphenicol acetyl transferase gene, and the polyadenylation signal from the soybean α' subunit gene. (B) Schematic representation of reporter constructs into which the -257 to -78 region from the α' subunit gene was introduced. Numbers (e.g. 6-36, 6-37A, etc.) are names given to the individual constructs. (C) Plot of CAT activity in developing seeds from transgenic tobacco plants transformed with the constructs indicated in (A) and (B). CAT activity is expressed relative to the level measured in seeds 8 days post-pollination (adapted from Chen et al., 1988).

stabilized by interaction with proteins that bind DNA sequences usually located upstream of the core promoter sequences. Some of these _trans_-acting factors are proteins which bear two functional surfaces. The first is a domain which specifically recognizes a sequence within the enhancer or UAS. The second surface interacts somehow with RNA polymerase or some other component thereby stabilizing formation of the transcription complex (see Struhl, 1987, and Ptashne, 1988, for reviews). For this reason, many researchers have begun to look for specific protein/DNA interactions in regulatory regions of these genes.

In plant systems, a number of groups have identified factors that bind to the upstream regions of regulated genes (Jofuku et al., 1987; Maier et al., 1987; Riggs et al., 1989) and, in a few cases, protein binding sites have been shown to correspond with sequences that have regulatory function (Green et al., 1987; Jensen et al., 1988; Green et al., 1989).

We recently identified a number of different factors which interact with upstream regions of the β-conglycinin genes (Allen et al., 1989). These factors have been termed Soybean Embryo Factors (SEF) 1 through 4, and can be distinguished by a number of different criteria. A common assay for the activity of DNA binding factors is the gel mobility shift assay. In this assay, radiolabelled DNA fragments representing specific sequences from the gene of interest are incubated in the presence of nuclear extracts. Upon electrophoresis through acrylamide or agarose gels, labelled DNAs which are bound by nuclear factors migrate more slowly than do unbound DNAs and appear as shifted bands after autoradiography of the gel. SEFs 1 through 4 each induce different shifts in such assays. A second distinction can be made between these factors when the sequences which they recognize are identified. This can be done in a number of different ways. The first and most direct method is a footprinting assay. In such a test, a specific end-labelled DNA fragment is incubated with nuclear extracts as they were for the mobility shift assays. Following this, DNase I or some other reagent is introduced into the incubation mixture. The reagent nicks the DNA fragment between exposed nucleotides, but is unable to nick bases in the region bound by the factor. When this reaction mix is electrophoresed through a sequencing gel, bands appear representing the position of each nick. A window or "footprint" will appear in the region recognized by the _trans_-acting factor as a series of bands missing in the sequencing gel. Other methods for identifying a factor's recognition sequence include Exonuclease III protection assays (e.g. see Allen et al., 1989) and mobility shift competition assays. In a competition assay, DNA fragments are incubated as they were for the mobility shift assays though a specific amount of "competitor" DNA is included in the mix. This competitor DNA is unlabelled and can represent different regions from the same or other genes. If the competitor used in the assay bears the same sequence which the _trans_-acting factor recognizes in the labelled DNA, it will compete for binding of that factor. Because factors bound to unlabelled DNAs will not appear after autoradiography, successful competition will decrease the intensity of the shifted bands in a mobility shift assay. By comparing the sequences and effectiveness of various labelled DNAs and competitors, one can identify the sequences recognized by the factor. While competition assays are simple to perform, they do not define the factors' binding sequences with the same resolution that can be obtained through footprinting.

Using such assays, we were able to determine that, of the four different binding factors, only SEF 3 had a unique interaction with the region from the α' subunit gene which was previously shown to bear enhancer-like activity (Allen et al., 1989). The sequences in the enhancer region which are recognized by SEF 3 center on two AACCCA repeats which had been postulated to be involved with function of the element (Chen et al., 1986). The next step in analysis of this DNA binding activity was to introduce into the apparent recognition sequence specific nucleotide changes, thereby altering the recognition sequence in an attempt to abolish binding of the factor. In this way the nucleotides responsible for binding of the factor could be identified. When a series of linkers (short fragments of heterologous DNA) were introduced into the region bearing the AACCCA repeats, a number of interesting results emerged. First, only those linkers which disrupted either of the AACCCA repeats themselves interfered with DNA binding activity. Other linkers introduced further upstream or downstream of the repeats or even linkers which disrupted the sixteen bases between the two repeats did not have significant effect upon binding activity (Allen et al., 1989; Allen and Beachy, unpublished results).

Following this, a synthetic oligonucleotide was made representing the 3′-most of these two AACCCA repeats along with 26 base pairs of surrounding sequence. In mobility shift assays it was found that monomers of this oligonucleotide were able to bind SEF 3 only very weakly. When two monomers were linked to reestablish the position and spacing of the second AACCCA repeat, full SEF 3 binding activity returned. Furthermore, when a dimer was constructed in which the first two C residues of each AACCCA repeat were changed to G's, SEF 3 binding activity was once again lost. These results suggest that binding of SEF 3 requires both of the AACCCA repeats and that (at least) the first two cytosine residues in the repeat are crucial for recognition (Allen et al., 1989).

An additional feature which makes involvement of SEF 3 in the regulation of the $\alpha'$ subunit gene likely is the change in SEF 3 activity over the course of soybean seed development. Transcription of the $\alpha'$ subunit gene begins in the developing embryo only as the seed approaches mid-maturation stages, and it declines as the seed approaches desiccation (Meinke et al., 1981; Naito et al., 1988). When we examined nuclear extracts prepared from soybean embryos from different stages of development, we found that SEF 3 activity closely parallelled transcriptional activity of the $\alpha'$ subunit gene: SEF 3 activity was undetectable in the youngest embryos, increased towards mid-maturation stages in development, and decreased as the embryos approached maturity (Allen et al., 1989; Crispino, Allen and Beachy, unpublished results).

## STUDIES IN TRANSGENIC PLANTS ARE NEEDED TO CONFIRM THE ROLE OF <u>TRANS</u>-ACTING FACTORS IN GENE EXPRESSION

The studies on DNA binding activity within the regulatory sequences of the $\alpha'$ gene provide good evidence that SEF 3 and perhaps other <u>trans</u>-acting factors are involved in the regulation of the $\beta$-conglycinin genes. While they are suggestive, these results do not tell us how the genes are controlled <u>in vivo</u>. One way to demonstrate the involvement of <u>trans</u>-acting factors in specific gene regulation has been to develop <u>in vitro</u> transcription systems. With such a system one can supply the factors normally required for transcription, i.e. RNA polymerase, template DNA, ribonucleotides, etc., and then test whether addition of a purified <u>trans</u>-acting factor increases transcription. This has been done in a number of animal systems (e.g. Freedman et al., 1989). As yet no reliable <u>in vitro</u> transcription system has been developed for a plant system. Such a system would be immensely useful to plant molecular biologists, but it would still provide a limited view of the complex regulatory and developmental processes that occur in whole plants. For this reason, each of the <u>in vitro</u> "discoveries" must be supported by <u>in vivo</u> tests. Once again, transgenic plants provide a tool for these studies.

In our lab, each of our hypotheses about the function of SEF 3 and the other soybean embryo factors is being tested by parallel experiments in transgenic plants. For example, each of the mutated $\alpha'$ subunit gene promoters which bear the linker insertions described previously have been placed before a reporter gene and introduced into tobacco plants. Likewise, specific base-pair alterations and rearrangements within the regulatory regions have been generated, linked to reporter genes and introduced into plants. Results from these experiments should help us to determine whether abolishing SEF 3 binding disrupts the function of the α' subunit enhancer, whether other as yet undiscovered regulatory sequences lie within this region, whether certain specific mutations cause the α' subunit promoter to lose its tissue specificity and what role if any SEFs 1, 2 and 4 play in the regulation of the β-conglycinin genes.

## CONCLUSIONS

The development of technologies for the transfer of foreign genes to plants has substantially helped many different areas of plant biology. In our research, the use of transgenic plants has been instrumental in the identification of some of the elements involved in the regulation of seed storage protein genes. In the absence of transgenic plants, it would have been difficult to determine the portion of the α' subunit gene of β-conglycinin that is responsible for tissue-specificity and developmental regulation. Furthermore, the identification of *cis*-acting sequence elements would have been much more difficult. Such results might have been obtained by examining scores of spontaneous mutations in whole, untransformed plants, as has been done for a number of genes in maize. But in soybeans, the generation of a sufficient number of useful mutations would be time-consuming, to say the least, and it would be especially difficult to identify any element with the type of precision that we have obtained in our studies using transgenic plants. The use of random mutations to study developmental and physiological processes in whole plants would likewise be difficult because there is very little detailed information available about the genetic organization of most plants, with only a few exceptions (e.g. maize and *Arabidopsis*). The use of transgenic plants overcomes some of these problems.

The use of transgenic plants in research has become more and more common in recent years. Transgenic plants will undoubtedly continue to prove useful in dissecting the mechanisms involved in gene transcription, protein synthesis and processing as well as a host of developmental processes. Transgenic plants will help researchers to improve the nutritional quality of existing crop plants and will help in the development of crop plants which can grow in increasingly diverse environments. Transgenic plants have already been used to overcome certain types of disease and predation among crop species and will continue to do so as the molecular biology of these problems becomes better understood.

## REFERENCES

Allen, R. D., Bernier, F., Lessard, P. A., and Beachy, R. N., 1989, Nuclear factors interact with a soybean β-conglycinin enhancer, *Plant Cell*, 1:623.

Barker, S. J., Harada, J. J., and Goldberg, R. B., 1988, Cellular localization of soybean storage protein mRNA in transformed tobacco seeds, *PNAS*, 85:458.

Beachy, R. N., Chen, Z.-L., Horsch, R. B., Rogers, S. G., Hoffmann, N. J., and Fraley, R. T., 1985, Accumulation and assembly of soybean β-conglycinin in seeds of transformed petunia plants, *EMBO J.*, 4:3047.

Beachy, R. N., Jarvis, N. P., and Barton, K. A., 1981, Biosynthesis of subunits of the soybean 7S storage protein, *J. Molecular Applied Genet.*, 1:19.

Beachy, R. N., Stark, D. M., Deom, C. M., Oliver, M. J., and Fraley, R. T., 1987, Expression of Sequences of Tobacco Mosaic Virus in Transgenic Plants and Their Role in Disease Resistance, *in* "Tailoring Genes for Crop Improvement," G. Bruening, J. Harada, T. Kosuge, and A. Hollaender, eds., Plenum Publishing Corporation, New York.

Bray, E. A., and Beachy, R. N., 1985, Regulation by ABA of β-conglycinin expression in cultured developing soybean cotyledons, Plant Physiol., 79:746.

Bray, E. A., Naito, S., Pan, N.-S., Anderson, E., Dube, P., and Beachy R. N., 1987, Expression of the β-subunit of β-conglycinin in seeds of transgenic plants, Planta, 172:364.

Broglie, K. E., Biddle, P., Cressman, R., and Broglie, R., 1989, Functional analysis of DNA sequences responsible for ethylene regulation of a bean chitinase gene in transgenic tobacco, Plant Cell, 1:599.

Castresana, C., Garcia-Luque, I., Alonso, E., Malik, V. S., and Cashmore, A.R., 1988, Both positive and negative regulatory elements mediate expression of a photoregulated CAB gene from Nicotiana plumbaginifolia, EMBO J., 7:1929.

Chen, Z.-L., Pan, N.-S., and Beachy, R. N., 1988, A DNA sequence element that confers seed-specific enhancement to a constitutive promoter, EMBO J., 2:297.

Chen, Z. L., Schuler, M. A., and Beachy, R. N., 1986, Functional analysis of regulatory elements in a plant embryo-specific gene, PNAS, 83:8560.

Coates, J. B., Medeiros, J. S., Thanh, V. H., and Nielsen, N. C., 1985, Characterization of the subunits of β-conglycinin, Arch. Biochem. Biophys., 243:184.

Colot, V., Robert, L. S., Kavanagh, T. A., Bevan, M. W., and Thompson, R. D., 1987, Localization of sequences in wheat endosperm protein genes which confer tissue-specific expression in tobacco, EMBO J., 6:3559.

de Wet, J. R., Wood, K. V., DeLuca, M., Helinski, D. R., and Subramani, S., 1987, Firefly luciferase gene: structure and expression in mammalian cells, Mol. Cell Biol., 7:725.

Dean, C., Favreau, M., Bedbrook, J., and Dunsmuir, P., 1989a, Sequences 5' to translation start regulate expression of petunia rbcS genes, Plant Cell, 1:209.

Dean, C., Favreau, M., Bond-Nutter, D., Bedbrook, J., and Dunsmuir, P., 1989b, Sequences downstream of translation start regulate quantitative expression of two petunia rbcS genes, Plant Cell, 1:201.

Elliott, R. C., Dickey, L. F., White, M. J., and Thompson, W. F., 1989, cis-acting elements for light regulation of pea ferredoxin I gene expression are located within transcribed sequences, Plant Cell, 1:691.

Forde, B. G, Day, H. M., Turton, J. F., Wen-jun, S., Cullimore, J. V., and Oliver, J. E., 1989, Two glutamine synthetase genes from Phaseolus vulgaris L. display contrasting developmental and spatial patterns of expression in transgenic Lotus corniculatas plants, Plant Cell, 1:391.

Freedman, L. P., Yoshinaga, S. K., Vanderbilt, J. N., and Yamamoto, K. R., 1989, In vitro transcription enhancement by purified derivatives of the glucocorticoid receptor, Science, 245:298.

Harada, J. J., Barker, S. J., and Goldberg, R. B., 1989, Soybean β-conglycinin genes are clustered in several DNA regions and are regulated by transcriptional and posttranscriptional processes, Plant Cell, 1:415.

Gallie, D. R., Lucas, W. J., and Walbot, V., 1989, Visualizing mRNA expression in plant protoplasts: Factors influencing efficient mRNA uptake and translation, Plant Cell, 1:301.

Gayler, K. R., and Sykes, G. E, 1985, Effects of nutritional stress on the storage proteins of soybeans, Plant Physiol., 78:582.

Giovannoni, J. J., Della Penna, D., Bennett, A. B., and Fischer, R. L., 1989, Expression of a chimeric polygalacturonase gene in transgenic rin (ripening inhibitor) tomato fruit results in polyuronide degradation but not fruit softening, Plant Cell, 1:53.

Green, P. J., Kay, S. A., and Chua, N.-H., 1987, Sequence-specific interactions of a pea nuclear factor with light-responsive elements upstream of the rbcS-3A gene, EMBO J., 6:2543.

Green, P. J., Yong, M.-H., Cuozzo, M., Kano-Murakami, Y., Silverstein, P., and Chua, N.-H., 1989, Binding site requirements for pea nuclear protein factor GT-1 correlate with sequences required for light-dependent transcriptional activation of the rbcS-3A gene, EMBO J., 7:4035.

Hoffman, L. M., Donaldson, D. D., Bookland, R., Rashka, K., and Herman, E. M., 1987, Synthesis and protein body deposition of maize 15-kd zein in transgenic tobacco seeds, EMBO J., 6:3213.
Hollowach, L. P., Thompson, J. F., and Madison, J. T., 1984, Effects of exogenous methionine on storage protein composition of soybean cotyledons cultured in vitro, Plant Physiol., 74:576.
Horsch, R. B., Fry, J. E., Hoffmann, N. L., Eicholtz, D., Rogers, S. G., and Fraley, R. T., 1985, A simple and general method for transferring genes into plants, Science, 227:119.
Jefferson, R. A., Burgess, S. M., and Hirsch, D., 1986, β-glucuronidase from E. coli as a gene-fusion marker, PNAS, 83:8447.
Jensen, E. O., Marker, K. A., Schell, J., and de Bruijn, F. J., 1988, Interaction of a nodule specific trans-acting factor with distinct DNA elements in the soybean leghaemoglobin IBC3 5' upstream region, EMBO J., 7:1265.
Jofuku, K. D., Okamuro, J. K., and Goldberg, R. B., 1987, Interaction of an embryo DNA binding protein with a soybean lectin gene upstream region, Nature, 328:734.
Kaulen, H., Schell, J., and Kreuzaler, F., 1986, Light-induced expression of the chimeric chalcone synthase-NPTII gene in tobacco cells, EMBO J., 5:1.
Kuhlemeier, C., Cuozzo, M., Green, P. J., Goyvaerts, E., Ward, K., and Chua, N.-H., 1988, Localization and conditional redundancy of regulatory elements in rbcS-3A, a pea gene encoding the small subunit of ribulose-bisphosphate carboxylase, PNAS, 85:4662.
Kuhlemeier, C., Strittmatter, G., Ward, K., and Chua, N.-H., 1989, The pea rbcS-3A promoter mediates light responsiveness but not organ specificity, Plant Cell, 1:471.
Ladin, B. F., Tierny, M. L., Meinke, D. W., Hosangadi, P., Veith, M., and Beachy, R. N., 1987, Developmental regulation of β-conglycinin in soybean axes and cotyledons, Plant Physiol., 84:35.
Linthorst, H. J. M., Meuwissen, R. L. J., Kauffmann, S., and Bol, J. F. 1989, Constitutive expression of pathogenesis-related proteins PR-1, GRP, and PR-S in tobacco has no effect on virus infection, Plant Cell, 1:285.
Logemann, J., Lipphardt, S., Lörz, H., Hauser, I., Willmitzer, L., and Schell, J., 1989, 5' upstream sequences from the wunl gene are responsible for gene activation by wounding in transgenic plants, Plant Cell, 1:151.
Maier, U.-G., Brown, J. W. S., Toloczyki, C., and Feix, G., 1987, Binding of a nuclear factor to a consensus sequence in the 5' flanking region of zein genes from maize, EMBO J., 6:17.
Medford, J. I., Horgan, R., El-Sawi, Z., and Klee, H. J., 1989, Alterations of endogenous cytokinins in transgenic plants using a chimeric isopentenyl transferase gene, Plant Cell, 1:403.
Meinke, D. W., Chen, J., and Beachy, R. N., 1981, Expression of storage-protein genes during soybean seed development, Planta, 153:130.
Meshi, T., Motoyoshi, F., Maeda, T., Yoshiwoka, S., Watanabe, H., and Okada, Y., 1989, Mutations in the tobacco mosaic virus 30-kD protein gene overcome Tm-2 resistance in tomato, Plant Cell, 1:515.
Nagy, F., Boutry, M., Hsu, M.-Y., Wang, M, and Chua, N.-H., 1987, The 5'-proximal region of the wheat Cab-1 gene contains a 268-bp enhancer-like sequence for phytochrome response, EMBO J., 6:2537.
Naito, S., Dube, P. H., and Beachy, R. N., 1988, Differential expression of developmentally regulated genes in transgenic plants, Plant Mol. Biol., 11:109.
Odell, J. T., Nagy, F., and Chua, N.-H., 1985, Identification of DNA sequences required for the activity of the cauliflower mosaic virus 35S promoter, Nature, 313:810.
Ptashne, M., 1988, How transcriptional activators work, Nature, 335:683.
Riggs, C. D., Voelker, T. A, and Chrispeels, M. J, 1989, Cotyledon nuclear

proteins bind to DNA fragments harboring regulatory elements of phytohemagglutinin genes, Plant Cell, 1:609.
Robert, L. S., Thompson, R. D., and Flavell, R. B., 1989, Tissue-specific expression of a wheat high molecular weight glutenin gene in transgenic tobacco, Plant Cell, 1:569.
Rosales, R., Vigneron, M., Macchi, M., Davidson, I., Xiao, J. H., and related motifs present in other promoters and enhancers, EMBO J., 6:3015.
Sanders, P. R., Winter, J. A., Zarnason, A. R., Rogers, S. G., and Fraley R. T., 1987, Comparison of cauliflower mosaic virus 35S and nopaline synthase promoters in transgenic plants, Nucl. Acids Res., 5:1543.
Sassone-Corsi, P., and Borelli, E., 1986, Transcriptional regulation by trans-acting factors, Trends Genet., 2:215.
Schell, J. S., 1987, Transgenic plants as tools to study the molecular organization of plant genes, Science, 237:1176.
Schernthaner, J. P., Matzke, M. A., and Matzke, A. J. M., 1988, Endosperm-specific activity of a zein gene promoter in transgenic tobacco plants, EMBO J., 7:1249.
Schmülling, T., Schell, J., and Spena, A., 1989, Promoters of the rolA, B, and C genes of Agrobacterium rhizogenes are differentially regulated in transgenic plants, Plant Cell, 1:665.
Sengupta-Gopalan, C., Reichert, N. A., Barker, R. F., Hall, T. C., and Kemp, J. D., 1985, Developmentally regulated expression of the bean $\beta$-phaseolin gene in tobacco seed, PNAS, 82:3320.
Serfling, E., Jasin, M., and Schaffner, W., 1985, Enhancers and eukaryotic gene transcription, Trends Genet., 1:224.
Shirsat, A., Wilford, N., Croy, R., and Boulter, D., 1989, Sequence responsible for the tissue specific promoter activity of a pea legumin gene in tobacco, Mol. Gen. Genet., 215:326.
Stockhaus, J., Eckes, P., Rocha-Sosa, M., Schell, J., and Willmitzer, L., 1987, Analysis of cis-active sequences involved in the leaf-specific expression of a potato gene in transgenic plants, PNAS, 84:7943.
Strittmatter, G., and Chua, N.-H., 1987, Artificial combination of two cis-regulatory elements generates a unique pattern of expression in transgenic plants, PNAS, 84:8986.
Struhl, K., 1987, Promoters, activator proteins, and the mechanism of transcriptional initiation in yeast, Cell, 49:295.
Thanh, V. H. and Shibasaki, K., 1978, Major proteins of soybean seeds: subunit structure of $\beta$-conglycinin, J. Agric. Food Chem., 26:692.
Tierney, M. L., Bray, E. A., Allen, R. D., Ma, Y., Drong, R. F., Slightom, J., and Beachy, R. N., 1987, Isolation and characterization of a genomic clone encoding the $\beta$-subunit of $\beta$-conglycinin, Planta, 172:356.
Timko, M. P., Kausch, A. P., Castresana, C., Fassler, J., Herrera-Estrella, L., Van den Broeck, G., Montagu, M., Schell, J., Cashmore, A. R., 1985, Light regulation of plant gene expression by an upstream enhancer-like element, Nature, 318:579.
Treisman, R., and Maniatis, T., 1985, Simian virus 40 enhancer increases number of RNA polymerase II molecules on linked DNA, Nature, 315:72.
Ueda, T., Pichersky, E., Malik, V. S., and Cashmore, A. R., 1989, Level of expression of the tomato rbcS-3A gene is modulated by a far upstream promoter element in a developmentally regulated manner, Plant Cell, 1:217.
Voelker, T. A., Herman, E. M., and Chrispeels, M. J., 1989, In vitro mutated phytohemagglutinin genes expressed in tobacco seeds: Role of glycans in protein targeting and stability, Plant Cell, 1:95.
Voelker, T. A., Sturm, A., and Chrispeels, M. J., 1987, Difference in expression between two seed lectin alleles obtained from normal and lectin-deficient beans are maintained in transgenic tobacco, EMBO J., 6:3571.
Weber, F., and Schaffner, W., 1985, Simian virus 40 enhancer increases RNA polymerase density within the linked gene, Nature, 315:75.
Witty, M., 1989, Thaumatin II: a simple marker gene for use in plants, Nucl. Acids Res., 17:3312.

# IMPACT OF SIMULATED DROUGHT STRESS ON PROTEIN BODY CYCLE IN RADICLES OF DEVELOPING AND GERMINATING COTTON SEEDS

Eugene L. Vigil

Climate Stress Laboratory
USDA/ARS, BARC-West
Beltsville, MD 20705

## INTRODUCTION

Changes in global climate, particularly reduced rain fall and elevated temperatures within the U.S.A., impact adversely on crop production. Because conditions of severe drought accompanied by elevated temperatures often result in production of low quality seed, research efforts to determine the effects of drought on seed filling are needed. Very little is known about the effects of drought stress on cellular and molecular regulation of organelle biogenesis related to storage of food reserves in ripening seeds. With this knowledge improvements in crop production can be tested either by employment of specific strategies involving use of limited water supplies during critical periods of seed development or by implementation of biotechnological methods to introduce drought and high temperature resistance into the genome of designated crop plants.

We have used cotton in our studies because it is an important agronomic plant grown widely in the world. It is a major source of natural fiber for textiles, oil for cooking and protein meal for animal feed. The source of these commercial products is the seed. Cotton is well suited for our studies on drought stress because it can be grown readily under greenhouse or field conditions.

We selected the radicle, because it is the first structure to emerge from the seed during germination and the cortical tissue therein is the site of major storage of food reserves, i.e. lipid and protein (Vigil, et al., 1984), which are utilized during germination (Vigil, et al, 1985a; Vigil and Frazier, 1989). Despite the abundance of information on the molecular biology of gene expression during cotton embryogenesis and seed ripening, there is virtually no information on the combined effects of drought and high temperature stress on cellular and molecular events of storage of protein in cotyledons. Even less is known about the radicle.

The general findings reported here deal with a simulated drought model, using whole plant defoliation, for studying effects of severe drought stress on protein body formation and filling. Our data show that there is a

*Recent Advances in the Development and Germination of Seeds*
Edited by R.B. Taylorson
Plenum Press, New York

correlation between protein body development and duration of exposure of maturing seeds to simulated drought stress. This effect resulted in reduced germination and seedling growth.

## COTTON EMBRYOGENESIS AND APPLICATION OF SIMULATED-DROUGHT STRESS

Once flowering commences the plant produces flowers continuously until inhibited by frost. During the period of active flowering, a single plant will have numerous bolls at different stages of development. Within each boll there are anywhere from three to five locules, containing approximately 9 ovules each. Except for the most apical ovules, which have limited placental vascularization, embryos in the remaining ovules develop in synchrony (Wade and Ramey, 1985). This is generally the case for locules of each boll on a plant. Extensive studies on cotton embryogenesis have established that development of the cotton embryo occurs during the first 25 to 30 days after anthesis (DAA) through active cell division. The next period from 30 to 50 DAA is marked by active protein and lipid accumulation (Dure, 1973). The storage proteins, namely vicilin (a glycoprotein) and a legumin-type, accumulate in cotton embryos through activation of several gene sets which are developmentally regulated (Dure et al., 1981; Galau and Dure, 1981). Cotton seeds also contain another group of proteins whose appearance during storage of protein and early seed desiccation appears to be regulated, in part, by abscisic acid. These proteins, referred to as late embryogenesis abundant (LEA proteins), appear to play a role in ameliorating the adverse affects of desiccation associated with seed maturation (Galau et al., 1987; Dure et al., 1989).

Seeds of the variety M-8, a double haploid, of cotton (Gossypium hirsutum L.) were used in all our experiments. Our general procedure for growth of seedlings involved using delinted, hot-water-treated (2 min at 80°C) seeds from the previous year's harvest. Plants were held in a 16 h day and 8 h night regime and selection for experimental use was based on uniformity in height and size and color of leaves. Flowers were tagged at anthesis and a daily record of flowering was kept for each plant.

Plant sampling and stress application were done based on availability of sufficient number of bolls per plant. Once cotton plants begin flowering and set fruit (bolls), there is a normal distribution over time of increase in number of flower/bolls per plant followed by decrease. We have observed this phenomenon for plants grown under both greenhouse and field conditions. Plants were selected which had bolls at 30, 40, 50, and 60 days after anthesis (DAA). These time periods represent the middle and late stages of seed maturation when the protein body population is being established and loading initiated (30 to 50 DAA) and time of final loading, seed maturation and boll opening (50 to 60 DAA) (Vigil et al., 1985a).

Because there is irregularity in development of embryos in the apical region due, in part, to incomplete vascularization the apical region due, in part, to incomplete vascularization to the placenta of the apical ovules (Wade and Ramey, 1985), we restricted our sampling to ovules in the middle to basal region of each locule. We previously confirmed this by examination with electron microscopy of radicles from embryos at the base and middle of each locule, finding that these embryos develop essentially in synchrony (Vigil, et al., unpublished data).

Simulated drought-stress by whole plant defoliation was performed on groups of plants grown under field conditions, using alternating replicates (split-plot design). All experimental plants to receive simulated drought stress were defoliated at the same time. Each of these plants had bolls at different ages of development. Sampling of bolls from control and stress plants during the experiment was at ten day intervals from the time of whole plant defoliation. There were three(3) samplings of the group of tagged bolls which were at 30 DAA at the start of the experiment, namely 40 DAA, 50 DAA and 60 DAA. These samples which received 10, 20 and 30 days of simulated drought stress were designated as 30-40 DAA, 30-50 DAA and 30-60 DAA, respectively. Correspondingly, there were two samplings of the group of tagged bolls which were at 40 DAA at the start of the experiment, namely, 50 DAA and 60 DAA; their designation being 40-50 DAA and 40-60 DAA, respectively. There was a single sample for the tagged bolls at 50 DAA at the start of the experiment, namely 60 DAA with a designation of 50-60 DAA. The net result of this sampling procedure was to provide us with developing seeds which received from 10 to 30 days of stress. In addition, it was possible to monitor changes at ten (10) day intervals for the bolls at 30 and 40 DAA at the start of the experiment to control seed maturity at 60 DAA.

For preparation of excised radicles for electron microscopy we used a modification of the osmium-potassium iodide procedure (Carrapico and Pais, 1981) which yielded consistently good preservation of membranes and proteinaceous material (Vigil et al., 1984, 1985b). The use of image analysis as a tool for such stereological analysis was effective in direct quantification of video images from electron microscopic negatives (Vigil et al., 1989). Random sampling for quantitative analysis of protein body development was done for cells in the apical and basal region of the cortex from at least ten different radicles, representing random samples from ten individual plants each for control and drought stressed plants. Since the apical and basal regions of the radicle become the meristematic region and region of elongation, respectively, in the primary root following germination, it was important to know what effect drought stress had on storage of protein in these two regions. We also wanted to know if there was any correlation in germination potential for radicles and duration of drought-stressed to developing seeds using an <u>in vitro</u> test system (Vigil and Frazier, 1989).

## PROTEIN BODY DEVELOPMENT IN COTTON RADICLES

Numerous cellular investigations have been conducted on the synthesis and loading process of protein reserves into storage organs (cotyledons) of oil seeds (See refs. in Chrispeels, 1985). Protein accumulates gradually in cotton, reaching 30 to 35% of total seed weight between 50 and 60 DAA (Bailey, 1948). In our initial fine structural studies on cotton radicles, we observed that protein bodies were primarily derivatives of the original vacuolar apparatus (and to a limited extent dilations of rough endoplasmic reticulum) that gradually fill with protein in parallel with appearance of numerous lipid bodies (Vigil et al., 1985a). Vacuole partitioning has also been observed in soybean cotyledons (Adams et al., 1985).

In cotton, protein loading begins at 30 to 35 DAA within small vacuoles derived by partitioning of the larger vacuoles observed at 20 to 25 DAA. By 40 DAA compact storage protein is present along the inner margin of the protein body membrane. The process of filling protein bodies is first noticeable as fine filamentous material in the vacuole lumen which aggregates as compact protein along the inner margin of the vacuolar membrane (Figs 1,3). This imparts a rigid appearance and smooth contour to the membrane in those areas. Between 40 and 50 DAA a significant amount of

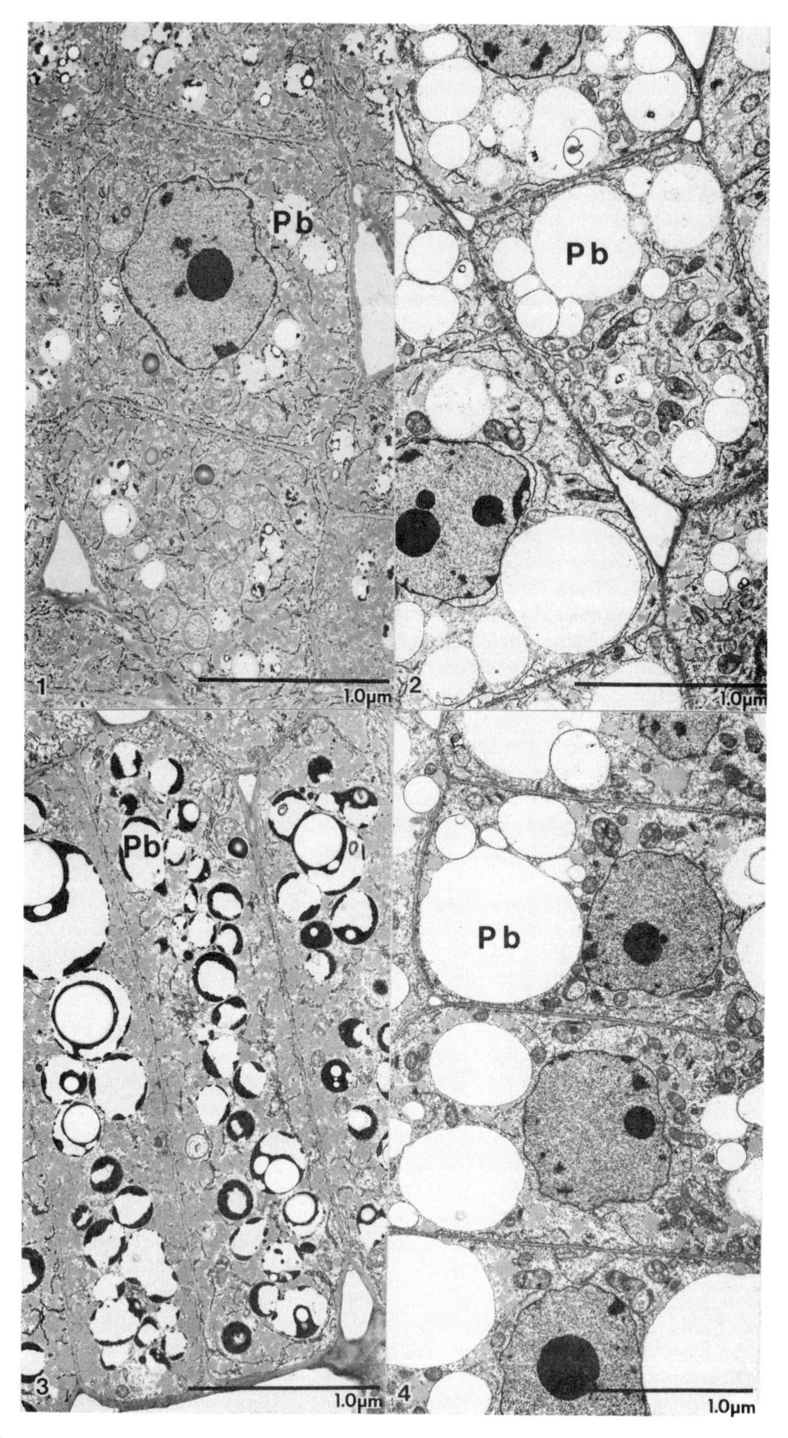


Figures 1-4. Longitudinal sections through several cells in apical (Figs. 1,2) and basal (Figs. 3,4) regions of the radicle cortex of seeds from control (Figs. 1,3) and defoliated (Figs. 2,4) plants. Age of control seeds is 40 DAA and defoliated is 30-40 DAA (10 days stress). PB - Protein body.

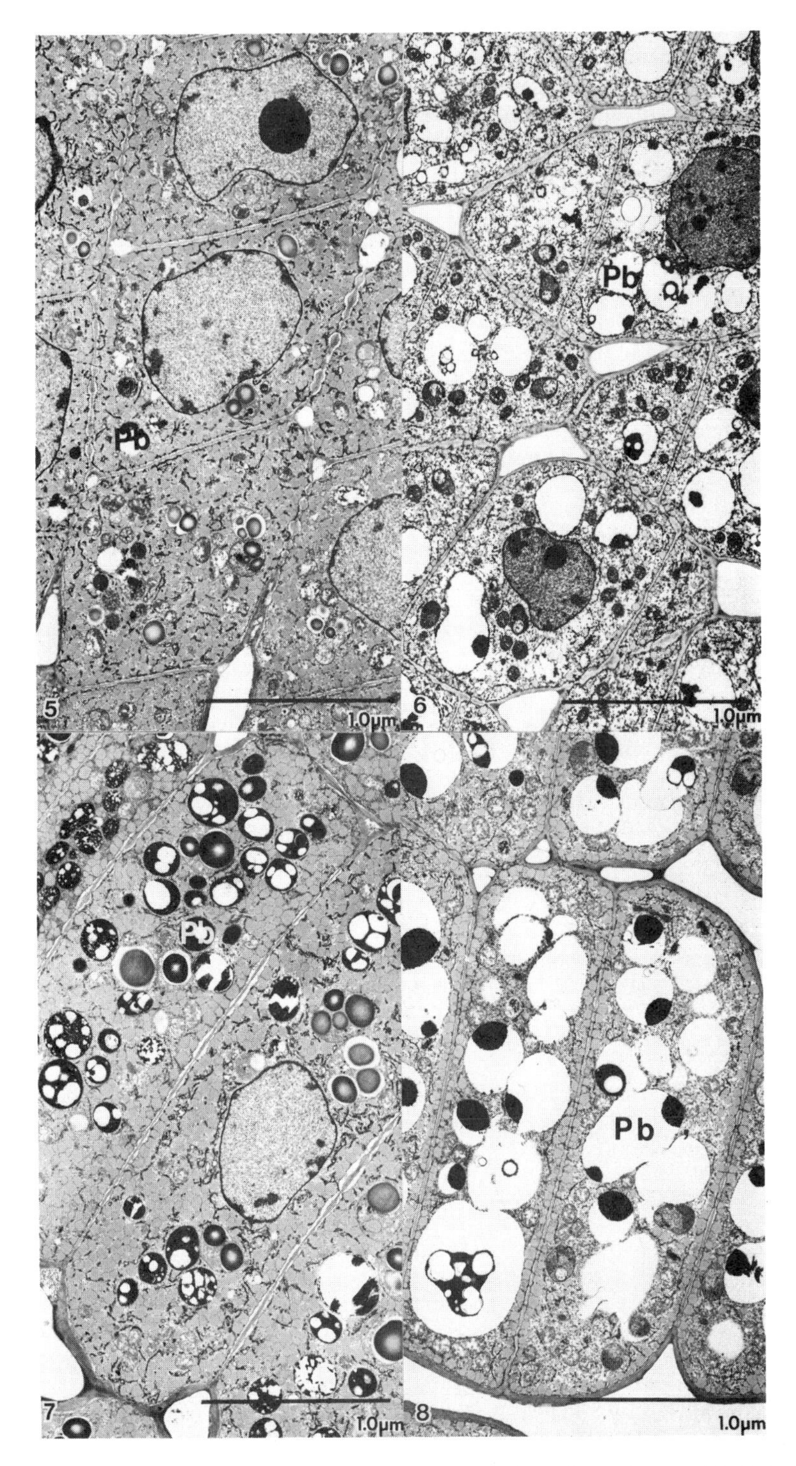


Figures 5-8. Protein body development in radicle of seeds at 50 DAA (Figs. 5,7 -control) and 30-50 DAA (Figs. 6,8 - defoliated). Note marked difference in amount of protein in protein bodies, particularly in the basal region of the cortex (Figs. 7,8).

protein accumulates within protein bodies, as evidenced by increased flocculent material in the lumen in parallel with an increase in area along the limiting membrane occupied by paracrystalline protein (Figs. 5,7). Phytin-containing globoids accumulate within protein bodies in parallel with protein. Completion of protein body loading occurs around 55 DAA. As protein body filling is completed under control conditions, the seed enters into a period of desiccation followed by boll opening at approximately 60 DAA (Figs. 9,11).

The use of whole plant defoliation, as a form of drought stress, has provided us with the first reproducible method for controlling protein body development. When plants containing bolls with ripening seeds varying in age from 30 to 50 DAA were defoliated and embryos sampled initially and at ten day intervals thereafter, we found that the impairment of protein body development reflected the age and amount of stress received (Vigil et al., 1989). Vacuole partitioning normally observed at 40 DAA was arrested when drought stress was applied to plants containing bolls at 30 DAA (30-40 DAA sample). This was true for protein body formation in apical (Figures 1,2) and basal (Figures 3,4) cells of cotton radicles. This pattern continued through 50 DAA (30-50 DAA sample) (Compare Figures 5,7 and 6,8) to 60 DAA(30-60 DAA sample), where differences in protein body structure and filling were quite dramatic for both apical and basal cells of the radicle cortex (Compare Figures 9,11 and 10,12). The amount of protein in vacuoles/protein bodies of drought-stressed plants, determined by quantitative image analysis (Vigil, et al., 1989), was only significantly different in cells of the basal region of the radicle cortex.

This was an important finding because radicle elongation during germination, which involves cell expansion in the basal region was impaired for seeds which received 30 days (30 DAA to 60 DAA) of drought stress. Data from germination tests of dry seeds or excised radicles from control and drought stressed (30 days stress) plants differed markedly; drought stressed seeds having only approximately 20% germination compared to nearly 100% for controls.

Examination of radicles of embryos which received 20 days of simulated drought stress starting at 40 DAA (40-50 DAA and 40-60 DAA samples) revealed a pattern of protein body development and filling similar to that of controls in both the apical and basal cells of the radicle cortex (Figures 13-16).

The cellular process of mineral deposition within protein bodies has not been easy to follow. Observations on phytin accumulation in protein bodies in developing castor bean endosperm (Greenwood et al., 1984) indicate that phytin accumulates within the cytoplasm as electron dense particles and which are ingested into the developing protein body, presumably by an endocytic process. In our observations of protein body development in cortical cells of cotton radicles, we found that phytin granules, i.e. globoids, appeared within protein bodies surrounded by protein. When we examined excised radicles from dry seeds of control and drought stressed plants (30-60 DAA) with energy dispersive X-ray analysis and scanning electron microscopy, we found that magnesium, potassium and phosphorous were selectively localized within protein bodies in both samples. Quantitative data from atomic absorption spectroscopy of excised radicles from these seeds and isolated protein bodies from radicles of control seeds confirmed that these minerals accumulate within protein bodies of the cotton radicle, and showed that drought stress limited the total amount of minerals entering the radicle (Vigil et al., unpublished data).

To answer the question of whether protein bodies in radicles were similar to those in cotyledons, we isolated protein bodies from each organ

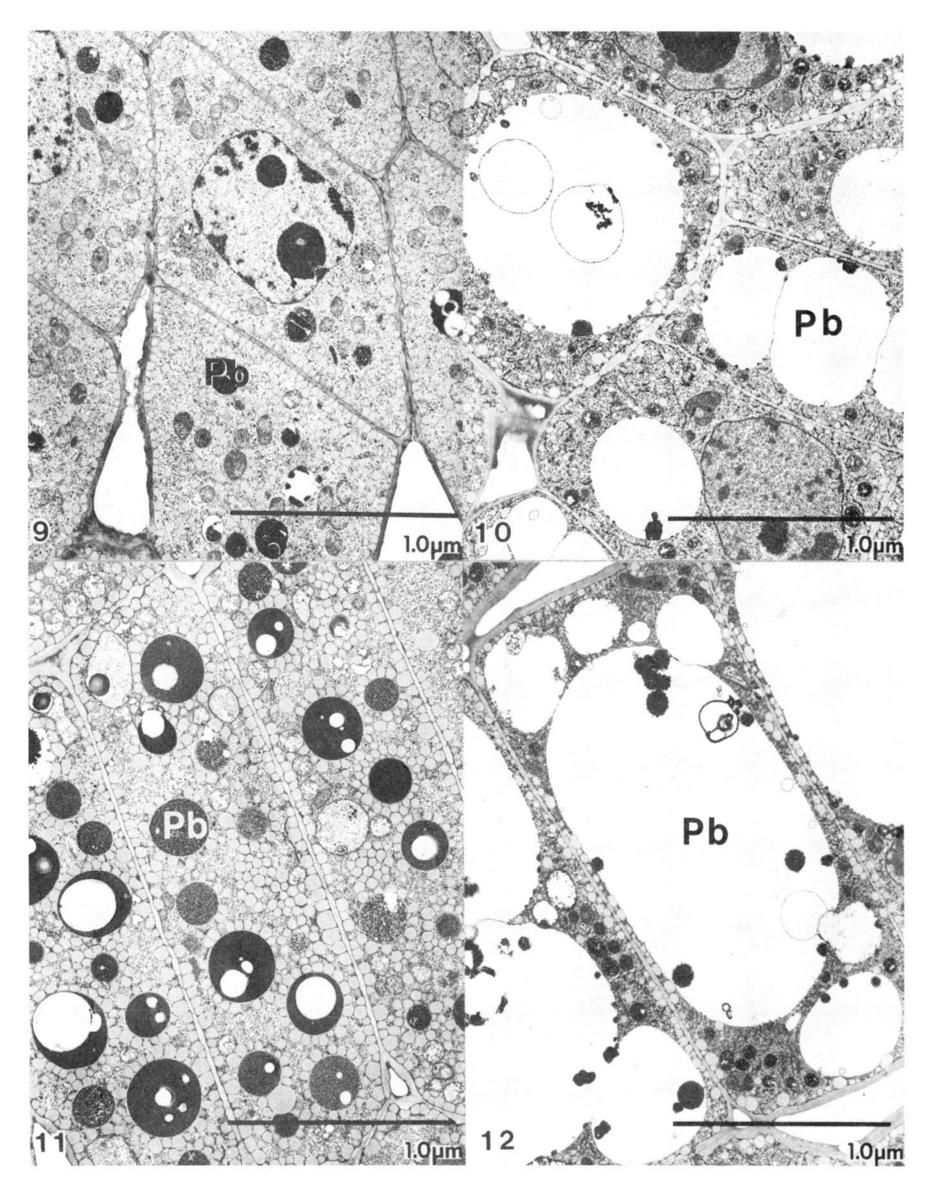


Figures 9-12. Figs. 9, 11 illustrate the appearance of protein bodies in cortical cells of the apical and basal regions, respectively, of radicles of mature seeds (60 DAA). The absence of significant vacuole partitioning and deposition of protein is most noticeable in the same regions of the radicle cortex for seeds from defoliated plants receiving 30 days of stress (30-60 DAA).

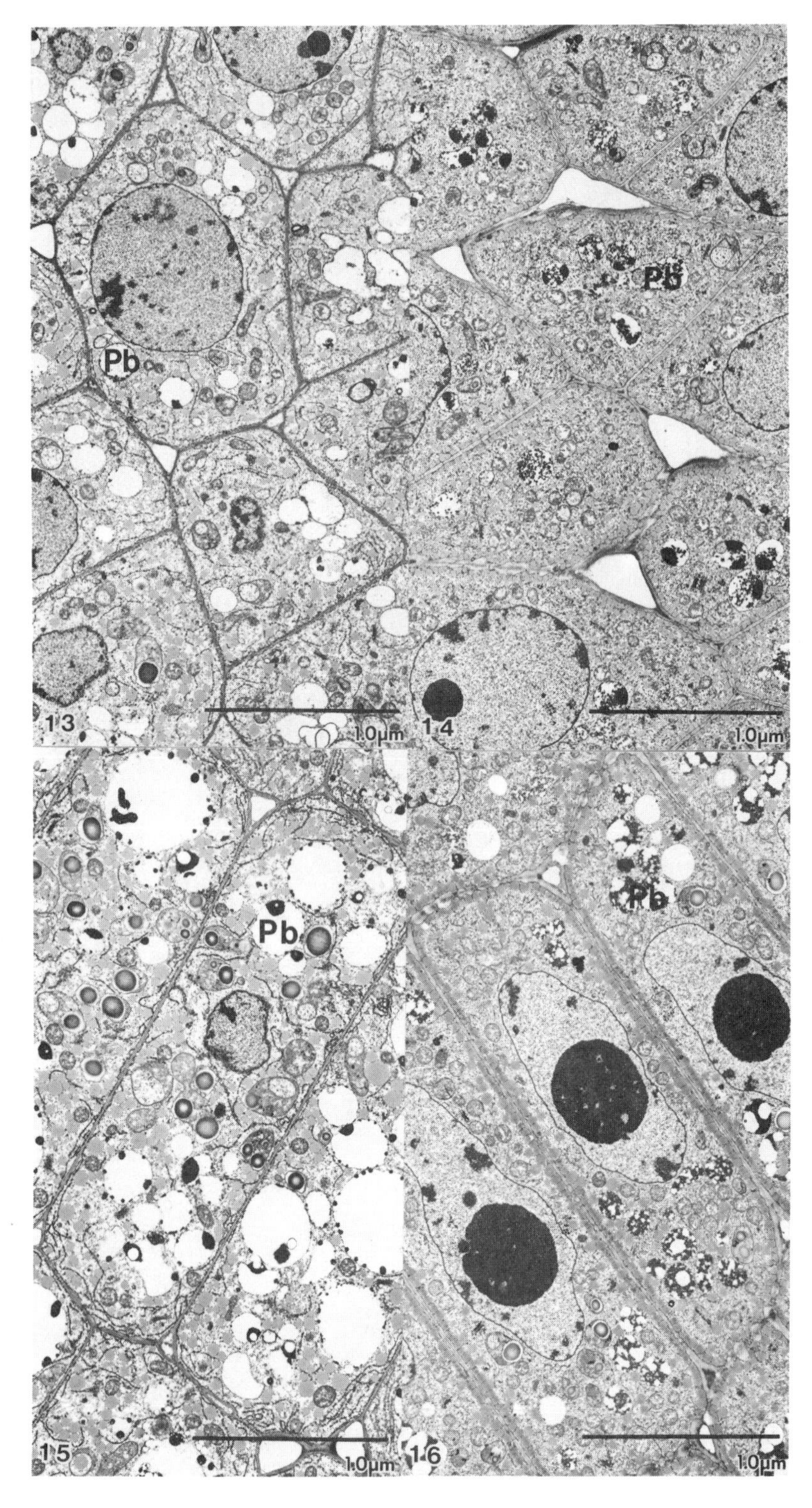


Figures 13-16. These images illustrate the effect of whole-plant defoliation at 40 DAA on protein body development in apical (Figs. 13,14) and basal (Figs. 15,16) cells of the radicle cortex over the next 10 (Figs. 14,15; 40-50 DAA) and 20 (Figs. 14,16; 40-60 DAA) days of seed ripening.

in anhydrous glycerol and resuspended the pellet in sodium dodecylsulfate. The samples were run on sodium dodecylsulfate polyacrylamide gel electrophoresis. Protein profiles were remarkably similar. The major differences were that protein bodies from radicles appear to have a lower concentration of storage protein polypeptides above 40 kd and between 18-20 kd and a slightly higher concentration of polypeptides in the 35 kd range (Vigil and Fang, unpublished data).

## SUMMARY

The cortex of cotton radicles contains numerous protein bodies whose stores are utilized during germination for cell expansion essential for radicle emergence. Formation of protein bodies occurs principally by vacuole partitioning between 30 and 40 DAA with major addition of protein during the next 15 to 20 days. Application of simulated drought stress by whole plant defoliation provided a simple, reproducible procedure to monitor the immediate impact of severe drought conditions on vacuole partitioning and storage of protein and minerals. The reduced photosynthate partitioning to the seed caused by leaf removal probably parallels the effect of severe drought on seed nutrition by abscission of wilted leaves. Quantitative image analysis provided data that indicated the period from 30 to 40 DAA is critical for vacuole partitioning, being arrested by simulated drought. Germination of drought-stressed seeds was reduced to approximately 20% for seeds receiving 30 days of stress (30 to 60 DAA). These findings direct attention to the importance of the initial period of vacuole partitioning and loading, as a critical early event for protein body formation and filling. The impact of simulated drought stress on this process is two fold; 1) Formation and filling of protein bodies is affected proportionately to the duration of stress imposed on ripening seeds; 2) The rate and per centage of germination, viz. radicle elongation, is similarly affected. This indicates that vacuole partitioning and initial filling are critical early events for protein body formation which, when arrested, affect subsequent germination, requiring utilization of stored food reserves.

## ACKNOWLEDGEMENTS

I wish to acknowledge gratefully the support and assistance of Randy Rowland for growing and tagging plants in greenhouse and field experiments, LC Frazier for preparing and photographing all tissue samples for electron microscopy and Christopher Pooley for processing electron micrographs with image analysis and preparing all figures for this paper.

## LITERATURE CITED

Adams, C. A., Norby, S. W., and Rinne, R. R., 1985, Production of multiple vacuoles as an early event in the ontogeny of protein bodies in developing soybean seeds, Crop Sci., 25:255.

Bailey, A. E., 1948, Cotton Seed and Cotton Seed Products. Interscience Publishing, New York.

Carrapico, F., and Pais, M. S., 1981, Iodure de potassium tetroxide d'osmium. Un melange d'impregnation pour la microscopie eletronique, C. R. Acad. Sc. Paris, t. 292:131.

Chrispeels, M. J., 1985, The role of the Golgi apparatus in the transport of and post-translational modification of vacuolar (protein body) proteins, Oxford Surveys of Plant Mol. and Cell Biol., 2:43.

Dure, L., III, 1973, Developmental regulation in cotton seed embryogenesis and germination. in: "Developmental Regulation: Aspects of Cell Differentiation," S. J. Coward, ed., Academic Press, New York.

Dure, L., III, Greenway, S. C., and Galau, G. A., 1981, Developmental biochemistry of cotton seed embryogenesis and germination: Changing messenger ribonucleic acid populations as shown by in vitro and in vivo protein synthesis. Biochem., 20:4162.

Dure, L., III, Crouch, M., Harada, J., Ho, T-H. D., Mundy, J., Quatrano, R., Thomas, T., and Sung, Z. R., 1989, Common amino acid sequence domains among the LEA proteins of higher plants. Plant Mol. Biol.. 12:475.

Galau, G. A. and Dure, L., III, 1981, Developmental biochemistry of cottonseed embryogenesis and germination: changing messenger ribonucleic acid populations as shown by reciprocal heterologous complementary deoxyribonucleic acid-messenger ribonucleic acid hybridization. Biochem., 20:4169.

Galau, G. A., Bijaisoradat, N., and Hughes, D. W., 1987, Accumulation kinetics of cotton late embryogenesis-abundant mRNAs and storage protein mRNAs; Coordinate regulation during embryogenesis and the role of abscisic acid. Devel. Biol., 123:198.

Greenwood, J. S., and Bewley, J. D., 1984, Subcellular distribution of phytin in the endosperm of developing castor bean: a possibility for its synthesis in the cytoplasm prior to deposition within protein bodies, Planta, 160:113.

Vigil, E. L. and Frazier, LC, 1989, In vitro germination test system for cotton seed radicles, Plant Physiol., 89:s130.

Vigil, E. L., Steere, R. L., Wergin, W. P., and Christiansen, M. N., 1984, Tissue preparation and fine structure of radicle apex from cotton seeds, Amer. J. Bot., 71:645.

Vigil, E. L., Steere, R. L., Christiansen, M. N. and Erbe, E. F., 1985a, Structural changes in protein bodies of cotton radicles during seed maturation and germination. in: "Botanical Microscopy," A. W. Robards, ed., Oxford University Press, Oxford.

Vigil, E. L., Steere, R. L., Wergin, W. P., and Christiansen, M. N., 1985b, Structure of plasma membrane in radicles from cotton seeds. Protoplasma, 129:168.

Vigil, E. L., Frazier, LC, and Pooley, C., 1989, Effect of drought stress on protein body development in radicles of cotton embryos: A quantitative study with image analysis. Cell Biol. Int. Rep., 13:47.

Wade, C. P. and Ramey, H. H., 1985, Potential biological NEP formation. 1. Distribution of motes, immature and mature seeds in locules. Proc. 39th Beltwide Cotton Conference, p.54.

# SYNTHETIC SEED FOR CLONAL PRODUCTION OF CROP PLANTS[1]

D.J. Gray

Central Florida Research and Education Center
Institute of Food and Agricultural Sciences
University of Florida
5336 University Avenue, Leesburg, FL 34748

## INTRODUCTION

Seed is the preferred planting vehicle for all crop plants due to ease of production and handling. For many crops (e.g. cereals, soybean etc.), genetically similar seed can be generated by self-pollination whereas, for others, uniform seed cannot be produced due to genetic self-incompatibilities as well as a variety of other reasons. In these latter instances, plants are propagated either vegetatively (e.g. fruit crops) or by genetically nonuniform seed (e.g. forage grasses). Even with self-pollinated crops, plants from different seeds are not genetically identical since they are subject to meiotic recombination during sexual reproduction (exceptions are crops that produce seed apomictically). Thus, elaborate and time consuming breeding systems have been developed in order to exploit hybrid vigor while maintaining seed uniformity.

Ability to propagate outstanding individuals vegetatively but at seed efficiencies would revolutionize production of both seed and vegetatively propagated crops (Gray, 1987b). For seed propagated crops, new hybrids could be used immediately without tedious development of parental inbreds and genetically identical plants would result. For vegetatively propagated crops, propagation and planting efficiencies would be dramatically increased. For both crop types, genetically engineered individuals could be increased and planted without sexual recombination of foreign genes.

Use of in vitro somatic embryogenesis to produce "synthetic seeds" offers the potential of suitably efficient vegetative propagation. Somatic embryos form continuously in cell cultures and production of several thousand embryos per gram of culture material has been achieved. Furthermore, somatic embryos are morphologically and, in most respects, developmentally analogous to zygotic embryos found in seeds. For example, somatic embryos possess typical embryonic organs and

[1]Florida Agricultural Experiment Station J. Series # R-00050.

*Recent Advances in the Development and Germination of Seeds*
Edited by R.B. Taylorson
Plenum Press, New York

can germinate bipolarly as do their zygotic counterparts. However, somatic embryogenesis has yet to gain utility as a means of plant propagation despite the fact that it has been recognized for over 35 years and has been documented for over 200 crop species. Only in recent years have attempts been made to confer seed-like storage and handling qualities on somatic embryos through development of synthetic seed technology.

Broadly defined, synthetic seed consists of somatic embryos processed to be of use in the production of a particular crop. For some crops, this could constitute hand manipulated, naked embryos reared in callus cultures. However, quiescent somatic embryos, encapsulated in synthetic seed coats and produced en masse by an automated process will be required for others. The purpose of this presentation is to discuss the applications and current developmental status of synthetic seed technology for crop production.

## APPLICATIONS OF SYNTHETIC SEED

Potential applications of synthetic seed will vary from crop to crop depending on the relative sophistication of existing production systems and the opportunities for improvement. Whether or not a cost advantage results from synthetic seed will ultimately determine its commercial use. Some potential applications for specific crops are presented in Table 1 and discussed below. They are selected to provide a spectrum of crop types in order to discuss different uses of and needs for synthetic seed technology.

### Seed Propagated, Self-incompatible Crops

Synthetic cultivars of seed propagated self-incompatible crops such as alfalfa (*Medicago sativa* L.)and orchardgrass (*Dactylis glomerata* L.) are laboriously developed by selecting phenotypically uniform but genetically distinct lines. These lines are then allowed to cross pollinate for seed production. Such seed is nonuniform and every resulting plant is a potentially distinct genotype. The nature of this breeding system makes it difficult to incorporate specific new genes into existing lines. Use of synthetic seed would allow single outstanding hybrids to be utilized as cultivars since self-fertilization would not be needed for seed increase. Such cultivars would be genetically uniform. Although excellent embryogenic culture systems exist for both alfalfa (eg. Stuart and Strickland, 1984) and orchardgrass (eg. Gray et al., 1984), a limitation to this application of synthetic seed technology is the low per plant value and the low cost of existing seed (Table 1). An intermediate use of synthetic seed for these crops may be for increase of parental lines prior to establishment in open crossing blocks.

### Vegetatively Propagated, Self-incompatible Crops

Planting efficiency of crops that are currently vegetatively propagated due to self-incompatibilities and long breeding cycles, such as fruits and nuts etc., could theoretically be increased by use of synthetic seed instead of cuttings. But, since existing methods tend to be cost effective, developmental costs of synthetic seed would likely not be justified. Use of

Table 1. Potential Applications of Synthetic Seed Technology (SST) for Selected Crop Species.

| Crop | Somatic Embryo Quality[a] | Relative Seed Cost[b] | Application[c] | Relative Need for SST[d] |
|---|---|---|---|---|
| Alfalfa | h | l | s | m |
| Corn | p | m | i | m |
| Cotton | p | m | h | m |
| Grape | h | na | s,g | m |
| Loblolly pine | p | h | c | h |
| Norway spruce | h | - | c | h |
| Orchardgrass | h | l | s | m |
| Soybean | p | m | h | m |
| Hybrid Tomato | n | v | d | h |
| Seedless Watermelon | n | v | d | h |

[a] Relative somatic embryo quality: h- highly developed embryos; p- poorly developed embryos; n- somatic embryos not obtained.
[b] Relative cost of seed: v- seed cost limits planting; h- seed is costly; m- moderate; l- relatively inexpensive; na- seed is not used.
[c] Application for synthetic seed: c- circumvent long breeding cycles; d- decrease hybrid seed cost; g- germplasm conservation; h- mass production of hybrids; i- eliminate need for inbreds.; s- circumvent self-incompatibility.
[d] Relative need: h- highly useful if implemented; m- existing methods are effective but implementation should yield improvements.

synthetic seed for germplasm conservation of these crops could be highly advantageous, however. For example, germplasm of grape (Vitis spp.) is currently maintained as living plants in vineyards. This method of conservation is expensive and subject to environmental disasters. Use of synthetic seed would allow clonal germplasm of grape to be conserved in seed repositories. More genotypes could be conserved since space problems would be eliminated. Moreover, development costs would be reduced since automated production equipment for mass production would not be needed for the relatively small number of synthetic seed required. This method of germplasm conservation would be particularly useful for tropical species where existing conservation is inadequate or nonexistent. Grape is a good experimental prospect since well developed somatic embryos have been obtained (eg. Gray, 1987c; 1989).

## Conifers

Forest conifers can be economically planted only by seed. Improvement via conventional breeding is extremely time consuming due to the long conifer life cycle. Furthermore, conifers are highly heterozygous so that seed from outstanding individuals does not necessarily result in improved progeny. Synthetic seed offers the possibility of cloning outstanding elite trees at reasonable costs, thus circumventing years of development. To date, well developed somatic embryos have been obtained only for Norway spruce (*Picea abies* L.) (eg. Becwar et al., 1989). Although somatic embryogenesis has been obtained for loblolly pine (*Pinus taeda* L.) (eg. Gupta and Durzan, 1987), well developed embryos have not been produced. The potentially great benefit of synthetic seed technology for the forest products industry provides compelling rationale for its continued development.

## Self-seeded Crops

Commercial quantities of hybrid seed are difficult to produce for certain seed propagated crops such as cotton (*Gossypium hirsutum* L.) and soybean (*Glycine max* Merrill.) due to cleistogamous flowers and/or problems with flower abscision. Thus, seed of most existing cultivars is derived from self-pollination. However, relatively small numbers of hybrids can laboriously be produced by extensive hand pollination and subsequently mass produced by use of synthetic seed. Hybrid vigor could then be exploited at the production level. Somatic embryogenesis has been obtained for both cotton (eg. Finer, 1988) and soybean (eg. Ranch et al., 1985),although the commercial potential of synthetic seed for such crops is unclear at this time, considering the moderate cost of existing seed (Table 1).

## High Value, Hybrid Seed Crops

For crops such as tomato (*Lycopersicon esculentum* Mill.) and seedless watermelon (*Citrullus lanatus* [Thunb.] Matsum. and Nakai), the high cost of producing hybrid seed is offset by the resulting crop value. For example, the cost per plant of establishing stands of seedless watermelon has exceeded $ 0.15 due to high seed cost and low germination rates (personal observation). In these instances, synthetic seed may actually be less expensive than conventionally-produced hybrid seed. Unfortunately, somatic embryogenesis has not been obtained for either of these species (Table 1) so that much development is still needed to gauge feasibility.

## Low Value, Hybrid Seed Crops

A final potential application of synthetic seed discussed here is to circumvent the need for inbred and male-sterile parental lines in hybrid seed production. For example, the hybrid corn (*Zea mays* L.) industry relies on inbred parentals to produce uniform hybrid seed. Mass hybridization is possible by use of male-sterile lines as females. Increased production costs over open pollinated seed that are incurred by use of inbred and male sterile lines are more than offset by the resulting yield and quality conferred by hybrid vigor.

However, development and maintenance of these parental lines consumes much of the time and resources of a breeding program and integration of new germplasm is slow. An intriguing possibility is the use of synthetic seed to propagate new hybrids and eliminate the need for parental inbreds and male-steriles altogether. This would facilitate commercialization of new hybrids and would probably stimulate competition since newcomers could produce cultivars without an existing stock of parental inbreds. Although somatic embryogenesis of corn is well described (eg. Kamo et al., 1985), it is unclear whether the successful hybrid corn industry would embrace this concept.

### Cost Estimates

The actual cost of producing synthetic seed is impossible to determine at this early stage of development since several key pieces of technology (ie. automated culture systems and hardened encapsulations as described below) have not yet been refined. Basing estimates on existing technology which involves much hand manipulation of cultures, greenhouse transplants of alfalfa would cost 3.3 cents each to produce by somatic embryogenesis (Redenbaugh et al., 1987). This cost is far in excess of conventional seed of most crops. However, expenses would undoubtedly be greatly reduced with a refined, automated system.

## SOMATIC VS. ZYGOTIC EMBRYO DEVELOPMENT

A primary difference between somatic and zygotic embryos is their method of initiation. Zygotic embryos develop only from fertilized eggs (zygotes) and resulting plants are hybrids, each containing a potentially different meiotic recombination of genes. Conversely, somatic embryos are produced from cells of only one individual and resulting plants are genetic duplicates or clones (Fig. 1). However, somatic and zygotic embryos share similar gross ontogenies, with both typically passing through globular, torpedo and cotyledonary stages for dicots (eg. Gray and Mortensen, 1987) and conifers (eg. Becwar et al., 1989), or globular, scutellar and coleoptilar stages for monocots (Gray and Conger, 1985).

Although somatic embryos are developmentally similar to zygotic embryos, significant differences exist which limit their use for propagation. For example, species that are monoembryonic and produce zygotic embryos via a suspensor (such as most seed propagated crops) often produce somatic embryos by a more evolutionarily primitive method whereby a mass of embryonic tissue (termed a proembryonal complex) is formed from which multiple somatic embryos develop (Fig. 1), (Haccius, 1978). The reason(s) for this reversion to a more primitive type of development are unknown. However, such somatic embryos tend to develop asynchronously so that multiple stages are present in cultures at any given time. Furthermore, embryos that develop over different time periods are subjected to changing nutrient regimes as medium becomes depleted then replenished between and during subcultures. Under such a fluctuating nutritional environment, embryos often bypass maturation altogether, becoming disorganized, forming new embryogenic cells and contributing to asynchrony (eg. Conger et al.,1989). Alternatively, the embryonic organs may develop at different rates, leading to precocious germination. Such embryos typically form only a root or a shoot but normal rapid germination and growth to a plant does not occur. Another related problem is that somatic embryos often exhibit structural anomalies such as extra cotyledons and poorly developed apical meristems (Ammirato, 1986). All of these problems are likely due to culture conditions and not factors intrinsic to somatic embryos since immature zygotic embryos often suffer similar irregularities when removed from seed and cultured in vitro (Norstog, 1967). The aforementioned problems with nutrition also appear to contribute to poor structural development. Thus, lack of synchronous cultures that produce uniformly mature somatic embryos is a serious obstacle to propagation.

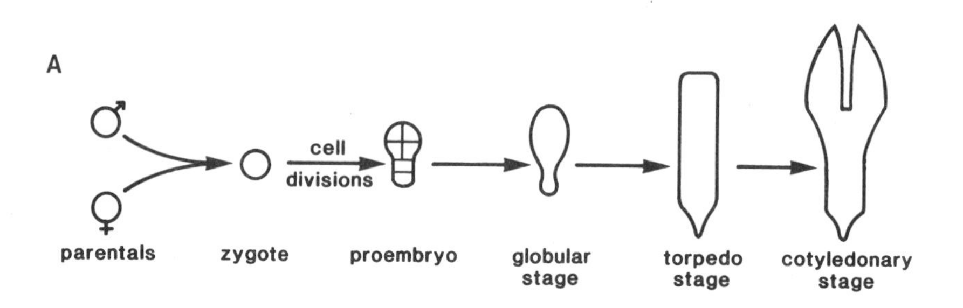


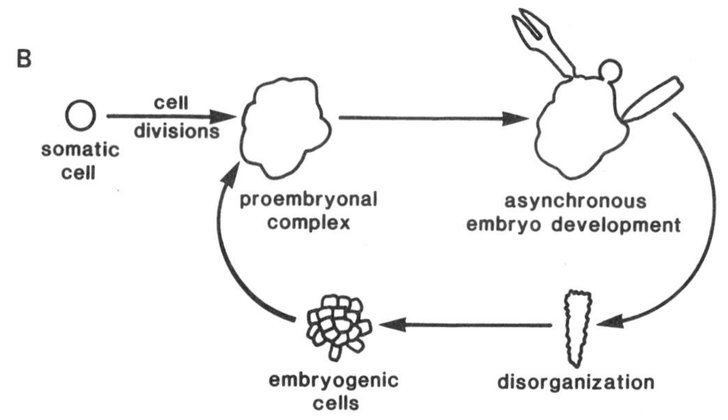


Fig. 1. Comparison of typical zygotic and somatic embryogenesis. A. Zygotic embryos form singly from fertilized eggs and typically possess a suspensor and pass sequentially through distinct developmental stages. B. Somatic embryos develop from a proembryonal complex and several embryos in various stages of development are often present. Maturation is often circumvented as embryos disorganize to form more proembryonal tissue.

Another significant difference between zygotic and somatic embryos is that somatic embryos typically lack a quiescent resting phase (Gray, 1987c). In orthodox seed, mature zygotic embryos typically become dehydrated and enter a resting period which is the major factor allowing seed to be conveniently stored. In contrast, somatic embryos continue to grow and typically either germinate, become disorganized into embryogenic tissue, or die. Lack of a quiescent resting phase in somatic embryos is a major drawback to their use as synthetic seed.

## REGULATION OF SOMATIC EMBRYO DEVELOPMENT

Attempts at regulating embryogenic cultures have taken two approaches: (i) physical separation of embryogenic tissues to produce uniform sized cell masses and favor more uniform embryo development, and (ii) use of growth regulators to physiologically synchronize development.

Embryogenic tissues are physically separated by passage through sieves of defined pore size. For example, cell masses from embryogenic alfalfa callus, passed through a 20 and then a 60 mesh sieve, measured 234 - 864 um in diameter and produced somatic embryos (Walker and Sato, 1981). These cell masses were analogous to isolated proembryonal complexes. Presumably, the proembryonal complexes were of a size capable of immediately producing just one embryo, resulting in overall uniformity. It is interesting that synchrony initiated at such an early developmental stage was maintained throughout subse-quent embryogenesis.

Physiological synchronization of embryogenesis has employed primarily one growth regulator, abscisic acid (ABA). ABA regulates development in some embryogenic culture systems including conifers (von Arnold and Hakman, 1988) and carrot (Ammirato, 1983). The mode of action of ABA in embryo development is unclear but it may interfere with the normal functions of endogenous growth regulators, causing a shift to conditions that favor embryogenesis. Alternatively, ABA may cause cell water content to decrease, leading to more normal development as described below. The role of ABA in embryo maturation appears to be related to its ability to suppress germination (Ammirato, 1983). With ABA, developing embryonic organs tend to mature without precociously germinating. ABA allows embryo development but inhibits plant development. Therefore, more complete embryo development occurs.

## QUIESCENCE AND DORMANCY IN SOMATIC EMBRYOS

Somatic embryos with reversible, arrested growth will be needed in order to mimic seed storage and handling characteristics. There are few examples of arrested growth being obtained in somatic embryos. For the purpose of this discussion, two types of arrested growth, quiescent and dormant, are distinguished based on definitions by Bewley and Black (1985). Quiescence is a resting phase that can be reversed solely by the addition of water. Dormancy is a form of quiescence that requires factors in addition to water, such as cold or heat treatments, for continuation of growth to occur. Although quiescence and dormancy have only recently been recognized in somatic embryos, it is possible that dormancy occurs in many instances where plants cannot be obtained from well developed somatic embryos. Thus, traditional methods for inducing and maintaining quiescence in seeds (eg. Barton, 1961; Owen, 1956) may be applicable for somatic embryos.

## Quiescence

The most logical means of inducing quiescence in rapidly growing somatic embryos is by controlled removal of water, since dehydration and rehydration cause arrest and resumption of growth in orthodox seeds. The first study that suggested quiescence could be induced in somatic embryos by dehydration utilized embryogenic carrot (Daucus carota L.) callus cultures (Nitzche, 1978). Dried callus was kept for seven days at room temperature. Regrowth of callus and, subsequently, plant regeneration occurred after imbibition; however, it was unclear as to whether the plants originated from preexisting embryos or those formed after imbibition. Survival was subsequently obtained from carrot somatic embryos dried for four days in a water soluble plastic resin, although recovery of plants was not reported (Kitto and Janick, 1985). Both of these studies utilized treatments with ABA and high concentrations of sucrose prior to dehydration, suggesting that these compounds were needed to produce mature, dehydration tolerant somatic embryos.

The first clear documentation of plant recovery from dehydrated quiescent somatic embryos utilized embryogenic callus cultures of orchardgrass (Gray and Conger, 1985b; Gray et al., 1987). Isolated somatic embryos were dehydrated by exposure to 70% relative humidity (RH) air and stored for up to 21 days at 23C. Under these conditions, the embryos became discolored, decreased in size and their outer cell walls collapsed (Fig. 2). An embryo water content of 13% was achieved within 24 hours, as determined by weight loss over time (Fig. 3), and was maintained over the entire storage period. This water content is similar to that of seeds maintained at 70% RH and is adequate for maintenance of viability during prolonged seed storage. The embryos were rehydrated by placement on solidified medium. During rehydration, the embryos swelled rapidly, and regained their normal white coloration. Only well developed, white, opaque somatic embryos were responsive. This somatic

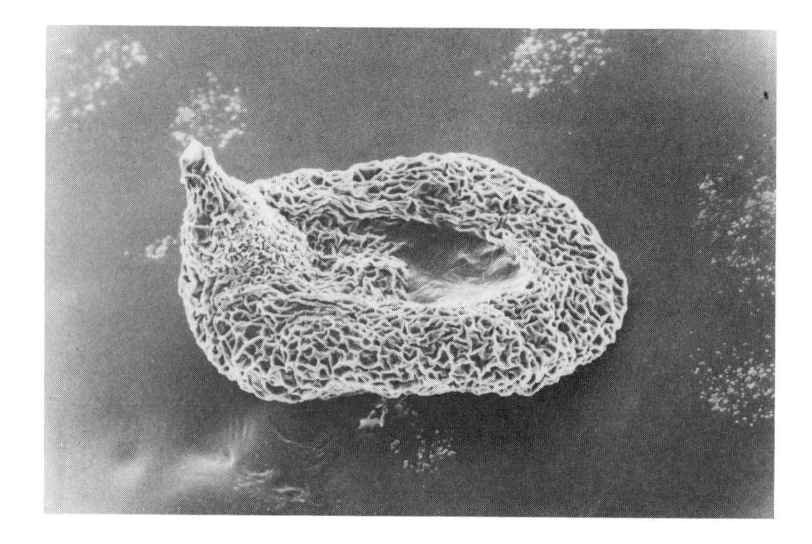

Fig. 2. SEM of dehydrated orchardgrass somatic embryo. Note collapsed epidermal cells. From Gray et al., 1987.

embryo type was structurally mature and contained starch and lipid storage compounds (Gray and Conger, 1985a). Four percent of the well developed dehydrated embryos that were stored for 21 days germinated and produced plants after imbibition (Table 2). This study showed that somatic embryos possessed the ability to become quiescent and suggested that higher germination percentages over longer storage periods would be possible with refinements to the system.

Quiescence was subsequently induced in somatic embryos of grape (Gray, 1987a; 1987c; 1989). Grape somatic embryos were dehydrated by the method described for orchardgrass. Embryos underwent similar morphological changes and equilibrated to 13% water content when stored at 70% RH. Genotypic differences in response were noted and those genotypes that produced relatively well developed somatic embryos were most responsive. After 21 days of dehydrated storage, 34% of embryos from one grape genotype produced plants following imbibition (Table 3). This study demonstrated that higher germination percentages were possible. Latest unpublished trials are achieving 100% germination after 28 days of dehydrated storage. Until recently, longer test storage periods were not necessary since viability declined over a relatively short time. With high germination percentages, longer storage periods are now needed to evaluate viability.

We have had similar success with somatic embryos of corn and soybean (unpublished). Recently, somatic embryos of alfalfa (McKersie et al., 1988) and wheat (Avena sativa L.) (Carman, 1988) were dehydrated and plants subsequently obtained. Alfalfa plants have now been recovered from somatic embryos after one year of dehydrated storage (Senaratna et al., 1989).

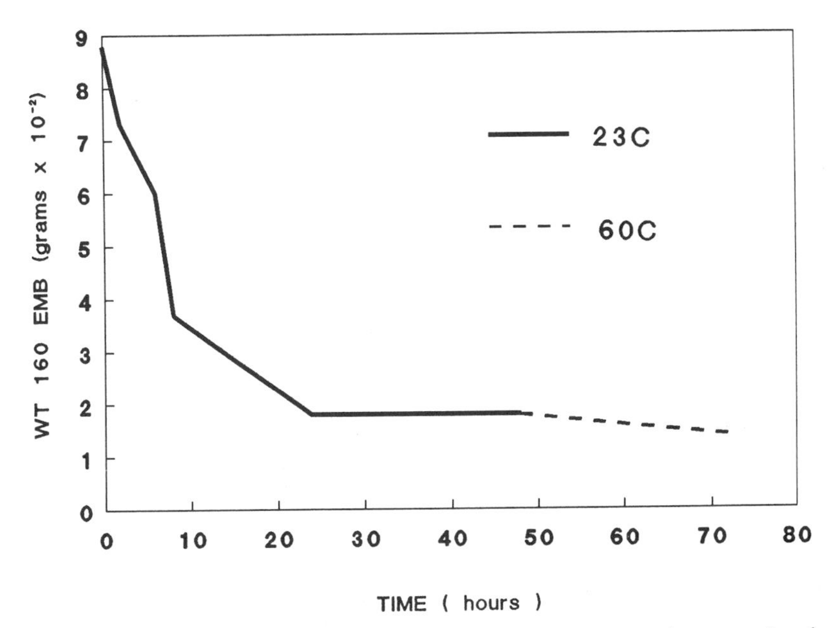


Fig. 3. Weight loss by orchardgrass somatic embryos during dehydration at 70% RH and 23C. Dotted line denotes additional drying at 60C. From Gray et al., 1987.

Table 2. Germination of dehydrated orchardgrass somatic embryos after storage at 23C for 0, 7,and 21 days[a]. From Gray et al., 1987.

| Response[b] | Days of Dehydrated Storage | | |
|---|---|---|---|
| | 0 | 7 | 21 |
| No germination | 126/28[c] | 333/74 | 396/88 |
| Germination-no further growth | 180/40 | 81/18 | 36/18 |
| Germination- viable plants | 144/32 | 36/8 | 18/4 |
| Total | 450/100 | 450/100 | 450/100 |

[a] Embryos were imbibed on solidified medium after test storage periods.
[b] Embryos that produced root hairs, roots, coleoptiles, and/or shoots but failed to develop further were scored as germinated-no further growth. Those that produced green leaves and continued to grow were considered to be viable.
[c] Number/percentage of total.

Table 3. Comparison of dehydration and benzyladenine (BA) for inducing germination in grape somatic embryos[a]. From Gray, 1989.

| Treatment | Percent Germination Response[b] | | | |
|---|---|---|---|---|
| | Hypocotyl | Root | Cotyledon | Shoot |
| Dehydration | 77 | 68 | 65 | 34 |
| 0.5 uM BA | 100 | 92 | 98 | 12 |
| Control | 76 | 88 | 36 | 0 |

[a] Well developed embryos were either dehydrated for 21 days at 70% RH and 27C, placed directly on medium with BA, or placed on basal medium (control).
[b] Germination response was based upon either enlargement and greening of hypocotyls and cotyledons or emergence of roots or shoots.

These results collectively demonstrate that quiescence can be induced in somatic embryos from a variety of crop species. Although the necessary combination of prolonged dehydrated storage with rapid germination under ex vitro conditions has not been achieved, the steady progress exhibited to date suggests that somatic embryos possess all of the abilities to fully function like their zygotic counterparts.

Dormancy

Although the occurrence of dormancy in somatic embryos has received little attention, it may be relatively common but overlooked. Poor germination that is typical of some embryogenic culture systems, instead, has been ascribed to inherent abnormalities of somatic embryo development. Yet, it is difficult to blame abnormalities for poor germination in culture systems where well developed embryos are formed. Unrecognized dormancy may function in these instances. There are two possible causes of dormancy in vitro: (i) dormancy that is typical for zygotic embryos of a particular species may also be expressed by somatic embryos in vitro, or (ii) dormancy may be artificially induced by aspects of the culture environment such as exogenously supplied growth regulators. For culture systems that produce well developed somatic embryos, the presence of dormancy should be readily detectable by application of standard dormancy breaking treatments such as exposure to temperature extremes, cytokinins, giberrellins, etc.

Although grape somatic embryos are well developed, they germinate poorly. Grape seed exhibits typical dormancy which is broken by cold stratification (Flemion, 1937). Grape somatic embryos also germinate after cold treatments (Gray and Mortensen, 1986). This suggests that the somatic embryos also are dormant. Studies of ABA concentration in embryogenic cultures showed a rapid increase during embryo development, which reached a peak at maturation (Rajasekaran et al., 1982). Cold stratification of somatic embryos resulted in a rapid decrease in ABA. Exogenously supplied ABA inhibited somatic embryo germination (Gray, 1989). Because ABA is implicated as a controlling factor of dormancy in many types of seeds (Walton, 1980), it is plausible to consider that ABA also functions in grape somatic embryos. In contrast to ABA, exogenously supplied gibberellin (GA) caused grape somatic embryos to germinate and the concentration of endogenous GA-like compounds increased during cold stratification (Pearce et al., 1987). This suggests a simple endogenous control of embryogenesis and germination whereby ABA inhibits precocious germination and thus promotes normal development while GA causes germination to occur. Although this is a relatively simplistic model, it allows hypotheses concerning dormancy and germination to be formulated and tested.

Dormancy of grape somatic embryos was also affected by dehydration (Gray, 1987, 1989). As described in the previous section, it was discovered that dehydrated embryos germinated to produce plants immediately after imbibition whereas plants were not recovered from nondehydrated controls. Because the dehydrated embryos did not require a typical dormancy breaking pretreatment, they were considered to be quiescent, not dormant. Conversely, dehydration could be considered to be a dormancy breaking treatment for somatic embryos. This latter conclusion is supported by studies of somatic embryogenesis in red oak (Quercus rubra L.) and soybean. Somatic embryos of red oak germinated and produced shoots only after a period of dehydration or treatment with osmotically active sugars (Gingas and Lineberger, 1988). Likewise, only a few soybean somatic embryos from several test genotypes germinated without dehydration pretreatments (Parrott et al., 1988). After dehydration, up to 100% of the embryos germinated. There are a number of embryogenic culture systems that exhibit poor embryo germination and may possess unrecognized dormancy. Perhaps similar dehydration treatments will lead to increased germination and plant recovery rates.

A significant difference between dehydrated grape somatic embryos and somatic embryos germinated by any other means relates to the early pattern of plant growth. Only dehydrated somatic embryos germinated in a "normal" seedling-like pattern with synchronous emergence of root and shoot (Fig. 4) (Gray, 1989). This was in contrast to germination of nondehydrated embryos after treatment with dormancy breaking growth regulators where embryonic organs became abnormally enlarged and emergence of roots and shoots were nonsynchronous. These results suggest a previously unrecognized effect of dehydration on somatic embryo maturation. Dehydration of seed to storage water levels is generally considered to be possible only after maturation and the acquisition of dehydration tolerance (Adams et al., 1983). Dehydration tolerance may occur by the stabilizing effect of increased sugar concentration on cell membranes (Koster and Leopold, 1988). However, seed water content drops continuously during most stages of development as relative dry weight increases. Furthermore, developing seed is subjected to diurnal fluxes in relative water content due to daily cycles of temperature and plant metabolic activity. Gradual dehydration as well as fluctuating cycles of dehydration may be key factors involved in regulation of seed (and zygotic embryo) maturation. These types of dehydration are missing or are much less pronounced in somatic embryos due to the fully hydrated nature of the in vitro environment. Dehydration treatments that mimic the water fluxes in developing seed may lead to better somatic embryo maturation and germination.

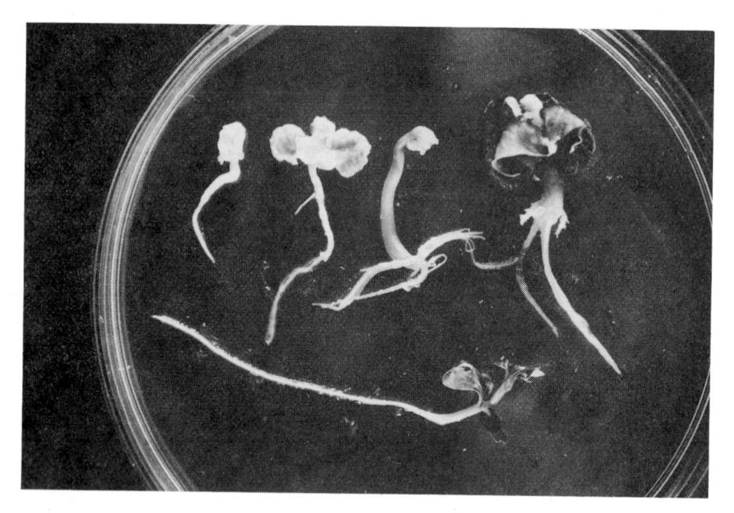

Fig. 4. Comparison of germination treatments for grape somatic embryos. From top left: control treatment; 1 uM gibberellic acid; 1 uM abscisic acid; 1uM benzyladenine. Bottom: dehydration at 70% RH for 21 days. From Gray, 1989.

## STRUCTURE OF SYNTHETIC SEED

By definition, synthetic seed will consist of a somatic embryo, but whether or not the embryo will require further processing, such as the induction of quiescence and/or addition of a protective encapsulation, will depend upon its specific application. For instance, use of naked somatic embryos plucked from callus cultures and germinated in soil plugs would be advantageous for certain ornamental crops that are now laboriously micropropagated by tissue culture. The relative ease of producing plants by somatic embryogenesis vs. micropropagation would, in itself, confer a cost advantage. Similarly, dehydrated quiescent somatic embryos without encapsulation may be useful for germplasm storage since they could be hand manipulated and carefully stored in protective containers. However, for mass propagation of field crops, a protective encapsulation will be necessary.

Encapsulations

The protective materials used to encapsulate somatic embryos are analogous to the seed coat of normal seed. Such synthetic seed coats may not only be useful for physical protection of the somatic embryo but could also carry nutrients, antibiotics, fungicides, etc. to assist in germination and plant survival. Encapsulated somatic embryos could conceivably be handled as seed using conventional planting equipment. Two types of encapsulations, hydrated and dry, have been envisioned.

For certain field crops that pass through a greenhouse transplant stage such as celery (*Apium graveolens* L.), non-quiescent somatic embryos placed in a hydrated encapsulation may be cost effective. Such an encapsulation would provide a measure of protection to the somatic embryo and would allow convenient handling as discrete planting units in the manner of seed. A hydrated encapsulation primarily composed of calcium alginate has been developed (Redenbaugh et al., 1987). Using this encapsulation, somatic embryos of several crop species have been successfully planted in soil; however, to date, plant recovery rates have been unacceptably low (Fujii et al., 1987).

The ultimate development of synthetic seed will involve use of dehydrated quiescent somatic embryos encased in dry, hardened synthetic seed coat material. This form of synthetic seed would be closely analogous to conventional seed in storage and handling characteristics and would be necessary for field planting using conventional technology. To date, only somatic embryos of carrot have been placed in a dry encapsulation (Kitto and Janick, 1985). Somatic embryos were dried en masse in a water soluble plastic resin and hardened wafers, each containing hundreds of embryos, resulted. Although some of the embryos "survived", plants were not obtained. Our increasing understanding of somatic embryo maturation and quiescence as discussed above may lead to renewed progress in development of improved encapsulations.

## PRODUCTION OF SYNTHETIC SEEDS

One advantage of somatic embryogenesis over other types of vegetative propagation is the potential ease of automation. Steps in synthetic seed production that are suitable for automation are: (i) the growth of cell cultures, (ii) sorting of embryos, and (iii) encapsulation and dehydration.

Because somatic embryos develop from cells, it is possible to scale up production by growing large liquid cultures in bioreactors. Large numbers of somatic embryos then can be induced to form once a suitable culture mass is reached. This is illustrated in Fig. 5 where results of an experiment with orchardgrass compared the effects of dicamba (a synthetic auxin) and casein hydrolysate on embryogenesis and showed over 2,300 somatic embryos produced from one gram of inoculum in six weeks. There are many similar examples of embryogenic suspension cultures, but few attempts to scale up production. Problems have arisen due to inherent difficulties of producing well developed embryos in liquid culture medium. For example, plant recovery from suspension culture of alfalfa was less than from callus and decreased still more when scale up to larger a bioreactor was attempted (Stuart et al., 1987). Improvements in bioreactor design are needed as well as a better understanding of somatic embryo development under liquid culture conditions in order to make progress in large scale production.

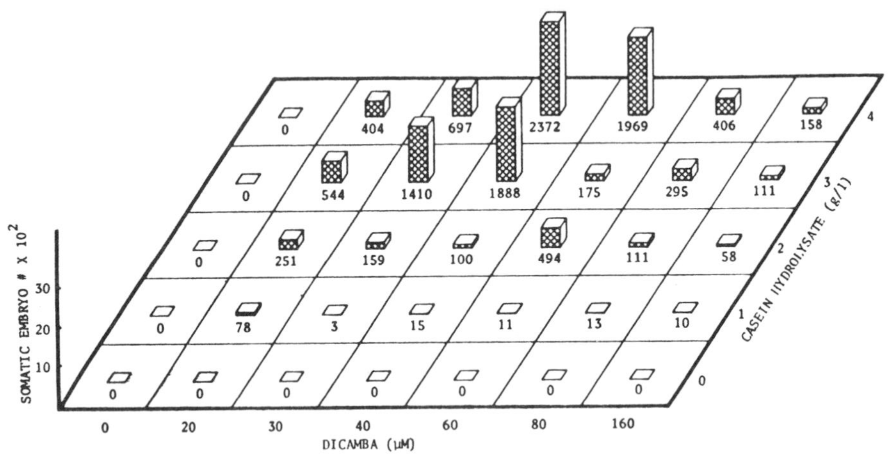


Fig. 5. Optimization of dicamba and casein hydrolysate levels for somatic embryogenesis from cell suspension cultures of orchardgrass. From Gray and Conger, 1985c.

Development of sorting equipment should be facilitated since somatic embryos fit within a discrete size and shape range. Recently, a machine that could sort several types of tissue culture propagules was developed (Levin et al., 1988). However, sufficient details were not given to gauge its use for sorting embryos.

Beyond somatic embryo production and sorting, methods to automate dehydration and encapsulation of embryos must be developed. It is possible that dehydration and encapsulation may be accomplished simultaneously by use of an osmotically active synthetic seed coat. Hydrated embryos placed in such a material would begin to lose water during the capsule hardening process. Final water content could then be controlled by composition of the encapsulation.

## CONCLUSIONS

Synthetic seed technology has potential application for a variety of crops. Use of synthetic seed must either reduce production costs or increase crop value in order to be effective. The relative benefit gained, when weighed against development costs, will determine whether its use is justified for a given crop. However, it is currently difficult to gauge the relative cost vs. benefit of synthetic seed due to the lack of commercial prototype systems. Additional research and development is needed in order to determine commercial feasibility. To date, promising progress has been made in producing mature, quiescent somatic embryos that functionally mimic seed. However, quiescent encapsulated somatic embryos with high soil germination rates have not yet been achieved for any crop. This is the challenge facing us as we attempt to advance the state of synthetic seed technology.

## REFERENCES


Adams, C. A., Fjerstad, M. C., and Rinne, R. W., 1983, Characteristics of soybean seed maturation: necessity for slow drying, Crop Sci. 23:265.

Ammirato, P. V., 1983, The regulation of somatic embryo development in plant cell cultures: suspension culture techniques and hormone requirements. Bio/Technology 1:68.

Ammirato, P. V., 1987, Organizational events during somatic embryogenesis in: "Plant Tissue and Cell Culture," Alan Liss, New York.

Barton, L. V., 1961, "Seed Preservation and Longevity," Leonard Hill, London.

Bewley, J. D., and Black, M., 1985, "Seeds: Physiology of Development and Germination," Plenum, New York.

Becwar, M. R., Noland, T. L., and Wycoff, J. L., 1989, Maturation, germination, and conversion of Norway spruce (Picea abies L.) somatic embryos to plants, In Vitro Cell. Dev. Biol. 25:575.

Carman J. G., 1988, Improved somatic embryogenesis in wheat by partial simulation of the in-ovulo oxygen, growth-regulator and desiccation environments, Planta 175:417.

Conger, B. V., Hovanesian, J. C., Trigiano, R. N.,and Gray, D. J., 1989, Somatic embryo ontogeny in suspension cultures of orchardgrass, Crop Sci. 29:448.

Finer, J. J., 1988, Plant regeneration from somatic embryogenic cultures of cotton (Gossypium hirsutum L.), Plant Cell Rpts. 7:399.

Flemion, F., 1937, After-ripening at 5C favours germination of grape seeds, Contr. Boyce Thompson Inst. 9:7.

Fujii, J. A., Slade, D. T., Redenbaugh, K., and Walker, K. A., 1987, Artificial seeds for plant propagation, Trends Biotechnol. 5:335.

Gingas, V. M., and Lineberger, R. D., Plantlet regeneration from asexual embryos of Quercus rubra L., HortScience 23:786.

Gray, D. J., 1987a, Effects of dehydration and other environmental factors on dormancy in grape somatic embryos, HortScience 22:1118.

Gray, D. J., 1987b, Introduction to the symposium, in: Proc. Symp. Synthetic Seed Technology for the Mass Cloning of Crop Plants: Problems and Perspectives, HortScience 22:796.

Gray, D. J., 1987c, Quiescence in monocotyledonous and dicotyledonous somatic embryos induced by dehydration, in: Proc. Symp. Synthetic Seed Technology for the Mass Cloning of Crop Plants; Problems and Perspectives, HortScience 22:810.

Gray, D. J., 1989, Effects of dehydration and exogenous growth regulators on dormancy, quiescence and germination of grape somatic embryos, In Vitro Cell. Dev. Biol. (in press).
Gray, D. J., and Conger, B. V., 1985a, Somatic embryo ontogeny in tissue cultures of orchardgrass, in: "Tissue Culture in Forestry and Agriculture," R. R. Henke, K. W. Hughes, M. J. Constantin, and A. Hollaender, eds., Plenum, New York.
Gray, D. J., and Conger, B. V., 1985b, Quiescence in somatic embryos of orchardgrass (Dactylis glomerata) induced by desiccation, Amer. J. Bot. 72:816.
Gray, D. J., and Conger, B. V., 1985c, Influence of dicamba and casein hydrolysate on somatic embryo number and culture quality in cell suspensions of Dactylis glomerata (Gramineae), Plant Cell Tissue Organ Cult. 4:123.
Gray, D. J., Conger, B. V., and Hanning, G. E., 1984, Somatic embryogenesis in suspension and suspension-derived callus cultures of Dactylis glomerata, Protoplasma, 122:196.
Gray, D. J., Conger, B. V., and Songstad, D. D., 1987, Desiccated quiescent somatic embryos of orchardgrass for use as synthetic seeds, In Vitro Cell. Dev. Biol. 23:29.
Gray, D. J., and Mortensen, J. A., 1987, Initiation and maintenance of long term somatic embryogenesis from anthers and ovaries of Vitis longii 'Microsperma', Plant Cell Tissue Organ Cult. 9:73.
Gupta, P. K., and Durzan, D. J., 1987, Biotechnology of somatic polyembryogenesis and plantlet regeneration in loblolly pine, Bio/Technology 5:147.
Haccius, B., 1978, Question of unicellular origin of non-zygotic embryos in callus cultures. Phytomorphology 28:74.
Kamo, K. K., Becwar, M. R., and Hodges, T. K., 1985, Regeneration of Zea mays L. from embryogenic callus, Bot. Gaz. 146:327.
Kitto, S. L., and Janick, J., 1985, Hardening treatments increase survival of synthetically coated asexual embryos of carrot, J. Amer. Soc. Hort. Sci. 110:282.
Koster, K. L., and Leopold, A. C., 1988, Sugars and desiccation tolerance in seeds, Plant Physiol. 88:829.
Levin, R., Gaba, V., Tal, B., Hirsch, S., DeNola, D., and Vasil, I. K., 1988, Automated plant tissue culture for mass propagation, Bio/Technology 6:1035.
McKersie, B. D., Bowley, S. R., Senaratna, T., Brown, D. C. W., and Bewley, J. D., 1988, Application of artificial seed technology in the production of alfalfa (Medicago sativa L.), In Vitro Cell. Dev. Biol. 24:71.
Nitzsche, W., 1978, Erhultung der Lebensfahigkeit in getrocknetem Kallus, Z. Pflanzenphysiol. 87:469.
Norstog, K., 1967, Studies on the survival of very small barley embryos in culture, Bull. Torrey Bot. Club 94:223.
Owen, E. B., 1956, "The Storage of Seeds for Maintenance of Viability," Commonw. Agr. Bur., Bucks, England.
Parrott, W. A., Dryden, G., Vogt, S., Hildebrand, D. F., Collins, G. B., and Williams, E. G., 1988, Optimization of somatic embryogenesis and embryo germination in soybean, In Vitro Cell. Dev. Biol. 24:817.
Pearce, D., Pharis, R. P., Rajasekaran, K., and Mullins, M. G., 1987, Effects of chilling and ABA on [$^3$H]Gibberellin $A_4$ metabolism in somatic embryos of grape (Vitis vinifera L. x V. rupestris Scheele), Plant Physiol. 80:381.
Rajasekaran, K., Vine, J., and Mullins, M. G., 1982, Dormancy in somatic embryos and seeds of Vitis: changes in endogenous

abscisic acid during embryogeny and germination, Planta 154:139.
Ranch, J. P., Oglesby, L., and Zielinsky, A. C., 1985, Plant regeneration from embryo-derived tissue cultures of soybean, In Vitro Cell. Dev. Biol. 21:653.
Redenbaugh, K., Slade, D., Viss, P., and Fujii, J. A., 1987, Encapsulation of somatic embryos in synthetic seed coats, in: Proc. Symp. Synthetic Seed Technol. for the Mass Cloning of Crop Plants: Problems and Perspectives, HortScience 22:803.
Senaratna, T., McKersie, B. D., and Brown, D. C. W., 1989, Artificial seeds: desiccated somatic embryos. In Vitro Cell. Dev. Biol. 25:39.
Stuart, D. A., and Strickland, S. G., 1984, Somatic embryogenesis from cell cultures of Medicago sativa L.: II. The interaction of amino acids with ammonium. Plant Sci. Lett. 34:175.
Stuart, D. A., Strickland, S. G., and Walker, K. A., 1987, Bioreactor production of alfalfa somatic embryos, in: Proc. Symp. Synthetic Seed Technol. for the Mass Cloning of Crop Plants: Problems and Perspectives, HortScience 22:800.
von Arnold, S., and Hakman, I., 1988, Regulation of somatic embryo development in Picea abies by abscisic acid (ABA), J. Plant Physiol. 132:164.
Walker, K. A., and Sato, S. J., 1981, Morphogenesis in callus tissue of Medicago sativa: the role of ammonium ion in somatic embryogenesis, Plant Cell Tissue Organ Culture 1:109.
Walton, D. C., 1980, Biochemistry and physiology of abscisic acid, Ann. Rev. Plant Physiol. 31:453.

# SYNTHESIS OF ACID-SOLUBLE, ABA-INDUCIBLE PROTEINS IN WHEAT EMBRYOS FROM DORMANT GRAIN

M. Walker-Simmons, K.E. Crane and S. Yao

USDA-ARS & Dept. of Agronomy & Soils
Washington State University
Pullman, WA 99164-6420

## INTRODUCTION

When dormant seeds are placed in water the seeds do not germinate even though environmental conditions are optimum for germination. What regulates dormancy in mature seeds is not yet known, but there is evidence that the plant hormone, ABA (abscisic acid) plays an important role, as illustrated in the following observations.

Plants incapable of producing ABA during seed development yield nondormant seeds. Arabidopsis mutants deficient in ABA production or in responsiveness to ABA during seed development yield seeds which lack dormancy and precociously germinate (Karssen et al., 1983). ABA-deficient mutants in maize, potato and tomato also precociously germinate.

Application of ABA to isolated embryos of immature wheat grains inhibits premature germination and stimulates the synthesis of late maturation proteins (Quatrano et al., 1983). Such responses to applied ABA may mimic the effect of endogenous ABA during grain development. Endogenous ABA levels increase during mid-maturation, and it is during this increase in ABA levels that seed dormancy is acquired.

Most seeds lose embryonic responsiveness to ABA as seeds mature. With maturation the concentration of exogenous ABA sufficient to block embryonic germination becomes abnormally high. In wheat, however, embryonic responsiveness to ABA only declines during seed maturation if grain dormancy decreases (Walker-Simmons, 1987; Morris et al., 1989). This results in ABA being relatively ineffective in blocking germination in embryos from mature, nondormant grain, while embryos from dormant grain remain highly responsive to ABA. For example, exogenous ABA is 10-100 fold more effective in suppressing germination of embryos from dormant compared to nondormant wheat grain (Walker-Simmons, 1988).

*Recent Advances in the Development and Germination of Seeds*
Edited by R.B. Taylorson
Plenum Press, New York

Having established that ABA is involved in dormancy regulation, questions remain as to the specific biochemical processes that respond to ABA. In this paper we examine the effects of ABA on new protein synthesis in the embryos from dormant, highly ABA-responsive grain. Specifically, we have determined effects of ABA on synthesis of acid-soluble proteins. Quatrano et al. (1983) demonstrated that a group of acid-soluble proteins could be prematurely induced by ABA in immature embryos. The same set of embryonic-specific proteins are normally synthesized in late maturation. mRNA for these proteins is present in mature dry grain but disappears during seed germination (Quatrano et al., 1983).

In this report we show that ABA-treated embryos from mature dormant grain continuously synthesized a large number of acid-soluble proteins for over 48 h post-imbibition. Control embryos incubated in water did not. First we demonstrate the responsiveness of the embryos to ABA as measured by the effects of increasing concentrations of ABA on synthesis of the acid-soluble proteins. Next we show the extent of synthesis of these proteins in embryos incubated in water or ABA for up to 48 h. Finally we present the results of two-dimensional protein analyses indicating that synthesis of a large number of acid-soluble proteins was regulated by ABA in embryos from dormant grain.

## TECHNIQUES EMPLOYED

Wheat (Triticum aestivum cultivar Brevor) was grown at Spillman Agronomy Farm near Pullman, WA in 1987. Grains were harvested at maturity and stored at -20°C. For germination assays grains were incubated in water for 2 h. Embryos were then dissected from the grains and incubated in water or 5 μM (±) ABA on blotting paper in petri dishes at 30°C. Embryonic germination was defined as an extension of the root or shoot axes.

For in vivo labeling embryos were incubated in distilled water (±)cis-trans ABA (Sigma Chemical Co.) at 30°C. One or 4 h prior to the end of the incubation period, embryos were transferred to fresh solutions containing 0.05 μCi/μl [$^{35}$S]methionine. [$^{35}$S]Methionine (1100 Ci/mmol) was from NEN. Ten embryos were incubated in 100 μl [$^{35}$S]methionine. At the end of the labeling period embryos were washed twice with 1 mM nonradioactive methionine and frozen in liquid nitrogen. Acid-soluble proteins were extracted by grinding the tissue in 50 mM HCl (Quatrano et., 1983). The extract was spun in a microfuge for 10 min, and then the supernatants were neutralized with NaOH and dried.

One-dimensional SDS-PAGE was performed as described by Laemmli (1970) with 12% (w/v) polyacrylamide. For two-dimensional electrophoresis acid-soluble proteins were solubilized in lysis buffer containing: 9 M urea, 2% Nonidet P-40, 1.6% ampholytes pH 5-7, 0.4% ampholytes, pH 3-10, and 5% mercaptoethanol. Two-dimensional PAGE gels were completed according to O'Farrell (1975) with an overlay buffer consisting of: 9 M urea, 5% Nonidet P-40, 0.8% ampholytes, pH 5-7, 0.2% ampholytes pH 3-10 and 5% mercaptoethanol. Ampholytes were from BioRad. Completed gels were dried and analyzed by fluorography. ABA was extracted and measured by indirect ELISA utilizing a monoclonal antibody as previously described (Walker-Simmons, 1987).

## EFFECTS OF ABA ON EMBRYONIC GERMINATION AND ACID-SOLUBLE PROTEIN SYNTHESIS

Embryos from dormant grain are highly responsive to ABA. This is demonstrated in Figure 1 by the strong inhibitory effect ABA has on embryonic germination. All the embryos placed in water germinated within 48 h, but embryonic germination was considerably slowed by the addition of ABA. Germination of embryos from nondormant mature grain is not affected by 5 µM ABA (Walker-Simmons, 1988).

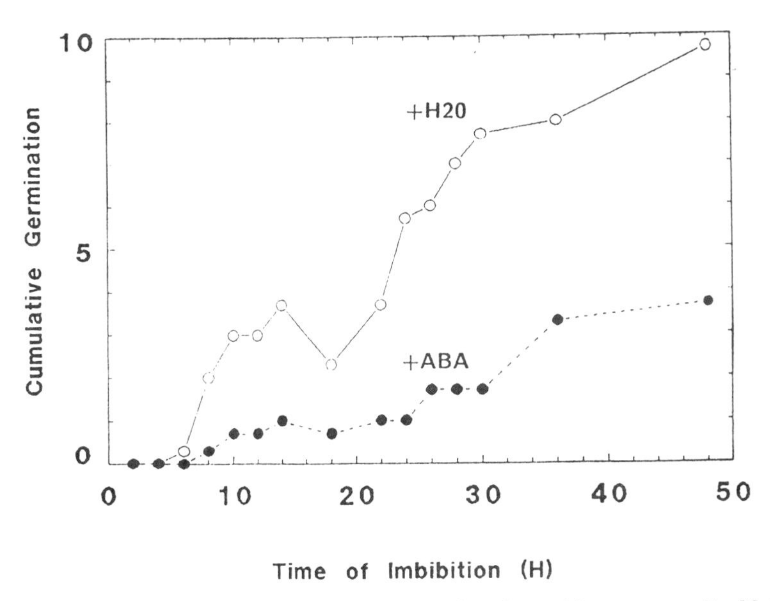


Fig. 1. Cumulative germination of embryos incubated in water ± 5 µM ABA. Data are the mean of 3 replicate assays of 10 embryos each.

The effect of ABA on acid-soluble protein synthesis in the embryos from dormant grain was measured using in vivo labeling techniques (Fig. 2); embryos were incubated in a wide range of ABA concentrations and synthesis at 8 h post-imbibition was measured. Significant acid-soluble protein synthesis was observed when the ABA concentration was raised to 0.5 µM. Maximum synthesis was observed at approximately 5 µM. Synthesis of many acid-soluble proteins with a wide range of molecular weights was enhanced with increasing ABA concentration.

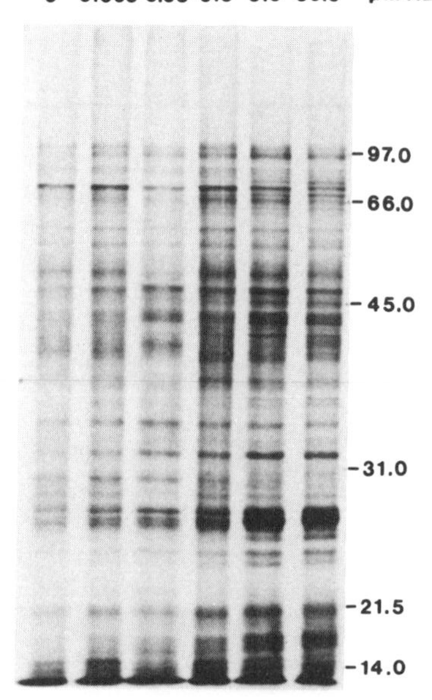


Figure 2. Dose response of ABA-responsive proteins. Embryos were incubated in water containing varying amounts of (±)ABA for 8 h. During the last h of the incubation embryos were pulse labeled with [$^{35}$S]methionine. Acid-soluble proteins were separated by SDS-PAGE and the fluorogram is shown. Bars indicate molecular weight markers.

The amount of synthesis of the ABA-inducible, acid-soluble proteins was measured during the first 8 h after imbibition in water or ABA. Synthesis was almost undetectable at 2 h (data not shown). As shown in Fig. 3 synthesis started increasing by 4 h with slight enhancement in the ABA-treated embryos. ABA enhancement of acid-soluble protein synthesis was readily apparent by 6 h and increased even more at 8 h.

Measurement for 16-48 h with 4 h pulse-labeling (Fig 4) showed that the ABA-treated embryos continuously synthesized the set of acid-soluble proteins. Far less synthesis of the acid-soluble proteins occurred in the control embryos and synthesis declined with incubation time.

Since long-term incubation in ABA compared to water has such a significant effect on acid-soluble protein synthesis it is of interest to determine whether incubation of the embryos in water ± 5 uM ABA affects endogenous ABA levels. Endogenous ABA levels were measured in embryos from the same samples as presented in Figure 4. Endogenous ABA concentration (Table 1) of the embryos incubated in water dropped from 4 μM to around 0.7 μM upon imbibition of water and gradually declined with long-term incubation. Endogenous ABA levels were higher in the embryos incubated in 5 μM ABA. Eight hours after imbibition the estimated endogenous ABA concentration was around 2 μM.

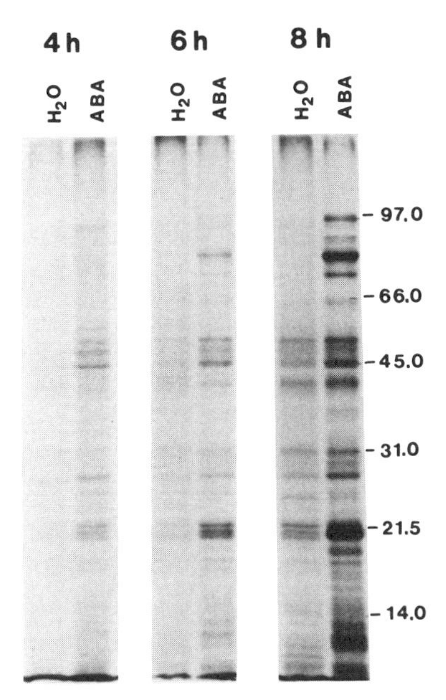


Figure 3. Time course of acid-soluble protein synthesis. Embryos were incubated in water ± 5 µM ABA. One hour prior to the times indicated embryos were transferred to fresh media containing [$^{35}$S]methionine. Acid-soluble proteins were separated by SDS-PAGE and the fluorogram of the gel is shown.

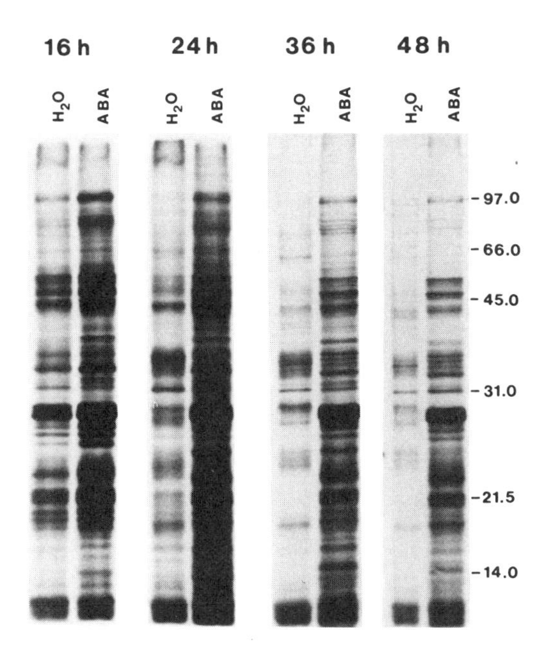


Figure 4. Long-term time course of acid-soluble protein synthesis. Embryos were incubated in water ± 5 µM ABA. Four hours prior to the times indicated embryos were transferred to fresh media containing [$^{35}$S]methionine.

Table 1. Endogenous ABA levels in embryos during long-term incubation in water $\pm$ 5 µM ABA

| Time of Incubation | ABA Levels (µM)* | |
|---|---|---|
| (h) | Water | ABA |
| 0 | 4.0 | |
| 8 | 0.65 | 2.0 |
| 16 | 0.76 | 1.2 |
| 24 | 0.55 | 1.7 |
| 36 | 0.57 | 0.92 |
| 48 | 0.59 | 1.1 |

*(+)ABA levels measured by immunoassay. ABA concentration was calculated based on corrections for $\pm$ ABA and assuming that all the water in the tissue was available to ABA.

## TWO-DIMENSIONAL GEL ANALYSES OF ACID-SOLUBLE PROTEINS

The one-dimensional gels indicate that synthesis of a large number of proteins are enhanced by ABA. In order to better evaluate how many proteins are affected by the hormone, two-dimensional analyses were conducted. A comparison of the ABA-treated and control embryos was made at 6 h post-imbibition. Results show that synthesis of many new polypeptides of molecular weights below 45 kD is induced or enhanced by ABA. Many of these polypeptides migrate to the basic region. Very similar results to those presented in Figure 5 were obtained with two-dimensional separations of the acid-soluble proteins at 8 and 12 h post imbibition (data not shown). Just as with the 6 h time point, continued incubation of the axes in 5 µM ABA at 8 and 12 h resulted in the induction or enhancement of a very large set of ABA-responsive proteins.

## COMPARISON TO FIELD CONDITIONS

In this work exogenously applied ABA (5 µM) caused the endogenous ABA to increase two-fold (Table 1). Enhanced synthesis of acid-soluble proteins resulted from the increase in ABA (Fig. 1). Endogenous ABA levels can increase two-fold without the application of external ABA. We have found that embryonic ABA levels can increase over two-fold when whole grains are wetted and then slowly dried (Walker-Simmons et al., 1989). Under these conditions actual ABA concentrations would increase even more due to water loss. Such wetting and slow drying situations could occur under field conditions when a light rain occurs followed by warm dry weather. Under such weather conditions ABA levels could increase and cause many changes including the enhancement of acid soluble protein synthesis. The final result could be suppression of embryonic germination.

How these ABA-responsive acid-soluble proteins are involved in ABA inhibition of embryonic germination is not known. The acid-soluble proteins are synthesized during seed development (Quatrano et al., 1983). At that time endogenous ABA levels peak and seeds undergo desiccation which produces quiescent seeds. During seed maturation embryos are very responsive to ABA as measured by the capability of ABA to block embryonic germination. As grains mature and increase in germinability, embryonic responsiveness to ABA is lost. However, embryos from grains which retain dormancy upon maturation continue to be highly responsive to ABA (Walker-Simmons, 1987), and it is in these embryos that we observe that ABA continues to block embryonic germination and cause continued synthesis of the acid-soluble proteins.

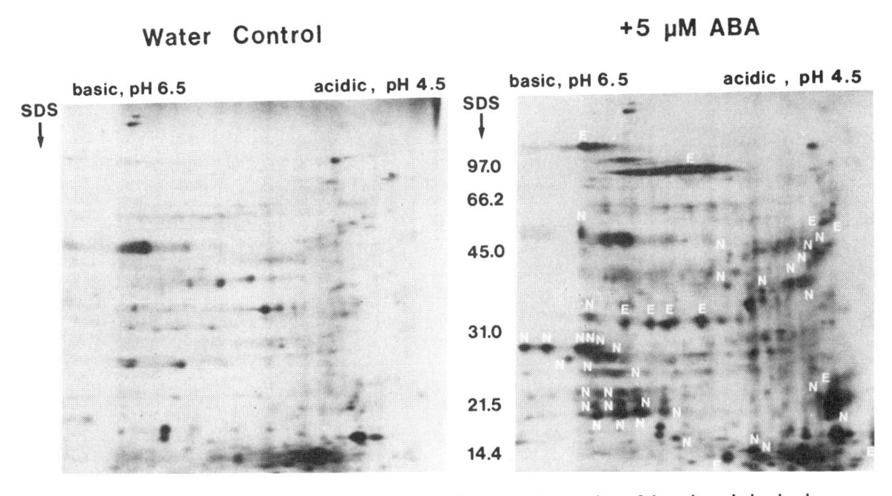


Figure 5. Two-dimensional gel electrophoresis of in vivo labeled proteins. Embryos were incubated with or without 5 μM ABA for 6 h. One h prior to the end of the incubation embryos were transferred to water ± ABA containing [$^{35}$S]methionine (50 μCi/ml). Fluorograms of the gels are shown with the molecular weight markers indicated. (N) new synthesis, (E) enhanced synthesis.

Recent reports have shown that ABA can induce the same proteins both during embryonic development as well as in water-stressed plant leaves (Gomez et al., 1988; Mundy and Chua, 1988; Chandler et al., 1988; Robertson et al., 1989). The ABA-inducible proteins may have a role in protecting developing seed embryos during grain desiccation and in leaves during water stress, as well as in regulating seed dormancy.

A large number of ABA-inducible proteins in the barley aleurone have been found to be heat stable (Jacobsen, 1988) probably because the proteins contain large amounts of hydrophilic amino acids. We have found that embryos from dormant grain also have a prolonged capacity for synthesis of the ABA-inducible, heat-stable proteins (Ried and Walker-Simmons, 1989). Mature dry embryonic axes from dormant or nondormant grain were incubated in water under optimum germination conditions. Embryonic axes from dormant grain exhibited prolonged synthesis of the heat stable proteins, while synthesis in the non-dormant axes decreased within a few hours. Many of the heat-stable proteins are also acid-soluble (J.L. Ried, unpublished data). Work is now underway in our laboratory to purify and characterize the ABA-inducible proteins associated with seed dormancy.

## CONCLUSIONS

In this report we have shown that embryos from dormant grain are highly responsive to exogenous ABA as measured by the inhibitory effects of ABA on embryonic germination. Incubation in 5 μM ABA blocks embryonic germination and causes a two-fold increase in endogenous ABA levels. Specific biochemical changes result from the ABA-treatment. ABA-treated embryos from mature dormant grain continuously synthesize acid-soluble proteins for over 48 h post-imbibition. Control embryos incubated in water do not. Synthesis of the acid-soluble proteins increases as the concentration of ABA is raised. Two-dimensional protein analyses indicate that synthesis of a large number of acid-soluble proteins is regulated by ABA in embryos from dormant grain.

## REFERENCES


Chandler, P.M., Walker-Simmons, M., King, R.W., Crouch, M., and Close, T.J., 1988, Expression of ABA-inducible genes in water stressed cereal seedlings, J. Cell. Biochem. (S 12C) p. 143.

Close, T.J., Kortt, A.A., and Chandler, P.M., 1989, A cDNA-based comparison of dehydration-induced proteins (dehydrins) in barley and corn, Plant Molec. Biol. 13:95.

Gomez, J., Sanchez-Martinez, D., Stiefel, V., Rigau, J., Puigdomenech, P., and Pages, M., 1988, A gene induced by the plant hormone abscisic acid in response to water stress encodes a glycine-rich protein, Nature 334: 262.

Jacobsen, J.V.,1988, ABA-Induced proteins in barley aleurone, 13th International Conference on Plant Growth Substances, Calgary, Canada. (Abstract).

Karssen, C.M., Brinkhorst-Van der Swan, D.L.C., Breekland, A.E. and Koorneef, M., 1983, Induction of dormancy during seed development by endogenous abscisic acid: studies on abscisic acid deficient genotypes of Arabidopsis thaliana (L.) Heynh, Planta 157: 158.

Laemmli, U.K., 1970, Cleavage of structural proteins during the assembly of the head of bacteriophage $T_4$, Nature 227: 680.

Morris, C.F., Moffatt, J.M., Sears, R.G., and Paulsen, G.M., 1989, Seed dormancy and responses of caryopses, embryos, and calli to abscisic acid in wheat, Plant Physiol. 90: 643.

Mundy, J. and Chua, N.-H., 1988, Abscisic acid and water-stress induce the expression of a novel rice gene, EMBO 7: 2279.

O'Farrell, P.H., 1975, High resolution two-dimensional electrophoresis of proteins, J. Biol. Chem. 250: 4007.

Quatrano, R.S., Ballo, B.L., Williamson, J.D., Hamblin, M.T., and Mansfield, M., 1983, ABA controlled expression of embryo-specific genes during wheat grain development, in: "Plant Molecular Biology", R. Goldburg, ed., Alan R. Liss, Inc., New York.

Ried, J.L., Walker-Simmons, M.K., 1989, Prolonged synthesis of ABA-inducible, heat stable proteins in embryonic axes from dormant wheat grain, Plant Physiol. 89(S): p. 171.

Robertson, M., Walker-Simmons, M., Munro, D., and Hill, R.D., 1989, The induction of alpha-amylase inhibitor synthesis in barley embryos and young seedlings by ABA and dehydration stress, Plant Physiol. In press.

Walker-Simmons, M., 1987, Embryonic ABA levels and sensitivity in developing wheat embryos of sprouting resistant and susceptible cultivars, Plant Physiol. 84: 61.

Walker-Simmons, M., 1988, Enhancement of ABA responsiveness in wheat embryos by high temperature, Plant, Cell and Environ. 11: 769.

Walker-Simmons, M., Chandler, P.M., Sesing, J., Ried, J.L., and Morris, C.F., 1989, Comparison of responses to ABA in embryos and dehydrating seedlings of sprouting-resistant and susceptible wheat cultivars, in: "Fifth International Symposium on Pre-Harvest Sprouting in Cereals", K. Ringlund, ed., Westview Press, Boulder, CO, In press.

# HORMONES, GENETIC MUTANTS AND SEED DEVELOPMENT

J. D. Smith[1], Franklin Fong[1], C. W. Magill[2], B. Greg Cobb[3] and D. G. Bai[1]

Department of Soil and Crop Sciences[1], Department of Plant Pathology[2] and Microbiology and Department of Horticultural Sciences[2], Texas A&M University, College Station, Texas

## INTRODUCTION

Genes unquestionably regulate phytohormone synthesis, and phytohormones are presumed to regulate the expression of specific genes at specific times in specific tissues. The regulatory functions of phytohormones appear to be concentration-dependent, and their rates of accumulation and/or degradation appear to be stimulated by changes in external or internal environmental conditions. To further confuse the situation, levels may change as a function of tissue age, and, since they can be translocated among tissues, their sites of synthesis.

Phinney and his colleagues (Phinney et al., 1982; Spray et al., 1984; Fujioka et al., 1988) have utilized andromonoecious dwarf mutants (_d1_, _d2_, _d3_, _d5_) in _Zea mays_ to elucidate the biosynthetic pathway of gibberellins, identify the specific enzymatic steps blocked by each mutant and, for _d1_, determine the biochemical reaction controlled by the normal allele. We have studied the viviparous mutants of maize in the belief that an understanding of their primary functions would provide information on the role of abscisic acid (ABA) in seed development regulation. Our discussion in this chapter will focus on our interpretations of the results of these investigations.

## THE VIVIPAROUS MUTANTS OF MAIZE

The viviparous mutants are all recessive and most have striking pleiotropic effects which facilitated their early discovery. The mutants we have worked with extensively include _vp1_, _vp2_, _vp5_, _vp7_(= _ps_), _vp8_, _vp9_(= _y7_), _w3_, _y3_ and _vp10_ (tentative designation). With the exception of _vp10_, they had been identified prior to 1955, and Robertson (1955; 1975) has described their pleiotropic effects in detail. Coe (1988) has presented additional information concerning these mutants including chromosome map locations and original references. The _vp10_ mutant was one of a number of ethylmethane sulfonate induced viviparous mutations generously given to us by M. G. Neuffer, Univ. of Missouri.

We have separated the viviparous mutants into three classes, based on their primary functions and genetic lesion sites. The only class I mutant thus far identified is _vp1_. It has normal carotenoids (Fong et al., 1983) and levels of ABA (Smith et al., 1978), it is relatively insensitive to exogenous ABA (Smith et al., 1978) and homozygous kernels carrying the necessary genes for anthocyanin accumulation in aleurone tissue are colorless (Robertson, 1955). McCarty et al. (1989) cloned the wild type _Vp1_ allele and determined that it encodes a 2500 b mRNA expressed in embryo and endosperm tissues of developing kernels. The

*Recent Advances in the Development and Germination of Seeds*
Edited by R.B. Taylorson
Plenum Press, New York

protein (VP) coded by this gene has been isolated (D.R. McCarty, Univ. of Florida, personal communication).

The class II mutants (vp2, vp5, vp7, vp9, w3, y3, y9) block specific steps in the central pathway for carotenoid biosynthesis (Fig. 1). Embryos of these mutants have reduced levels of ABA and are viviparous. Growth of cultured embryos is inhibited by exogenous ABA (Smith et al., 1978), and class II mutants, except y3 and y9, produce lethal, albino seedlings (Robertson, 1975) due to photooxidation of chlorophyll in the absence of carotenoids. The y3 mutant appears to be temperature sensitive, producing green/albino sectorial plants, while y9 appears to be 'leaky', producing pale green seedlings. Some plants of both mutants survive to maturity.

Class III mutants have normal carotenoids but reduced levels of ABA, and we believe they represent lesions in the biosynthetic pathway between an unidentified carotenoid precursor and ABA. Both vp8 and vp10 appear to belong to this class, although vp10 inclusion is based on preliminary data and is tentative. Rescued embryos of vp8 produce small seedlings with narrow, pointed leaves which grow more slowly with age and eventually die. They appear to accumulate a weak growth inhibitor, which suggests that the genetic lesion is quite close to ABA. Efforts to rescue vp10 seedlings and grow them to maturity are in progress. Typically, seedling leaves do not unfold, but, when mechanically separated, seedling growth rate is comparable to vp1. Leaf tips tend to be necrotic, and seedlings appear to wilt more readily than vp1.

## Primary Functions of Wild Type Alleles

Table 1 summarizes the primary functions of the wild type alleles at these loci as we interpret the genetic and biochemical data. Our designation of the Vp1 gene product as a regulatory protein is based upon: (1) the insensitivity of vp1 embryos and seedlings to exogenous ABA; (2) the absolute requirement of the Vp1 product for induction of embryo dormancy, suppression of lipase activity in aleurone and scutellum tissues (Bai, 1989) and induction of anthocyanin synthesis in aleurone tissue; and (3) the recent isolation of the protein coded by Vp1.

The identification of enzymes associated with the Class II mutants is based upon TLC/MS determination of intermediate carotenoids which accumulate when specific mutants are homozygous. With the exception of y3 (temperature sensitive) and y9 (leaky), none of the carotenoids beyond the ones which accumulated in specific homozygous mutants were found. Our assignment of trivial names and subunit designations is

Table 1. Primary functions of polypeptide products coded by the wild type alleles of the viviparous mutants of maize. Trivial names assigned to specific enzymes and A or B subunit designations are arbitrary.

| Wild type allele | Primary function of gene product |
|---|---|
| Vp1 | Regulatory protein (VP) |
| Vp2 | Phytoene dehydrogenase; subunit A |
| Vp5 | Phytoene dehydrogenase; subunit B |
| Vp7 (= Ps) | Beta-cyclase II |
| Vp8 | Unidentified enzyme between carotenoid precursor and ABA |
| Vp9 (= y7) | Zeta-carotene dehydrogenase; subunit A |
| Vp10 | Unidentified enzyme between carotenoid precursor and vp8 |
| W3 | Phytofluene dehydrogenase |
| Y3 | Alpha- and Beta-carotene hydroxylase |
| Y9 | Zeta-carotene dehydrogenase; subunit B |

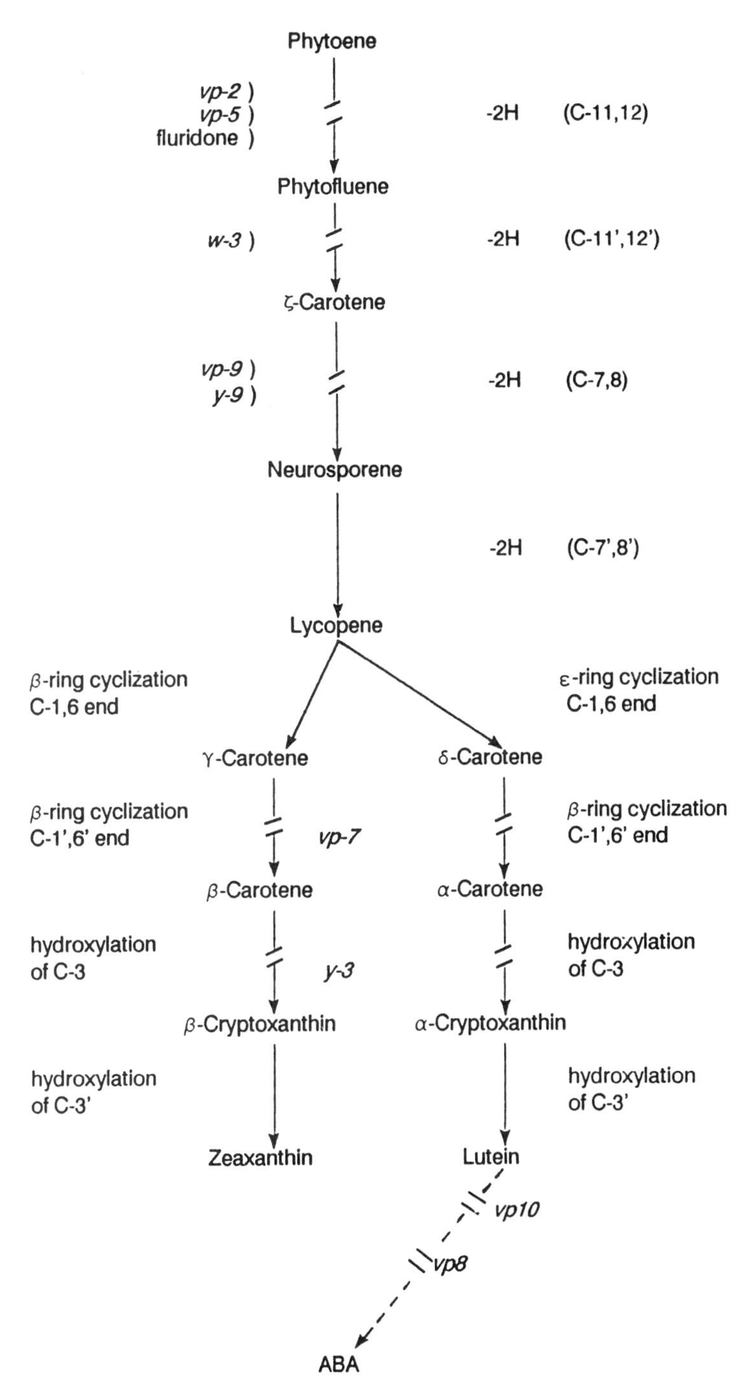


Fig. 1. Primary functions of ABA-deficient viviparous mutants. Solid arrows (————>) indicate verified pathways. Dashed arrow (– – – –>) indicates postulated pathway.

descriptive but arbitrary. The fact that different mutations block specific desaturation reactions indicates that different enzymes catalyze these desaturations. Similarly, when two non-complementing mutations both block a single desaturation step, this suggests that the holo-enzyme consists of at least two subunits. We designated vp7 as a beta-cyclase II mutation because gamma- and delta-carotene accumulate, but no beta- or alpha-carotene, or subsequent dicyclic carotenoids, are found.

Although class III mutants have reduced ABA, we have not yet identified intermediate metabolites which accumulate in vp8 and vp10. Thus, presumed functions (Table 1) and genetic lesion sites (Fig. 1) are based on interpretation of genetic and morphological data.

SOURCES OF ABA IN MAIZE EMBRYOS

ABA levels in normal maize embryos vary with age. They tend to be low (ca. 0.15 pmol ABA·embryo$^{-1}$) until about 15 days after pollination (DAP) and increase rapidly until about 26 DAP (ca. 14.0 pmol ABA·embryo$^{-1}$). ABA levels remain high for about 10 days, after which they decline to relatively low levels in fully mature kernels. Similar age-dependent variation in ABA levels occurs in viviparous mutant embryos, but there is always significantly less ABA in homozygous mutant embryos, except vp1, than in wild type embryos segregating on the same ears.

These observations are consistent with the possibility that more than one pathway for ABA synthesis operates in maize embryos. However, since homozygous mutant kernels are produced by self pollinating heterozygous normal plants, it is also possible that some of the ABA found in developing embryos is translocated from the maternal plant, as Karssen, et al. (1983) reported occurs in *Arabidopsis thaliana*.

The latter possibility was testable. Fluridone (1-methyl-3-phenyl-5-(3-[trifluoromethyl]phenyl)-4-(1H)-pyridinone) inhibits carotenoid synthesis by blocking phytoene desaturation (Bartels and Watson, 1978). When applied to developing kernels before 10 DAP, fluridone treated kernels, and all class II mutant kernels except y3 and y9, contain no detectable xanthophylls, and they are usually all viviparous. When fluridone is added to the medium for in vitro kernel culture (Gengenbach, 1977; Cobb and Hannah, 1983), xanthophyll synthesis is completely inhibited in the kernels and in the cob sector of the kernel blocks after they are placed in culture (Hole, 1988; Hole et al., 1989). Kernels removed from the plant at 3-5 DAP and cultured for a period of time are essentially free of inputs from the maternal plant.

Data showing relative amounts of ABA recovered from 15 DAP embryos grown in vivo and in vitro (Table 2) provide approximate comparisons of the ABA contributions from different sources at this age. The in vivo values for wild type, fluridone-treated (FT) wild type and w3 embryos suggest that about half the ABA had been translocated from maternal tissue. The comparison of in vivo and in vitro grown w3 embryos suggests that a substantial portion of maternal ABA, perhaps 30%, originates in cob tissues, and we subsequently determined that ABA is synthesized in cultured cob tissue from unpollinated ears (Hole et al., 1989). The ABA levels in embryos cultured with fluridone are very low (ca. 10 fmol ABA·embryo$^{-1}$). It seems likely that this residual ABA originates in cob tissue, since ABA levels are not reduced to zero by 15 DAP in cob sectors cultured with fluridone (Table 3).

While not unequivocal, these data show a strong causal relationship between the presence of xanthophylls and ABA levels, and they support the growing body of evidence (recently reviewed by Zeevaart and Creelman, 1988) which indicates that ABA is derived from a xanthophyll precursor in higher plants.

REGULATION OF ABA LEVELS IN MAIZE EMBRYOS

The mechanisms which regulate ABA synthesis and degradation in maize embryos are unknown. However, observations of field grown

Table 2. ABA levels, expressed as percentage of appropriate wild type control, found in 15 DAP embryos grown under different conditions. In vivo materials were harvested at 15 DAP, while in vitro materials were harvested at 5 DAP and kernel blocks were cultured on synthetic medium for 10 days. Fluridone treatments (FT) were accomplished by spraying in vivo ears with a fluridone solution (100 $mg \cdot l^{-1}$) at 9 and 11 DAP, while in vitro kernel blocks were cultured on medium containing fluridone (100 $mg \cdot l^{-1}$)

| Genotype | ABA level | |
|---|---|---|
| | In vivo | In vitro |
| Wild type | 100% | 100% |
| w3 | 54 | 15 |
| FT-wild type | 47 | 2 |

Table 3. Abscisic acid levels ± SE of isolated cob tissue cultured from 5 DAP with and without fluridone (100 $mg \cdot l^{-1}$).

| Tissue Age | ABA ($ng \cdot g^{-1}FW$) | |
|---|---|---|
| (DAP) | Without Fluridone | With Fluridone |
| 12 | 30 ± 1 | 26 ± 1 |
| 15 | 51 ± 3 | 11 ± 1 |
| 18 | 55 ± 4 | 10 ± 1 |

materials are beginning to shed some light on what is happening, at least in a general sense. Our field plots require irrigation, and daytime maximum/night minimum temperatures average about 36°C/24°C during the seed development period. Seed develop quite rapidly under these conditions, with distinct black layer apparent by 35 DAP.

Embryo ABA levels increase about 100-fold between 15 and 25 DAP. Comparisons of ABA-deficient and wild type embryos indicate that both *in situ* and maternal ABA components increase during this period, but the maternal component comprises about 60 to 70% of the ABA at 25 DAP. We have also observed rather large fluctuations in ABA levels in embryos of the same age harvested on different days, and ABA levels are consistently higher when ears are harvested in late afternoon rather than early morning.

Homozygous ABA-deficient mutants do not always germinate prematurely (Robertson, 1955), and we recovered relatively large amounts of dormant kernels for all mutants except *vp1* in 1983, which was an exceptionally hot, dry season. Subsequently, we have determined that the frequency of such dormant kernels is appreciably enhanced if plants are subjected to water stress between 10 and 15 DAP. Conversely, growth of viviparous seedlings on the ear is usually limited to eruption of the coleoptile, but in cool, wet seasons it is not unusual to find seedlings with 2 or 3 leaves and roots that are several inches long. Collectively, these observations suggest that accumulation of ABA synthesized in the kernel, *in situ* ABA, may be genetically regulated, but the ABA contribution from the maternal plant is largely determined by environmental conditions.

Presuming that ABA is derived from a xanthophyll precursor, an understanding of how ABA synthesis is regulated seems to reside in class III mutants, which interrupt the biosynthetic pathway between xanthophyll and ABA. The only maize mutant in this class that has been studied is vp8, and it appears to be close to ABA. Although vp10 and several other induced mutants appear to fall in this class, we have very little information concerning them at this time. However, several class III mutants have been identified in other species (Zeevaart and Creelman, 1988). Investigations of the ability of the tomato mutants flc, sit and not to convert xanthoxin to ABA (Perry et al., 1988) indicate that the genetic lesions for flc and sit are between xanthoxin and ABA, while not interrupts ABA synthesis prior to xanthoxin.

Any discussion of the mechanism of ABA synthesis regulation is purely speculative at this time, but the key gene probably codes for an unidentified enzyme which cleaves the xanthophyll precursor. Rapid increases in ABA levels induced by water stress suggest that the initial trigger may be related to changes in turgor pressure, osmotic potential or water potential.

## ABA REGULATION OF SPECIFIC EVENTS DURING KERNEL DEVELOPMENT

### Vivipary

Viviparous sprouting is similar, but not identical, to germination. Light and electron microscopic comparisons of tissues from normal and viviparous kernels (Bai, 1989) at 16, 25 and 35 DAP revealed that events observed in aleurone, scutellum and embryo tissues of viviparous kernels were similar to those seen during germination. However, no obvious differences in endosperm development were detected during this period.

In normal kernels, protein and lipid bodies (spherosomes) are present in aleurone and scutellum by 16 DAP. These enlarge with age and the protein bodies appear to be surrounded by spherosomes. Aleurone and scutellum tissue in vp5 is similar to wild type at 16 DAP, but degradation of spherosomes in the aleurone is apparent by 25 DAP and by 35 DAP degradation of both spherosomes and protein bodies is obvious. The same events were observed in vp1 aleurone, but they were visible earlier and were much more dramatic at comparable ages. In vp1, spherosome degradation was detected at 16 DAP, both spherosomes and protein bodies were obviously being catabolized by 25 DAP and by 35 DAP the spherosomes had been largely consumed and protein bodies were seriously degraded. Similar events were observed in scutellum tissues, but they lagged about 10 days behind comparable stages of degradation observed in the aleurone. In vp5, spherosome appearance at 35 DAP was similar to that observed in 25 DAP aleurone, while 35 and 25 DAP scutellum appeared similar to 25 and 16 DAP aleurone, respectively, in vp1.

The first developmental differences we observed between normal and vp5 embryos were related to the radicle (vp1 embryos were not analyzed). Cell division and enlargement in the radicle virtually cease at 15 DAP in wild type kernels, but vp5 radicles were visibly larger at 17 DAP and lignified xylem elements were identifiable at 19 DAP. Using the Abbe and Stein (1954) system for identifying stages during embryogeny according to the number of true leaf primordia differentiated, the coleoptilar stage is evident at 9 DAP, stage 1 at 11 DAP, stage 4 at 15 DAP and stage 6 at 21 DAP in the inbred TX5855 (Fong and Smith, 1987). No additional leaf primordia are initiated in this line, and maximum dry weight of the embryo axis is reached about 26 DAP. In vp5 plumules, the rate of differentiation of leaf primordia was only slightly accelerated, but the plumules were visibly larger at 21 DAP. By 25 DAP the elongated plumules are frequently visible through the pericarp of vp5 kernels, and they have often ruptured the pericarp by this age in vp1 kernels.

We have conducted a number of experiments to inhibit dormancy induction in normal kernels by inhibiting carotenogenesis with fluridone and to induce dormancy in viviparous kernels by treating them with exogenous ABA. In field grown materials the induction of dormancy can be inhibited by fluridone treatments made at 13 DAP and earlier, with maximum vivipary (ca. 70%) observed when treatments were made at 9 and 11 DAP (Fong et al., 1983). Applications of fluridone at later dates resulted in increased phytoene levels, but the kernels were all dormant and subsequent germination rates of mature seed was about 95%. Application of ABA to ears segregating for ABA-deficient mutants, or previously treated with fluridone, have been partially successful, but only when ABA treatments were made at 14 or 15 DAP (Fong et al., 1983). Up to 75% of the viviparous kernels treated with ABA ($10^{-4}$mol) at 15 DAP appear to be dormant, but less than half of these will germinate. Efforts to induce dormancy in vp1 embryos have been completely unsuccessful.

Experiments utilizing kernel blocks cultured on media containing fluridone with and without ABA addition at various ages have given similar but more definitive results. Kernels cultured with fluridone are 100% viviparous. When ABA was added to the medium at 13, 14 and 15 DAP, 57, 63 and 80% of the kernels were dormant and these were 100% viable. The addition of ABA later than 15 DAP failed to induce dormancy in any kernels (Hole, 1988; Hole et al., 1989).

Sussex (1978) has pointed out that embryo dormancy is an inducible rather than obligatory stage of seed development. It is now apparent that the induction of dormancy requires both ABA and the VP protein. The regulatory effect of ABA is concentration-dependent, but the source of the ABA does not appear to be important. Normal kernels cultured in vitro are dormant, which indicates that the kernel and its associated cob tissue produce sufficient ABA to induce embryo dormancy. The fact that ABA-deficient mutants are usually viviparous indicates that the contribution of ABA from maternal plant tissues is usually not sufficient to induce dormancy at the time dormancy must be induced. However, the occasional recovery of viable mutant kernels, and the ability to increase the frequency of such kernels by stressing the plants, indicates that the maternal contribution can be above the threshold level for dormancy induction. Using composited data from different experiments we estimate that the threshold level for induction of embryo dormancy in the maize inbred TX5855 lies between 9 and 12 ng ABA$\cdot g^{-1}$ fresh weight of 15 DAP embryo tissue.

It appears that dormancy must be induced during a rather narrow time window at or near stage 4 of embryogenesis. The failure to induce dormancy at this time results in kernels which are either viviparous or non-viable even though they may appear to be normal. Our anatomical observations and data from experiments to inhibit or induce dormancy in both field grown and cultured kernels are quite consistent in identifying stage 4 as the sensitive developmental stage.

In similar experiments with ABA-deficient and ABA-insensitive mutants in *Arabidopsis thaliana*, Koornneef et al. (1989) obtained somewhat different results. Although they found that abnormal seed development associated with homozygous mutants was largely alleviated in seed from heterozygous plants or by application of ABA to the plant roots during seed development, dormancy was not induced by either maternal or exogenous ABA. However, this apparent difference in dormancy induction response may simply be a function of experimental techniques applicable to the different species. In maize, we were able to spray ABA ($10^{-4}$ mol) directly on the naked caryopses and exposed cob tissue or culture the kernels on medium containing ABA, which probably increased the uptake of exogenous ABA during kernel development.

## Regulation of Lipase Activity

Ultrastructural studies of aleurone and scutellum tissues from normal and viviparous kernels suggested ABA might be involved in the regulation of lipase. We assayed crude homogenates of aleurone and scutellum tissues from wild type and mutant kernels for lipase activity (Bai, 1989). Activity was measured as total fatty acid released per gram fresh weight of tissue per minute (Lin et al., 1983) (Table 4).

Table 4. Total fatty acid ($nmol \cdot gFW^{-1} \cdot min^{-1}$)[a] released in each tissue homogenate. TX5855, vp1 and vp5 materials were grown in the field and harvested at ages indicated. FT-TX5855 kernels were harvested at 5 DAP and cultured on medium containing fluridone.

| Genotype | Age (DAP) | Aleurone | Scutellum |
|---|---|---|---|
| FT-TX5855 | 13 | 170 | 25 |
| | 15 | 84 | 38 |
| | 17 | nd | 145 |
| vp1 | 16 | 171 | 117 |
| | 25 | 158 | 4,235 |
| | 35 | 143 | 4,811 |
| vp5 | 16 | 80 | 57 |
| | 25 | 101 | 76 |
| | 35 | 356 | 1,813 |
| TX5855 | 16 | 31 | nd |
| | 25 | 50 | 29 |
| | 35 | 30 | 29 |

nd = not determined

[a]values less than 50 were not significantly different from background.

Lipase is suppressed in wild type tissues during kernel development. In both the ABA-insensitive (vp1) and ABA-deficient (vp5) genotypes, active lipase was present in aleurone and scutellum tissues. The relative activities determined at different ages were consistent with the apparent rates of lipid body degradation at comparable ages.

Lipase is active much earlier in vp1 than in vp5, and the decline in activity with age shown for vp1 aleurone is probably related to the rapid degradation of lipid bodies in this tissue. Wild type kernels cultured on fluridone sprout prior to 25 DAP, and the same age series could not be used for this material. However, the 13, 15 and 17 DAP kernels were comparable to the 16 DAP field grown materials, and lipase activity in FT-TX5855 was similar to vp1 for both aleurone and scutellum.

These data suggest that the gene for lipase is repressed during normal kernel development, and that both the VP protein and ABA are required for repression. Since ABA levels in vp5 embryos are about 5 times greater than in FT-TX5855 embryos at 15 DAP, it appears that the regulatory effects of ABA are concentration-dependent. Since VP appears to be an on/off component of this regulatory system and the Vp1 allele is always present in normal maize kernels, this suggests that ABA is the effective regulator of expression of the gene for lipase. The high lipase activity detected in 13 DAP FT-TX5855 aleurone also suggests that repression of lipase is an early part of a cascade of events which result in the induction of embryo dormancy.

Regulation of Anthocyanin Synthesis

The red and blue colors associated with anthocyanin pigments in the aleurone tissue of maize kernels, and other plant parts, have attracted a great deal of attention, and studies of color segregations of maize kernels predate the rediscovery of Mendelian genetics (East and Hayes, 1911). Due to the ease with which discrete color variations can be identified, the genetic control of anthocyanin pigmentation in maize is more completely defined than that of any other character in higher plants. At least eight dominant genes (A, A2, Bz, Bz2, C C2 R and Vp1) (Coe et al., 1988) are required for anthocyanin synthesis in the aleurone, which is colorless if any of these alleles is absent. In addition, homozygous in is required for full color expression. Given the appropriate genotype, mature kernels are deep purple, verging on dark blue, in color.

The vp1 mutant is unique among the viviparous mutants in several ways. It has normal levels of ABA, it is insensitive to exogenous ABA and it is required for anthocyanin synthesis in the aleurone. Mature kernels of the ABA-deficient mutants we have tested in full color (ACR) genetic backgrounds have colored aleurone. These include vp2, vp5, vp7 and vp9. However, anthocyanin accumulation in the viviparous segregates lags behind that in normal kernels, and when vivipary is partially suppressed the viviparous kernels are visibly lighter colored at maturity. This could be due to metabolic differences in the aleurones of viviparous and normal kernels, but it also seemed possible that these differences in color intensity were related to differences in ABA levels.

Normal ACR TX5855 kernels grown in the field have yellow endosperm, dark purple aleurone and dormant embryos. Fluridone treatment (100 $mgl^{-1}$) at 9 DAP blocks carotenoid and *in situ* ABA synthesis in ACR TX5855 kernels, and mature kernels have white endosperm, purple aleurone and viviparous embryos.

ACR TX5855 ears also were harvested at 5 DAP and cultured on 4 different media: (1) standard; (2) standard + fluridone; (3) standard + ABA; and (4) standard + fluridone + ABA. The results of this experiment are summarized in Table 5.

Anthocyanin synthesis in aleurone tissue is not induced in the kernels cultured on fluridone. The intense pigmentation in kernels cultured with both fluridone and ABA in the medium indicates that the absence of anthocyanins was due to the near absence of ABA rather than fluridone inhibition of anthocyanin synthesis. Both the intensity and

Table 5. Progression of anthocyanin accumulation in field grown and cultured ACR TX5855 kernels over time. Fluridone (100 $mg \cdot l^{-1}$) and ABA ($10^{-4}$ mol) treatments modulated *in situ* ABA levels and maternal ABA contribution is largely removed from *in vitro* cultured kerenls. Pigmentation in aleurone and endosperm tissues and the embryo status are described.[a]

| | Kernel Age (DAP) | | | | | |
|---|---|---|---|---|---|---|
| | 15 | 17 | 19 | 21 | 23 | 25 |
| Field Grown | | | | | | |
| ACR TX5855 | LP,Y | LP,Y | P,Y | P,Y | P,Y,D | DP,Y,D |
| FT-ACRTX5855 | C,W | LP,W | LP,W | P,W | P,W,V | P,W,V |
| In Vitro Cultured | | | | | | |
| Medium | | | | | | |
| Standard | C,Y | LP,Y | P,Y | P,Y,D | P,Y,D | DP,Y,D |
| Fluridone | C,W | C,W | C,W | C,W,V | C,W,V | C,W,V |
| ABA | LP,Y | P,V | P,Y | P,Y,D | DP,Y,D | DP,Y,D |
| Fluridone + ABA | LP,W | P,W | P,W | P,W,D | DP,W,D | DP,W,D |

[a]Aleurone: C = colorless; LP = light purple; P = purple; DP = dark purple
Endosperm: Y = yellow; W = white
Embryo: D = dormant; V = viviparous

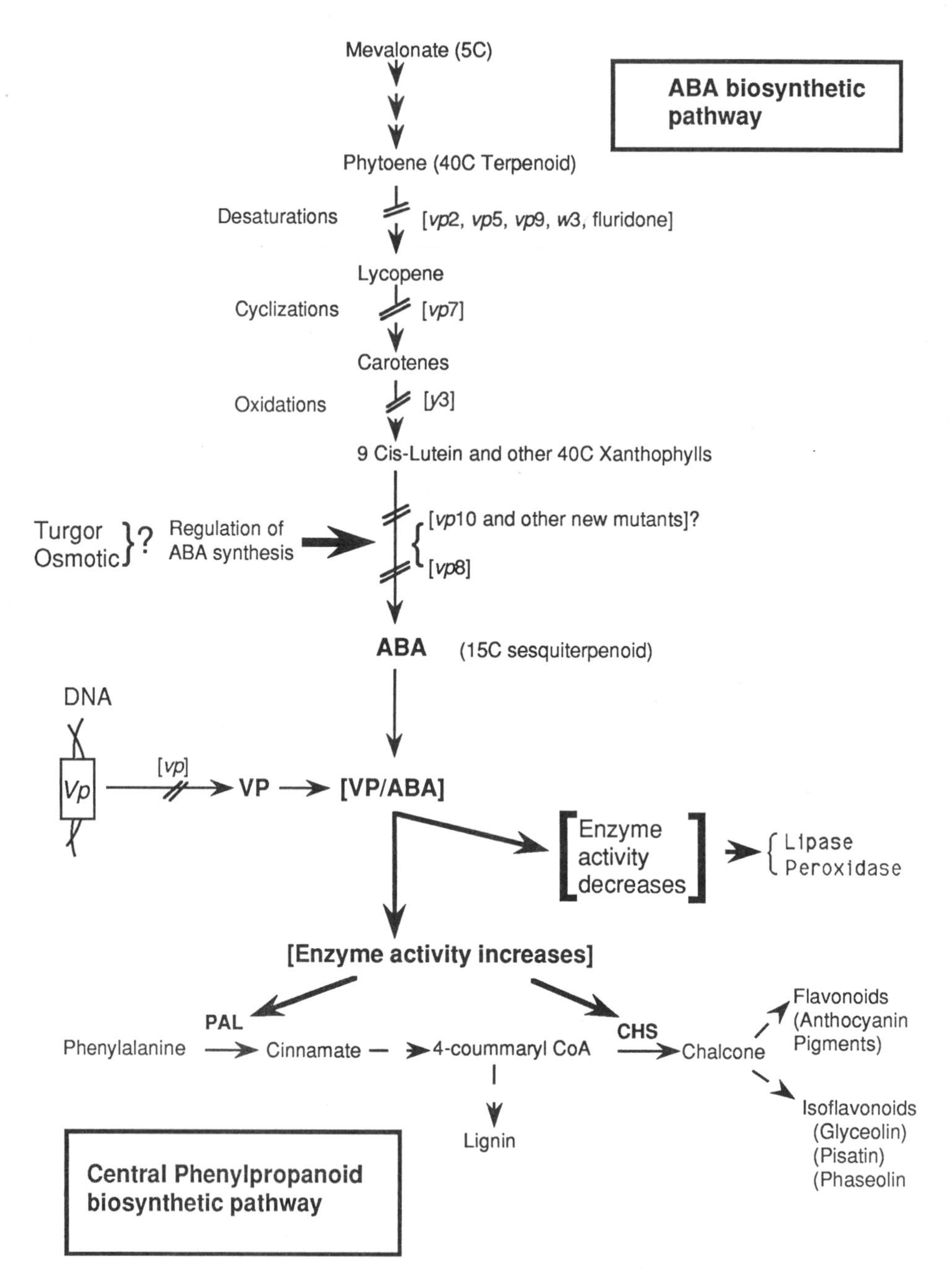


Fig. 2. Hypothetical model for synthesis and known functions of the presumed ABA/VP aggregate. Suppression and induction of enzyme activities is presumably accomplished by regulating transcription of the appropriate genes. VP acts as an on/off regulator. ABA effects on enzyme activities are concentration dependent.

time of visible expression of anthocyanins are enhanced by exogenous ABA.

The induction of anthocyanin synthesis appears to be dependent upon ABA, and this function of ABA is also concentration-dependent. It seems probable that all of the regulatory functions ascribed to the Vp1 gene (Smith et al., 1978; Smith et al., 1980; Dooner, 1985; McCarty, 1989) will eventually be shown to require ABA as a co-factor.

## CONCLUDING REMARKS

ABA and the VP protein are intimately involved in controlling maize embryo dormancy as well as lipase and anthocyanin expression. ABA also is known to induce or inhibit the synthesis of a variety of proteins in various species of higher plants (Kuhlemeier et al., 1987). In maize, Dooner (1985) has reported that three phenylpropanoid pathway enzymes and two unrelated enzymes are not expressed in homozygous vp1, and we have shown that ABA is required to induce anthocyanin and to repress activity of lipase and three peroxidase isozymes (F. Fong, unpublished data). Yet, virtually nothing is known about the mechanisms involved in the regulation of gene expression. Thus, we have taken the liberty to propose a hypothetical model which may partially elucidate the regulatory functions of ABA and VP (Fig. 2).

This model incorporates the known and some presumed primary functions of the viviparous mutants of maize. It presumes, on the basis of molecular simplicity and functional complexity, that ABA does not bind to DNA. Vp1 is the structural gene for the VP protein, and we presume that VP is a regulatory protein. Because of the chromosomal locations of genes involved we have further assumed that VP regulates transcription of specific genes. The ability of VP to induce some genes and repress others may or may not indicate bifunctional characteristics for VP, but this is not important to the model. Similarly, VP may be a DNA binding protein, or it could be an activating protein, but this, again, is of no importance to the model.

Since both ABA and VP appear to be absolute requirements for the systems we have investigated, our critical assumption is that ABA does bind to VP and that specific induction/repression events depend upon the state of this presumed ABA/VP aggregate. Although it suggests that individual phenylpropanoid enzymes are ABA/VP regulated, it is quite possible that induction of enzymes distal to phenylalanine ammonia lyase (PAL) in the pathway is precursor driven.

The model also provides a possible mechanism for time- and tissue-specific regulation of genes. ABA could not act as a regulator prior to the time that the VP protein was synthesized or in tissues in which Vp1 was repressed. However, if VP is present, regulatory functions would depend upon ABA concentration.

We believe this model is consistent with present information concerning the known regulatory functions of ABA and the Vp1 gene in maize kernels, but it is also highly speculative. It is presented at this time in the belief that other investigators will either verify or correct the model.

## LITERATURE CITED

Abbe, E. C., and Stein, O. L., 1954, The growth of the short apex in maize embryogeny, Am. J. Bot. 41:285.

Bai, D. G., 1989, Regulation of lipase activity by Vp1 and abscisic acid during development of the caryopsis of Zea mays L. Ph.D. Dissertation, Texas A&M University Library, College Station.

Bartels, P. G., and Watson, C. W., 1978, Inhibition of carotenoid synthesis by fluridone and norflurazon, Weed Sci. 26:198.

Cobb, B. G., and Hannah, L. C., 1983, Development of wild type, shrunken-1 and shrunken-2 maize kernels grown in vitro, Theor. Appl. Genet. 65:47.

Coe, E. H., Jr., Neuffer, M. G., and Hoisington, D. A., 1988, The genetics of corn, in: "Corn and Corn Improvement," G. F. Sprague and J. W. Dudley, eds., American Society of Agronomy, Inc., Madison.

Dooner, H. K., 1985, Viviparous-1 mutation in maize conditions pleiotropic enzyme deficiencies in the aleurone, Plant Physiol. 77:486,

East, E. M., and Hayes, H. K., 1911, Inheritance in maize, Conn. Agric. Exp. Sta. Bull. 167.

Fong, F., and Smith, J. D., 1987, Morphological stages in embryo development in TX5855, Maize Genet. News Lett. 61:40.

Fong, F., Koehler, D. E., and Smith, J. D., 1983, Fluridone induction of vivipary during maize seed development, in: "Third International Symposium on Pre-harvest Sprouting in Cereals," J. E. Kruger and D. E. Laberge, eds., Westview Press, Boulder.

Fong, F., Smith, J. D., and Koehler, D. E., 1983, Early events in maize seed development, Plant Physiol. 73:899.

Fujioka, S., Yamane, H, Spray, C. R., Gaskin, P., MacMillan, J., Phinney, B. O., and Takahaski, N., 1988, Qualitative and quantitative analysis of gibberellins in vegetative shoots of normal, dwarf-1, dwarf-2, dwarf-3 and dwarf-5 seedlings of Zea mays L., Plant Physiol. 88:1367.

Gengenbach, B. G., 1977, Development of maize caryopses resulting from in vitro pollination, Planta 134:91.

Hole, D. J., 1988, Development of Zea mays L. caryopses: Regulation of embryo dormancy and differences in protein patterns associated with manipulation of ABA levels in intact kernels cultured in vitro, Ph.D. Dissertation, Texas A&M University Library, College Station.

Hole, D. J., Smith, J. D., and Cobb, B. G., 1989, Regulation of embryo dormancy by manipulations of abscisic acid in kernels and associated cob tissue of Zea mays L. cultured in vitro, Plant Physiol. 90:(in press).

Karssen, C. M., Brinkhorst-van der Swan, D. L. C., Breekland, A. E., and Koorneef, M., 1983, Induction of dormancy during seed development by endogenous abscisic acid: Studies on abscisic acid deficient genotypes of Arabidopsis thaliana (L.) Heynh., Planta 157:158.

Koornneef, M., Hanhart, C. J., Hilhorst, H. W. M., and Karssen, C. M., 1989, In vivo inhibition of seed development and reserve protein accumulation in recombinants of abscisic acid biosynthesis and responsiveness mutants in Arabidopsis thaliana, Plant Physiol. 90:(In press).

Kuhlemeier, C., Green, P. J., and Chua, N-H., 1987, Regulation of gene expression in higher plants, in: "Annual Review of Plant Physiology," W. R. Briggs, R. J. Jones and V. Walbot, eds., Annual Reviews Inc., Palo Alto.

Lin, Y-H., Wimer, L. T., and Huang, A. H. C., 1983, Lipase in the lipid bodies of corn scutella during seedling growth, Plant Physiol. 73:460.

McCarty, D. R., Carson, C. B., Stinard, P. S., and Robertson, D. S. 1989. Molecular analysis of viviparous-1: an abscisic acid-insensitive mutant of maize. Plant Cell 1:523.

Neill, S. J., Horgan R., and Parry, A. D., 1986, The carotenoid and abscisic acid content of viviparous kernels and seedlings of Zea mays L., Planta 169:87.

Perry, A. D., Neill S. J., and Horgan, R., 1988, Xanthoxin levels and metabolism in wild-type and wilty mutants of tomato, Planta 173:397.

Phinney, B. O., and Spray, C., 1982, Chemical genetics and the gibberellin pathway in Zea mays L., in: "Plant Growth Substances 1982," P. F. Wareing, ed., Academic Press, New York.

Robertson, D. S., 1955, The genetics of vivipary in maize, Genetics 40:745.

Robertson, D. S., 1975, Survey of the albino and white-endosperm mutants of maize, J. Hered. 66:67.

Smith, J. D., Koehler D. E., and Fong, F., 1980, Genetic studies utilizing viviparous mutants of Zea mays indicate that abscisic acid may play a critical role in the induction of seed dormancy, Genetics 94:S100.

Smith, J. D., McDaniel, S., and Lively, S., 1978, Regulation of embryo growth by abscisic acid in vitro, Maize Genet. News Lett. 52:107.

Spray, C., Phinney, B. O., Gaskin, P., Gilmore, S. J., and MacMillan, J. 1984, Internode length in Zea mays L. The dwarf-1 mutation controls the 3-beta-hydroxylation of gibberellin $A_{20}$ to gibberellin A, Planta 160:464.

Sussex, I. M., 1978, Dormancy and development, in: "Dormancy and Developmental Arrest," M. E. Clutter, ed., Academic Press, New York.

Zeevaart, J. A. D., and Creelman, R. A., 1988, Metabolism and physiology of abscisic acid, in: "Annual Review of Plant Physiology and Plant Molecular Biology," W. R. Briggs, R. L. Jones and V. Walbot, eds., Annual Reviews, Inc., Palo Alto.

# MEMBRANE BEHAVIOR IN DROUGHT AND ITS PHYSIOLOGICAL SIGNIFICANCE

Folkert A. Hoekstra[1], John H. Crowe[2] and Lois M. Crowe[2]

[1]Department of Plant Physiology
Agricultural University
Wageningen, The Netherlands

[2]Department of Zoology
University of California
Davis, CA 95616, USA

## INTRODUCTION

During the last decade numerous efforts have been made to understand the mechanism of desiccation tolerance in plants and animals, particularly in those that exhibit the extreme dehydration known as anhydrobiosis (Crowe and Clegg 1978; Leopold, 1986). Research has focused on the following areas: metabolic changes during drought stress, particularly carbohydrates; the biochemistry and biophysics of membrane lipids; changes in DNA/RNA and protein synthesis; hormonal regulation of desiccation tolerance; and also more applied aspects such as the influence of environmental factors during seed and pollen rehydration.

In this paper we focus on the behavior of membranes in pollen and seeds, particularly their phospholipids during dehydration and rehydration stresses. Phase behavior of model lipid systems will be shown to be similar to that seen in membranes of intact dry cells. Also described is a Fourier transform infrared spectroscopic method that enables measurement of phase changes in membranes of intact cells. This method has considerably increased our understanding of the biophysical basis of imbibitional damage.

## DESICCATION AND LEAKAGE OF ENDOGENOUS SOLUTES

In higher plants, seeds and pollen are generally much more tolerant to dehydration than are flowers, leaves or roots. During the development of seeds and pollen, a shift from a desiccation sensitive to a tolerant state is often seen. Mustard seeds, for instance, become desiccation tolerant after about half of the 2 months' developmental period has elapsed (Fischer et al., 1988). A much shorter development has been established in *Papaver* pollen (Hoekstra and Van Roekel, 1988). Maximal germination capacity is reached at one day prior to flower opening (Fig. 1). The germination capacity increases during the last 3 days of maturation in the flower bud, and at the same time the pollen becomes desiccation tolerant. In Fig. 1 desiccation tolerance is expressed as the capacity of the rehydrated grains to retain the fluorescein deesterified in the cytosol from added fluorescein diacetate (FDA), and also as the extent to which intracellular $K^+$ leaks into the medium during the first 3 minutes of imbibition. These



*Recent Advances in the Development and Germination of Seeds*
Edited by R.B. Taylorson
Plenum Press, New York

data illustrate a common theme in studies on dehydration tolerance – the integrity of permeability barriers during and after dehydration – a theme that has been the focus of our work for some years. Although we will emphasize imbibitional events here, it is likely that at least part of the injury that results in leakage shown in Fig. 1 originates during the dehydration step (Hoekstra et al., 1988).

Injury at the membrane level may occur during re-imbibition of dry organs. In this case cells remain essentially viable during the dehydration step. Injury can be prevented by carefully selecting the proper conditions during rehydration. Fig. 2A shows germination and $K^+$-leakage data from *Narcissus* pollen having different initial moisture contents. A high and a low temperature of imbibition was selected. Leakage of $K^+$ during the first three minutes of imbibition is particularly serious at the low imbibition temperature and at low moisture contents. Leakage is negatively correlated with germination. Fig. 2A further shows that imbibition at 29°C considerably improves germination of pollen of relatively low moisture contents. Whereas we call the damage in the 0°C-imbibed grains "imbibitional chilling damage" and that at 29°C "imbibitional damage", these two forms of damage do not essentially differ as to the mechanism involved (Hoekstra and Van der Wal, 1988). Although pollen seems an ideal material for such imbibition studies due to its relatively small size and simple organization lacking differentiated tissues, strikingly similar patterns were obtained with a more complex system such as seeds (Fig. 2B). As mentioned before, leakage points to problems at the membrane level. The beneficial effect of elevated temperatures during imbibition suggests involvement of a phase transition, as does the interaction with initial moisture content. This involvement may become more understandable after the following section on phospholipid behavior at low water activities.

## BEHAVIOR OF PHOSPHOLIPID BILAYERS AT LOW WATER ACTIVITIES

More than 20 years ago Chapman et al. (1967) published a differential scanning calorimetry study on the behavior of dipalmitoylphosphatidylcholine-water systems (di-C16:0-PC; DPPC). They showed that the gel to the

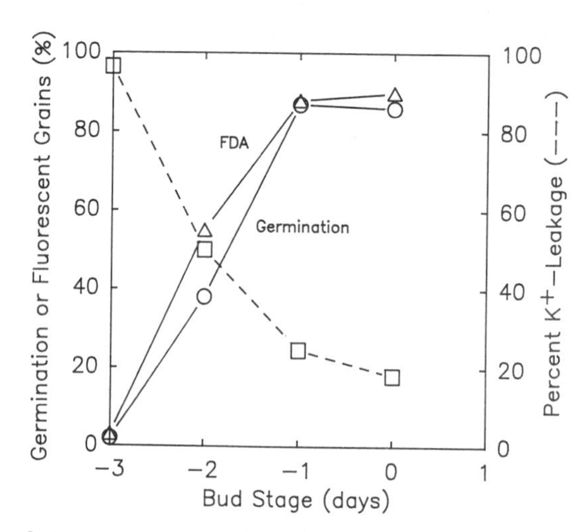


Fig.1. Effect of drying on germination in vitro, fluorescein retention (FDA test), and $K^+$-leakage of *Papaver* pollen during its development in the flower bud. Anthesis is at day zero. Dried pollen was rehydrated in humid air for 1 h prior to the tests.

liquid crystalline phase change is centered around 41 °C. Put more simply, during heating DPPC melts at 41 °C. The transition temperature, $T_m$, slowly rises when the water hydrogen bonded around the polar head group is removed. Removal of the last 10-12 mol of water causes a steep rise in $T_m$ of 70 °C in total (Fig. 3). This behavior is completely reversible upon re-addition of $H_2O$. The amount of 12 mol $H_2O$ per mol phospholipid represents between 0.2-0.3 g $H_2O$ per g phospholipid, which is approximately 0.2-0.3 g $H_2O$ per g dry weight in a living cell.

The increase in transition temperature with drying can be explained as follows. Due to the removal of water, the lateral spacing of the polar head groups decreases, leading to increased opportunities for van der Waals' interactions between the hydrocarbon chains (reviewed in Crowe and Crowe, 1988), resulting in an increase in $T_m$. In other words, a hydrated lipid that is in liquid crystalline phase at physiological temperatures will often be in gel phase when it is dry (Crowe et al., 1988).

Because phospholipids of plants have more (cis)unsaturated acyl

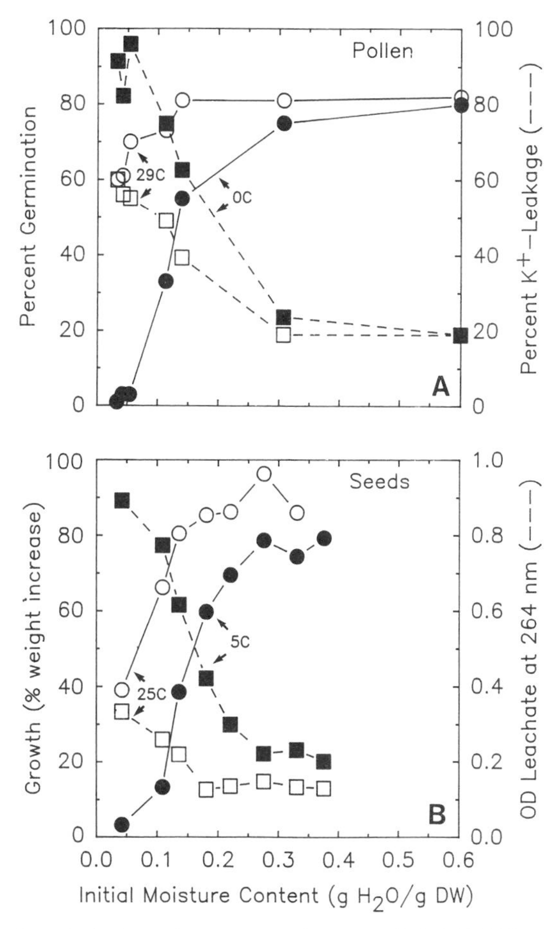


Fig. 2. Effect of initial moisture content and imbibition temperature on performance and leakage. A: Narcissus pollen imbibed for 3 minutes at 0 or 29°C and subsequently cultivated at 22°C. B: Replotted data of Pollock (1969) on Lima bean seeds imbibed for 30 minutes at 5 or 25°C, and subsequently grown at 25°C. The different initial moisture levels were obtained by different techniques of humidification from the vapor phase.

chains, their transition temperatures are low due to an increased curvature of the acyl chains, causing the opportunities for van der Waals' interactions to decrease. For instance, dilinolenoylphosphatidylcholine (di-C18:3-PC) has a hydrated $T_m$ slightly less than -60 $^{o}C$ (D.V. Lynch, personal communication). The increase in $T_m$ due to dehydration, also of other phosphatidylcholines tested, is in the order of 60-70 $^{o}C$ (Crowe et al., 1989d; D.V. Lynch, personal communication).

During the thermotropic phase transition, one may expect the temporary presence of both gel and liquid crystalline phases. The permeability properties of such phase separating vesicles have been studied in the presence of water. Fig. 4 shows that when dimyristoylphosphatidylcholine (di-C14:0-PC; DMPC) vesicles are heated through their phase transition, around 23-24 $^{o}C$, all the dye (carboxyfluorescein) trapped in the aqueous interior leaks into the medium. The midpoint of the leakage was calculated to occur simultaneously with the calorimetric transition (Fig. 4). If the vesicles had been made at temperatures well above 24 $^{o}C$, complete leakage would then occur only upon cooling through the transition temperature. Currently such leakage is interpreted as indicating packing defects existing at boundaries between liquid crystalline and gel phase domains.

Returning to the behavior of phospholipid vesicles during desiccation, one might expect leakage when the rising transition temperature reaches room temperature. However, this leakage cannot be observed during desiccation due to the limited availability of $H_2O$ as a carrier for entrapped compounds. In contrast, during rehydration in excess water, leakage is observed when the phospholipid melts again (Crowe et al., 1985).

With other phospholipids such as the phosphatidylethanolamines (PE's), the situation is more complex. The polar head group of PE possesses a positively charged N on the amino group that can form a bond of strong ionic character with the negatively charged phosphate of an adjacent PE molecule. This may cause such a close contact that PE molecules are forced into a curved structure, leading to the formation of lipid tubes, with the acyl chains directed to the outside and the polar head group oriented toward the center of the hydrated core. When such disturbance occurs in a plant membrane composed of different types of phospholipids, complete loss

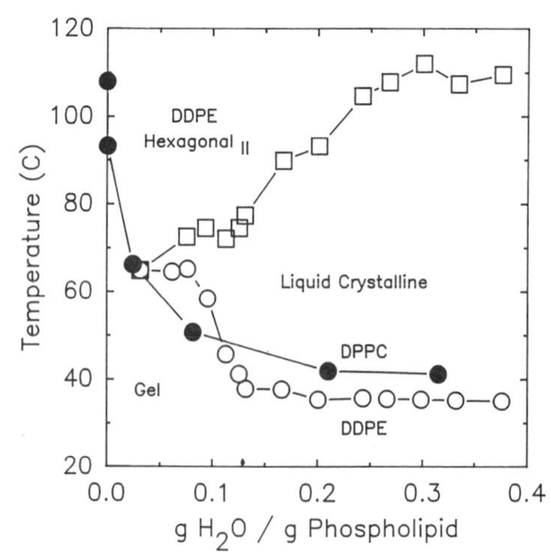


Fig. 3. Phase diagram of dipalmitoylphosphatidylcholine (closed symbols; DPPC) (data from Chapman et al., 1967, and from Crowe et al., 1988) shown together with that for didodecylphosphatidylethanolamine (open symbols; DDPE)(data from Seddon et al., 1983) at various water contents (Crowe and Crowe, 1988).

of semipermeability is observed, as, for instance, upon freezing (Crowe and Crowe, 1982). The chance for the formation of the so-called $hexagonal_{II}$ phase (inverted hexagonal) is increased by any treatment that increases the volume occupied by the acyl chains, or that reduces the distance between the polar head groups. Among such treatments are heating and an increase in cis-unsaturation of the acyl chains, and dehydration, respectively (Cullis and DeKruijff, 1979). Fig. 3 shows the phase diagram of didodecyl-phosphatidylethanolamine (di-C12:0-PE; DDPE). Under hydrated condition DDPE melts around 34 $^{o}C$ and goes from the liquid crystalline into the $hexagonal_{II}$ phase at about 110 $^{o}C$ ($T_h$). Dehydration reduces $T_h$ and increases $T_m$ which might allow for a merge into one single transition (Seddon, 1983). This means that at low water activity a direct shift might occur from the gel into the $hexagonal_{II}$ phase.

Desiccation tolerant seeds and pollen are apparently able to circumvent injury associated with gel and $hexagonal_{II}$ phase formation. It is likely that they are able to do so by chemical changes that lead to increased lateral spacing of the polar head groups or decreased van der Waals' interactions of the hydrocarbon chains, and thereby depress $T_m$ and increase $T_h$. This can be achieved by increased unsaturation, or, as we will discuss later in more detail, by certain disaccharides that can replace the water around the polar head groups.

## CHEMICAL CHANGES WITH DROUGHT STRESS

### Phospholipids

The dramatic shift from desiccation sensitivity to tolerance during the last 3 days of Papaver pollen development is particularly suitable for a study of possible changes in chemical composition. We analysed phospholipid composition and the degree of unsaturation of the acyl chains (Hoekstra and van Roekel, 1988). Fig. 5 shows that, much to our surprise, we could detect no changes in phospholipid composition. The expected decrease in mol percent of PE, which may be effective at preventing $hexagonal_{II}$ lipid, as observed during drought stress in wheat seedlings

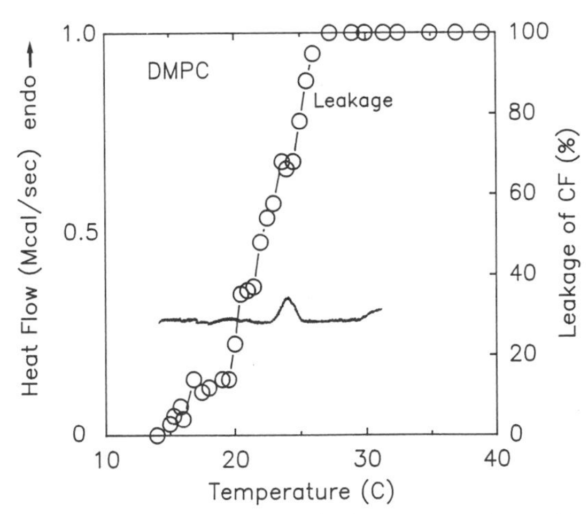


Fig.4. Leakage of carboxyfluorescein (CF) from unilamellar dimyristoyl-phosphatidylcholine (DMPC) vesicles as they are heated through their transition temperature (data from Crowe et al., 1989d). Also shown is a differential scanning calorimetry plot (data from Rudolph et al., 1986).

(Vigh et al., 1986), apparently did not occur here. Papaver pollen phospholipids contain palmitic and linolenic acid as the most prominent esterified fatty acids (more than 80% of the total). Contrary to our expectation, the already highly unsaturated phospholipids became more saturated during the shift to desiccation tolerance, particularly in the case of PE. As we have seen earlier, this may reduce the chance for the formation of hexagonal$_{II}$ phase lipid (Cullis and DeKruijff, 1979).

Soybean seeds are sensitive to imbibition in the cold, whereas seeds of a related family member, the pea, are not. In an attempt to find a chemical explanation for this difference, Priestley and Leopold (1980) analysed phospholipid composition and degree of unsaturation, and several

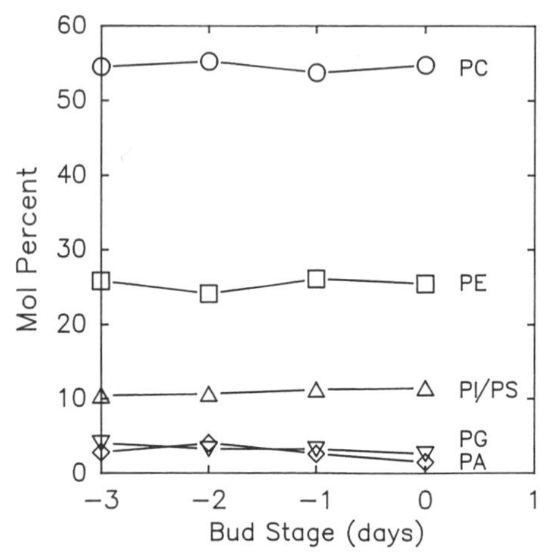


Fig. 5. Changes in phospholipid composition of fresh Papaver pollen during maturation. Pollen was liberated mechanically from anthers and not previously dried.

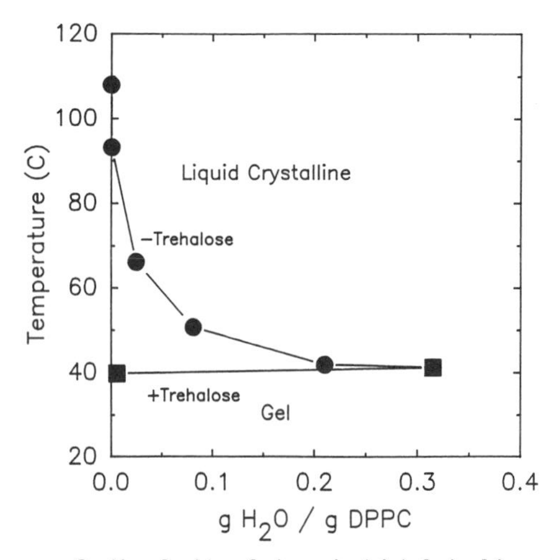


Fig. 6. Phase diagram of dipalmitoylphosphatidylcholine (DPPC) (data from Chapman et al., 1967, and from Crowe et al., 1988) in the absence or presence of trehalose (1.5 g trehalose/g DPPC) (data from Crowe et al., 1985) at various water contents.

properties of liposomes prepared from the isolated phospholipids. There were no differences that were significant enough to explain the differential sensitivity to imbibition in the cold between the two species. Removal of the seed coat from pea seeds, however, renders them sensitive to cold imbibition (see Herner, 1986). This brings a morphological component to the mechanisms of imbibitional damage.

**Soluble carbohydrates**

In animal systems and also in yeasts and bacterial spores, desiccation tolerance has been linked to substantial increases in the disaccharide, trehalose, constituting up to 20% of the DW (Crowe et al., 1984). Too rapid a desiccation causes leakage and death, correlated with relatively low trehalose levels (Madin and Crowe, 1975). Liposome studies indicate that trehalose is very effective at preventing fusion and leakage during freeze drying (Crowe et al., 1986) or air drying (Hoekstra et al., 1989) by replacing the water around the polar head groups, thus providing sufficient lateral spacing between them (Crowe et al., 1989d). Fig. 6 shows that, in

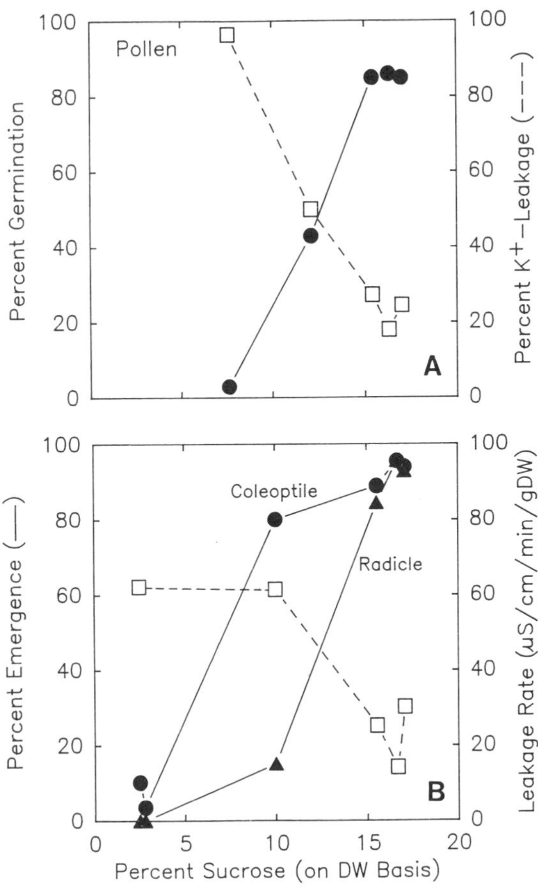


Fig. 7. Correlations of sucrose content and the capacities for growth and retention of solutes after drying and rehydration. A: during the last 3 days of <u>Papaver</u> pollen development (data from Hoekstra and van Roekel, 1988). B: after imbibition of corn seeds over a 60 h period (data from Koster and Leopold, 1988).

the presence of trehalose, dry DPPC vesicles do not exhibit an increase in their transition temperature. This may be the priciple by which anhydrobiotic organisms escape from desiccation stress.

Trehalose does not seem to occur in seeds (Koster and Leopold, 1988) and pollen (Hoekstra, 1986). Instead, sucrose may act as the protective agent in higher plants (Hoekstra et al., 1989). Sucrose has also been shown to stabilize liposomes during drying and rehydration, although slightly less effectively than trehalose.

During the last 3 days of maturation of *Papaver* pollen, starch is degraded and sucrose formed. Sucrose, the major soluble carbohydrate present in this pollen ( 96%), rises from 6% of the dry weight in dehydration sensitive pollen to about 17% of the DW in the tolerant ones. Immature, desiccation sensitive pollen, isolated mechanically from the anthers and kept humid, independently degrades starch, doubles its sucrose content and becomes desiccation tolerant. A similar breakdown of starch occurs during the development of desiccation tolerance in Mustard seeds (Fischer et al., 1988). When $K^+$-leakage and germination capacity of *Papaver* pollen are plotted against the sucrose content, one can observe an increased resistance to desiccation with elevated sucrose content (Fig. 7A). This is a plot during the acquisition of desiccation tolerance. A strikingly similar plot is shown in Fig. 7B during the loss of desiccation tolerance in corn seeds when sucrose levels decrease. Dependent on the species, seeds may also contain other sugars in considerable quantities such as raffinose and stachyose (Koster and Leopold, 1988). Raffinose also stabilizes liposomes to a certain extent (Crowe et al., 1985), whereas monosaccharides are generally less effective. Stachyose has not been tested in this regard. Although the relationship between high sucrose levels and desiccation tolerance as shown in Fig. 7 is only correlative, we suggest that these factors are causally linked. As evidence, we point to the data from Fig. 8. These data show that liposomes prepared from phospholipids isolated from *Papaver* pollen may be dehydrated and rehydrated without loss

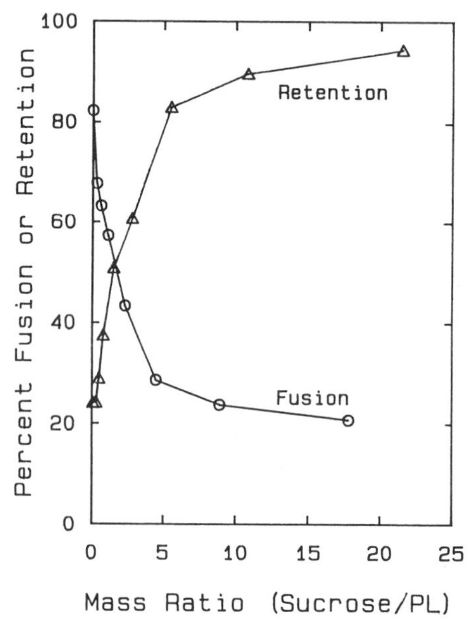


Fig. 8. Effects of a series of concentrations of sucrose on fusion and solute retention of liposomes prepared from *Papaver* pollen phospholipids, upon a cycle of dehydration and rehydration (data from Hoekstra et al., 1988).

of contents or fusion between the vesicles, provided they are dried in the presence of sucrose. It is intriguing that the liposomes are stabilized at similar sucrose levels as encountered in the intact pollen grain (a ratio of sucrose to phospholipid of about 4:1).

## OCCURRENCE OF HEXAGONAL$_{II}$ LIPIDS IN CELLS DURING A DRYING CYCLE

Working with dry yeast cells, Van Steveninck and Ledeboer (1974) found a dramatic improvement of survival and decrease of $K^+$-leakage at a high imbibition temperature. These authors suggested that the influence of temperature on the reconstitution process involves a phase transition of the membrane lipids from the gel into the liquid crystalline state during cold shock.

Simon (1974) has presented another explanation for the leakage and subsequent injury that occur during the initial stages of imbibition. Because of the known phase behavior of brain phospholipids he suggested that membranes in dry seeds are not arranged in the normal bilayer, but rather in hexagonal$_{II}$ arrays, with the remaining water in the core of these

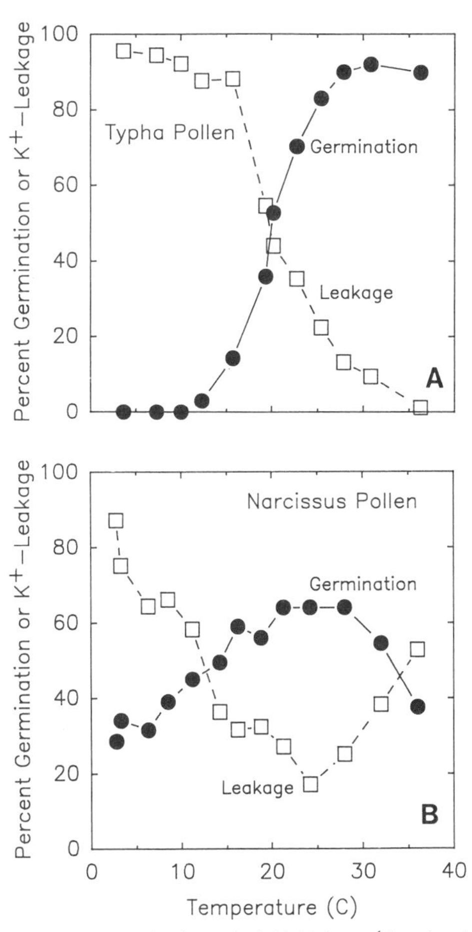


Fig. 9. Effect of temperature during imbibition (3 minutes) of dry pollen species (water content approximately 0.06 g $H_2O$/g DW) on germination and $K^+$-leakage. Growth temperature was 22°C. A: Typha pollen; B: Narcissus pollen.

lipid tubes. Upon reimbibition the porous state would become reorganized into the typical bilayer. Prehumidification of the seeds with water vapor would therefore be beneficial, because it would give the disorganized membranes time to assume the bilayer configuration before bulk water is present. Too rapid a movement of water into dry seeds, particularly at low temperature would interfere with this reorganization. Simons hypothesis was quickly embraced by Shivanna and Heslop-Harrison (1981) to explain imbibitional injury in dry pollen species, and the suggestion of Van Steveninck and Ledeboer (1974) was temporarily forgotten.

However, evidence against the occurrence of the $hexagonal_{II}$ configuration in dry seeds is accumulating through X-ray diffraction (McKersie and Stinson 1980; Seewaldt et al., 1981), and in pollen through $^{31}P$-NMR data, which indicate that the bilayer structure is exclusively present down to hydration levels of at least 11% (Priestley and DeKruijff, 1982). From later work on the same pollen species it can be inferred that cold hydration at 11% moisture content is fatal (Hoekstra, 1984). From freeze fracture electron micrographs of dry pollen, Platt-Aloia et al. (1986) were similarly unable to detect signs of this non-bilayer lipid phase. A much more simple argument against the presence of $hexagonal_{II}$ phase is provided by the widely observed phenomenon in both dry seeds (Herner, 1986) and pollen (Hoekstra and Van der Wal, 1988), that imbibition at elevated temperatures prevents leakage and allows for excellent germination. On the basis of the phase diagram of a PE (DDPE in Fig. 3) one would assume that the occurrence of $hexagonal_{II}$ phase lipid ought to be promoted at higher temperatures, which is contradictory with the results.

An example of the beneficial effect of imbibition of pollen at elevated temperatures is given in Fig. 9A. Leakage of endogenous $K^+$ in the first 3 minutes of imbibition is reduced and viability preserved. However, there is scattered evidence that the theory of Simon (1974) may hold true in certain pollen species under conditions of high temperature and extreme dehydration. Fig. 9B, for instance, shows that Narcissus pollen, after an initial improvement of its performance with temperature, starts to become leaky again at temperatures over 25°C. This would be compatible with a much reduced $T_h$ of phosphatidylethanolamines at low water activity. Prehydration of Narcissus pollen from the vapor phase prior to imbibition abolishes the high temperature sensitivity, which is in line with the much higher $T_h$ of the hydrated lipid.

## IN SITU MEASUREMENT OF GEL TO LIQUID CRYSTALLINE PHASE CHANGES

### Measurement of phase transitions with Fourier Transform Infrared Spectroscopy (FTIR)

We have seen in Fig. 6 that trehalose has a profound effect on the $T_m$ of dry DPPC vesicles. If one wishes to measure phase changes of membranes in dehydrating cells, a method is required that leaves the cells essentially intact, in order to be able to observe the impact of cytosolic disaccharides that accumulate during desiccation. Employing FTIR spectroscopy we have recorded IR spectra of $CH_2$ stretching vibrations in various pollen species (see Crowe et al., 1989a, b, c, d) and seed parts of cauliflower. When pure lipids pass from the gel into the liquid crystalline phase, the absorption peaks shift a few wave numbers to a higher frequency. This can be observed in both the symmetric $CH_2$ stretch centered around 2853 $cm^{-1}$ and the asymmetric $CH_2$ stretch around 2925 $cm^{-1}$ (Fig. 10). Because the majority of $CH_2$ chains in a cell are encountered in lipids, the spectra of pollen and seed parts refer to lipids. The spectra of dry radicle tips in Fig. 10, taken as an example, were recorded at -17°C when the bulk of the lipid was in gel phase, and at +19°C when it had melted. Pollen spectra look very similar (Crowe et al., 1989b). By plotting the frequency of the absorption peaks of the $CH_2$ symmetric stretch versus the temperature one can estimate the temperature at which half of the lipid has melted ($T_m$ in Fig. 11). The seemingly small changes in peak position in Fig. 10 are large enough to be measured by the equipment with an accuracy of about 0.1 wave number. The range over which the lipid melts is rather broad (15-20 $^{o}C$). This is understandable as FTIR spectroscopy cumulatively records the melting of gel phase domains of different molecular species of lipids at their different locations in the cell. Fig. 11 further shows that depending on the moisture content of the pollen, the frequency shift occurred at a different temperature range. This points to the interaction of the lipidic component with water.

## Assignment of peaks to phospholipids

As lipids in pollen and seeds are composed of a polar phospholipid fraction in the membranes and a neutral fraction of storage fats in lipid bodies or small droplets, the frequency shift may be caused by a combination of the two. However, storage lipids do not interact with water because of their lack of hydrophilic groups, and, therefore, will not cause frequency shifts with the water content. We are rather confident that the curves of Fig. 11 represent phase shifts of phospholipids in membranes from gel (lower frequency) to liquid crystalline (higher frequency) . Although the lipid of Typha pollen consists of at least 50% neutral lipids, we did not observe a considerable interference of these lipids with the wave number plots of Fig. 11. It may be that the shape of the oil bodies is responsible for their lack of signal.

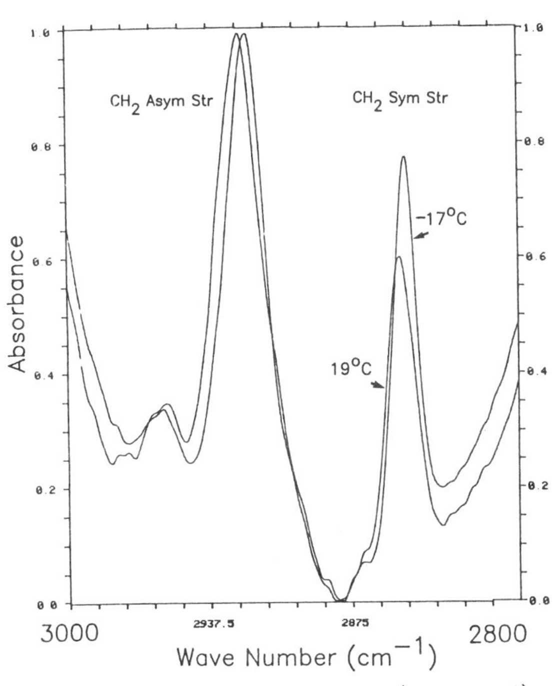


Fig. 10. Infrared spectra for "intact" dry (squashed) radicle tips of cauliflower seeds containing about 0.05 g of water per g of dry weight. As temperature is increased from -17$^{o}$C (gel phase) to 19$^{o}$C (liquid crystalline phase) the vibrational frequency (wave number) for the $CH_2$ bands increases.

**Phase diagram for an intact cell**

When physiological data such as germination capacity and $K^+$-leakage are also plotted versus temperature , they closely match the shift in wave number for a given initial moisture content (Fig. 12). Similar to $T_m$ for the lipid, one can calculate $G_m$, the temperature at which half of the maximal germination capacity is reached. By doing this for a range of initial moisture contents, sufficient values of $T_m$ and $G_m$ were collected to construct a hydration dependent phase diagram for intact Typha pollen (Fig. 13). This is the first such diagram for a living cell. The diagram has predictive value in that at all combinations of moisture content and temperature below the curve, the bulk of phospholipids are in gel phase, and above the curve in liquid crystalline phase. Imbibition at the given combinations of initial moisture content and temperature below the curve leads to leakage and loss of viability, and above the curve to retention of solutes and germination.

The shape of the $T_m$ curve in an intact cell is remarkably similar to that of a phospholipid such as DPPC (see Fig. 3). However, the absolute difference between $T_m$ values in hydrated and dry intact pollen seems to be 30°C less than usually found for a phospholipid. This may be the result of the moderating effect of sucrose which in Typha pollen represents about 23% of the dry weight. Sucrose makes up 96% of soluble carbohydrates present in this species. The data in Table I on the transition temperature of microsomal membranes isolated from Typha pollen support this concept. Whereas the hydrated transition is below 0°C, as in hydrated intact pollen, the dry $T_m$ is much higher than that of intact dry pollen. Adding sucrose to the microsomes before freeze drying considerably depressed $T_m$ to just under the $T_m$ of dry pollen.

## MECHANISM OF IMBIBITIONAL LEAKAGE: AN HYPOTHESIS

FTIR work has provided evidence that desiccation leads to an increase in the gel to liquid crystalline transition temperature of phospholipids in intact cells. Depending on the interaction with sugars, this increase may be more or less extensive. The similarities of the $T_m$ and $G_m$ plots in Fig.

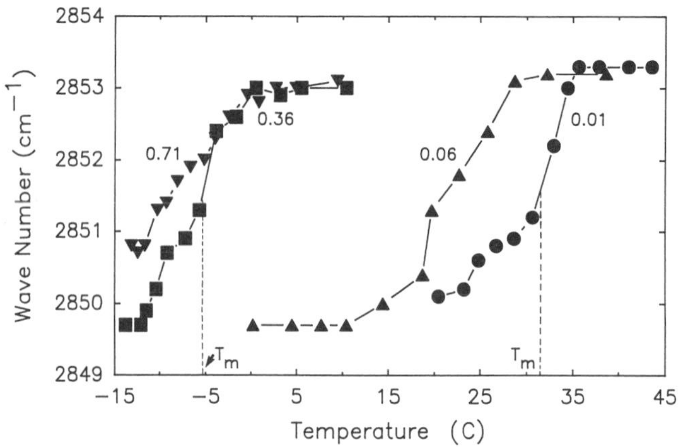


Fig. 11. Vibrational frequencies for $CH_2$ symmetric stretch as a function of temperature in samples of Typha pollen with different moisture contents. Water contents (as g $H_2O$/ g DW) are shown in the numbers alongside each curve. The midpoint of the transition ($T_m$) is indicated for two of the curves.

13 suggest that germination capacity of pollen is closely linked to the status of the membranes at imbibition. Presence of gel phase lipid leads to failure of germination.

The protective effect of prehydration in humid air might be explained as follows. The slow re-occupation by water molecules of the polar head groups slowly drives the transition temperature down. The mixed phases that

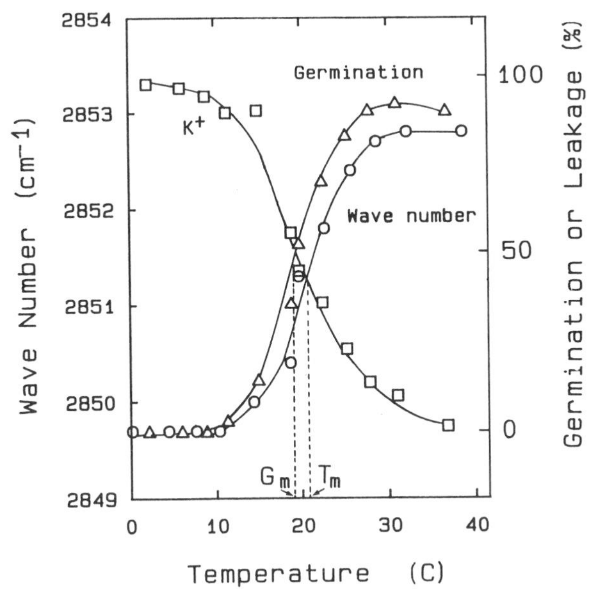


Fig. 12. Plot of the frequency of the absorbance maxima of the $CH_2$ symmetric stretch versus temperature for pollen of Typha (water content about 0.06 g $H_2O$/g DW). Curves of germination and $K^+$-leakage versus temperature are also plotted. $T_m$ and $G_m$ are "mid" values.

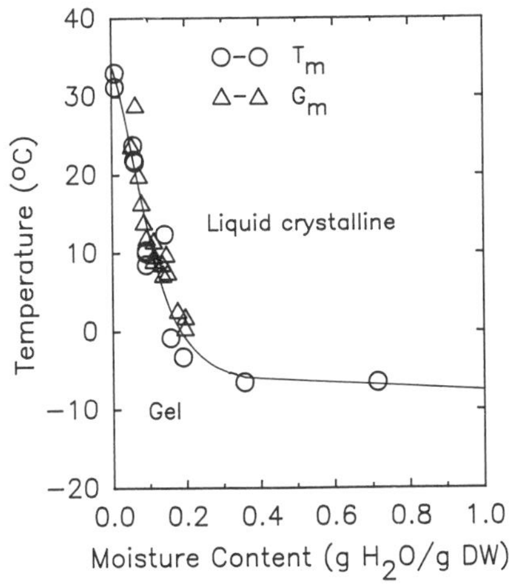


Fig. 13. Phase diagram of membrane polar lipids in intact Typha pollen. $G_m$ and $T_m$ values were determined as in Fig. 12.

Table I. Transition temperatures ($T_m$) of Typha pollen material, measured by FTIR spectroscopy.

| Pollen Material | $T_m$ ($^oC$) |
|---|---|
| **Intact Pollen** | |
| Dry (less than 0.01 g/g) | 33 |
| Hydrated (0.7 g/g) | -7 |
| **Microsomal Membranes** | |
| Dry | 59 |
| Hydrated | -6 |
| Dry + Sucrose | 25 |

will occur when the $T_m$ passes room temperature, might permit leakage. However, there is not enough free water for the transport of solutes out of the cell. When, eventually, free water becomes available, the transition temperature has already dropped below $0^oC$. In contrast, the situation during addition of liquid water is entirely different. During the short period that mixed phases exist, excess water is present, thus leakage can occur. These ideas are supported by examples from the seed literature on the beneficial effect of slowing down the rate of water uptake during cold

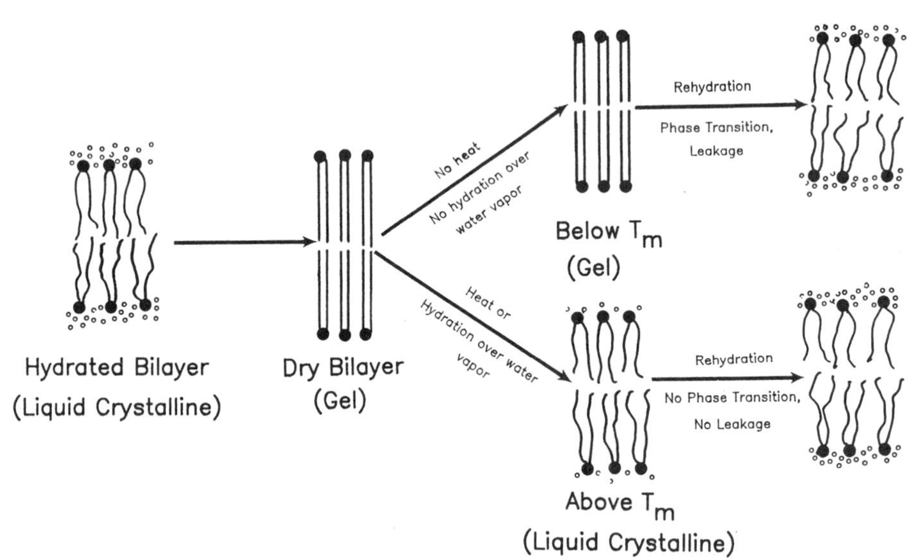


Fig. 14. Diagrammatic representation of the proposed mechanism for imbibitional leakage. Depending on the temperature a bilayer may enter the gel phase during dehydration. When this bilayer is not heated or prehumidified by exposure to water vapor (upper pathway) addition of bulk water will cause a phase change to liquid crystalline, and leakage and death are the result. When the dry bilayer is heated or exposed to water vapor (lower pathway), it can pass through the phase transition in the absence of bulk water. Under these better conditions addition of liquid water will not cause a phase transition, and leakage is prevented. Reproduced from Crowe et al. (1989b).

imbibition (reviewed by Herner, 1986). Thus a humid wave may precede liquid water, which can depress $T_m$ prior to the arrival of the liquid. A high rate of water uptake at elevated temperatures does not harm dry cells, because the warming up *per se* melts gel phase lipid at imbibition. These ideas are summarized in Fig. 14. Any treatment that leads to avoidance of a phase transition during the entry of bulk water preserves vitality and reduces leakage. It must be emphasized that in comparison to pollen, seeds are much more complex in their behavior toward imbibition stress, because of the presence of a seed coat or testa that impedes water uptake and hampers leakage. The presence of differentiated tissues may permit a certain extent of injury, while not necessarily causing loss of viability for the entire seed.

It should be expected that structural aspects are involved in the injury. Gel phase lipid in membranes may not be flexible enough to withstand the forces of penetrating masses of liquid water (Hoekstra and van der Wal, 1988). The extensive vesiculation of lipids as observed in imbibitionally damaged *Typha* pollen grains (Sack et al., 1988) may be an indication for this lack of flexibility.

## LIMITATIONS OF THE FTIR METHOD

*Typha* pollen gave reasonable plots of frequency versus temperature over a 15-20°C range, from which a $T_m$ value could be calculated. However, one might expect problems when membranes in cells have widely different compositions. In another pollen species we observed very broad transitions, over 40°C, with clearly several subtransitions of particular classes of membranes. It will be extremely difficult to derive a $T_m$ value with a reasonable degree of confidence from such plots. Alternatively, if a plasma membrane, consisting of only a few percent of the total phospholipid content in a cell, deviates strongly in composition from the bulk phospholipid, physiological data may not closely match $T_m$ values. One such complicated plot is shown in Fig. 15, representing dry and hydrated radicle tips of cauliflower seeds. The steep part around -10°C of the dry material may be linked to neutral lipid. What can be established with certainty here is that desiccation leads to an increase in $T_m$ of membrane lipids, since hydration shifts the curve to the lower temperature range (left). Problems

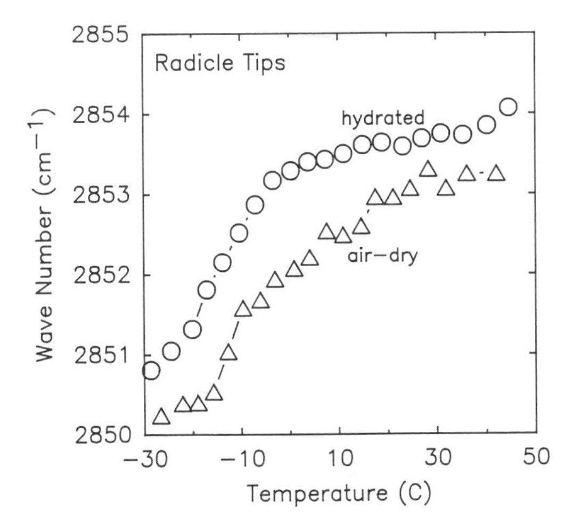


Fig. 15. Frequency change of the absorbance maxima of the $CH_2$ symmetric stretch of hydrated and dry radicle tips of cauliflower seeds versus temperature.

with the possible application of the FTIR spectroscopic technique to the study of desiccation sensitivity may arise from the fact that drought may degrade cells to a certain extent. This renders it extremely difficult to compare such cells with dry, desiccation tolerant cells.

## CONCLUDING REMARKS AND OUTLOOK

Desiccation and rehydration have profound effects on membrane integrity of seeds and pollen. Based on the behavior of model lipid systems exposed to conditions of low water activity, we have tried to predict what will occur in desiccation sensitive and tolerant cells. Fourier transform infrared spectroscopy enables the analysis of lipid behavior in intact dry cells, and shows that even in anhydrobiotic organisms the gel to liquid crystalline transition temperature rises with dehydration. We developed a hypothesis for the mechanism of imbibitional (chilling) injury which explains the involvement of initial moisture content and temperature of imbibition in the phenomenon.

Sucrose depresses the phase transition temperature during desiccation, and most likely enables pollen and seeds to withstand drought and, to a certain extent, circumvent imbibitional injury. We did not review the behavior of cytosolic and membrane proteins during desiccation. They can also undergo transitions due to water loss, sometimes irreversibly. Several disaccharides such as trehalose and sucrose are also effective at preserving proteins against the adverse effects of water loss (Carpenter et al., 1987a and b).

Whereas abscisic acid seems to play a regulatory role during seed maturation with regard to prevention of precocious germination and development of desiccation tolerance, data on pollen are scarce. However, in an abstract by Lipp and Kusters (1988) the occurrence of peaks in ABA levels in pollen several days before anthesis is mentioned. This is an indication that the regulation of drought tolerance may be similar in seeds and in pollen.

Abscisic acid and early desiccation induce considerable changes in protein synthesis in seeds (Quatrano, 1986; Kermode et al., 1989). Stress induced trehalose accumulation was shown to require de novo RNA synthesis (Attfield, 1987). It is of great importance that in future research the identity and function of de novo proteins associated with the development of desiccation tolerance are elucidated. This will enable a connection between biochemical research and the biophysical approach described in this paper. It may then be possible to more fully appreciate the tactics used by anhydrobiotic organisms to escape from the adverse effects of desiccation and utilize it to the benefit of their survival.

**Acknowledgements:** We gratefully acknowledge the help of Dr. Tony Haigh in providing large quantities of the tiny axes and radicle tips of dry cauliflower ("Bloemkool") seeds, and the improvements he made in the text. This work was supported by grant 88-37264-4068 from the U.S. Department of Agriculture Competitive Grants Program.

## REFERENCES

Attfield, P. V. , 1987, Trehalose accumulates in Saccharomyces cerevisiae during exposure to agents that induce heat shock response, FEBS Letters, **225:** 259.

Carpenter, J. F., Crowe, L. M. and Crowe, J.H., 1987a, Stabilization of phosphofructokinase with sugars during freeze-drying: character-

ization of enhanced protection in the presence of divalent cations, Biochim. Biophys. Acta, **923**:109.
Carpenter, J. F., Martin, B., Crowe, L. M. and Crowe, J.H., 1987b, Stabilization of phosphofructokinase during air-drying with sugars and sugar/transition metal mixtures, Cryobiol., **24**:455.
Chapman, D., Williams, R. M. and Ladbrooke, B. D., 1967, Physical studies of phospholipids VI: thermotropic and lyotropic mesomorphism of some 1,2-diacyl-phosphatidylcholines (lecithins), Chem. Phys. Lipids, **1**:445.
Crowe, J. H. and Clegg, J. S., 1978, Dry Biological Systems, Acad. Press, New York.
Crowe, J. H., Crowe, L. M. and Chapman, D., 1984, Preservation of membranes in anhydrobiotic organisms: the role of trehalose, Science, **223**:701.
Crowe, J. H., Crowe, L. M., Carpenter, J. F. and Aurell Wistrom, C., 1987, Stabilization of dry phospholipid bilayers and proteins by sugars, Biochem. J., **242**:1.
Crowe, J. H., Crowe, L. M., Carpenter, J. F., Rudolph, A. S., Aurell-Wistrom, C., Spargo, B. J. and Anchordoguy, T. J., 1988, Interactions of sugars with membranes, Biochim. Biophys. Acta, **947**:367.
Crowe, J. H., Crowe, L. M., Hoekstra, F. A. and Aurell Wistrom, C., 1989a, Effects of water on the stability of phospholipid bilayers: the problem of imbibition damage in dry organisms, Crop Sci. Soc. Am. Special Publ. no. **14**:1.
Crowe, J. H., Hoekstra, F. A. and Crowe, L. M., 1989b, Membrane phase transitions are responsible for imbibitional damage in dry organisms, Proc. Natl. Acad. Sci. USA, **86**:520.
Crowe, J. H., Hoekstra, F. A., Crowe, L. M., Anchordoguy, T. J. and Drobnis, E., 1989c, Lipid phase transitions measured in intact cells with Fourier Transform infrared spectroscopy, Cryobiol., **26**:76.
Crowe, J. H., Crowe, L. M., and Hoekstra, F. A., 1989d, Phase transitions and permeability changes in dry membranes during rehydration, J. Bioenerg. Biomembranes, **21**:77.
Crowe, L. M. and Crowe J. H., 1982, Hydration-dependent hexagonal phase lipid in a biological membrane, Arch. Biochem. Biophys., **217**:582.
Crowe, L. M., Crowe, J. H. and Chapman, D., 1985, Interaction of carbohydrates with dry dipalmitoylphosphatidylcholine, Arch. Biochem. Biophys., **236**:289.
Crowe, L. M., Womersley, C., Crowe, J. H., Reid, D., Appel, L. and Rudolph, A., 1986, Prevention of fusion and leakage in freeze-dried liposomes by carbohydrates, Biochim. Biophys. Acta **861**:131.
Crowe L. M. and Crowe J. H., 1988, Effects of water and carbohydrates on membrane fluidity, in: "Physiological regulation of membrane fluidity", R. C. Aloia, C. C. Curtain and L. M. Gordon, eds., pp 75-99, Alan R. Liss Inc., New York.
Cullis, P. R. and De Kruijff, B., 1979, Lipid polymorphism and the functional roles of lipids in biological membranes, Biochim. Biophys. Acta, **559**:399.
Fischer, W., Bergfeld, R., Planchy, C., Schafer, R. and Schopfer, P., 1988, Accumulation of storage materials, precocious germination and development of desiccation tolerance during seed maturation in mustard (Sinapis alba L.), Bot. Acta, **101**:344.
Herner, R. C., 1986, Germination under cold soil conditions, HortScience, **21**:1118.
Hoekstra F. A., 1984, Imbibitional chilling injury in pollen: involvement of the respiratory chain, Plant Physiol., **74**:815.
Hoekstra, F. A., 1986, Water content in relation to stress in pollen, in: "Membranes metabolism and dry organisms", pp. 102-122, A. C. Leopold, ed., Comstock Publ. Ass., Ithaca/London.
Hoekstra, F. A. and Van der Wal, E. G., 1988, Initial moisture content and temperature of imbibition determine extent of imbibitional injury in pollen, J. Plant Physiol., **133**:257.

Hoekstra, F. A. and Van Roekel, T., 1988, Desiccation tolerance of Papaver dubium L. pollen during its development in the anther: possible role of phospholipid composition and sucrose content, Plant Physiol. **88**:626.

Hoekstra, F. A., Van Roekel, T and Ten Pas, N., 1988, Pollen maturation and desiccation tolerance, in: "Sexual reproduction in higher plants", M. Cresti, P. Gori and E. Pacini, eds., Springer Verlag, Berlin/New York.

Hoekstra, F. A., Crowe, L. M. and Crowe, J.H., 1989, Differential desiccation sensitivity of corn and Pennisetum pollen linked to their sucrose contents, Plant Cell Environment, **12**:83.

Kermode, A. R., Oishi, M. Y. and Bewley, J. D., 1989, Regulatory roles for desiccation and abscisic acid in seed development: A comparison of the evidence from whole seeds and isolated embryos, Crop Sci. Soc. Am. special publ. #14:23.

Koster, K. L. and Leopold, A. C., 1988, Sugars and desiccation tolerance in seeds, Plant Physiol., **88**:829.

Leopold, A. C., 1986, Membranes, Metabolism, and Dry Organisms, Comstock Publ. Ass., Ithaca/London.

Lipp, J. and Kusters, A., 1988, ABA, IAA and proline in pollen of different developmental stages, Abstract # 734, 6th Congress of FESPP, Split, Yugoslavia.

Madin, K. A. C. and Crowe, J. H., 1975, Anhydrobiosis in nematodes: carbohydrate and lipid metabolism during dehydration, J. Exp. Zool., **193**:335.

McKersie, B. D. and Stinson, R. H., 1980, Effect of dehydration on leakage and membrane structure in Lotus corniculatus L. seeds, Plant Physiol., **66**:316.

Platt-Aloia, K. A., Lord, E. M., DeMason, D. A. and Thomson, W. W., 1986, Freeze-fracture observations on membranes of dry and hydrated pollen from Collomia, Phoenix and Zea, Planta, **168**:291.

Pollock, B. M., 1969, Imbibition temperature sensitivity of lima bean seeds controlled by initial seed moisture, Plant Physiol., **44**:907.

Priestley, D. A. and Leopold, A. C., 1980, The relevance of seed membrane lipids to imbibitional chilling effects, Physiol. Plant., **49**:198.

Priestley, D. A. and DeKruijff, B, 1982, Phospholipid motional characteristics in a dry biological system, Plant Physiol., **70**:1075.

Quatrano, R. S., 1986, Regulation of gene expression by abscisic acid during angiosperm embryo development., Oxford Surv. Plant Mol. Cell Biol., **3**:467.

Rudolph, A. S., Crowe, J. H. and Crowe, L. M., 1986, Effects of three stabilizing agents-proline, betaine, and trehalose- on membrane phospholipids, Arch. Biochem. Biophys., **245**:134.

Sack, F. D., Leopold, A. C. and Hoekstra, F. A., 1988, Structural correlates of imbibitional injury in Typha pollen, Amer. J. Bot., **75**:570.

Seddon, J. M., Cevc, G. and Marsh, D., 1983, Calorimetric studies of the gel-fluid and lamellar-inverted hexagonal phase transitions in dialkyl- and diacylphosphatidylethanolamines, Biochemistry, **22**:1280.

Seewaldt, V., Priestley, D. A., Leopold, A. C., Feigenson, G. W. and Goodsaid-Zalduondo, F., 1981, Membrane organization in soybean seeds during hydration, Planta, **152**:19.

Shivanna, K. R. and Heslop-Harrison, J., 1981, Membrane state and pollen viability, Ann. Bot. **47**:759.

Simon, E. W., 1974, Phospholipids and plant membrane permeability, New Phytol., **73**:377.

Van Steveninck, J. and Ledeboer, A. M., 1974, Phase transitions in the yeast cell membrane: the influence of temperature on the reconstitution of active dry yeast, Biochim. Biophys. Acta, **352**:64.

Vigh, L., Huitema, H. , Woltjes, J. and Van Hasselt, P. R., 1986, Drought stress-induced changes in the composition and physical state of phospholipids in wheat, Physiol. Plant., **67**:92.

# THE BASIS OF RECALCITRANT SEED BEHAVIOUR

## Cell Biology of the Homoiohydrous Seed Condition

Patricia Berjak, Jill M. Farrant and N.W. Pammenter

Plant Cell Biology Research Group, Department of Biology
University of Natal, King George V Avenue
Durban, 4001 South Africa

## INTRODUCTION

The term 'recalcitrance', defined as obstinate disobedience, refers to seeds that undergo no maturation drying as the final phase of development, tolerate very little post-shedding desiccation and are often chilling-sensitive. Such seeds are unstorable by any of the methods used for air-dry orthodox seeds. Since these terms were introduced by Roberts in 1973, much of the widely-disseminated literature has been systematically collated to afford an overview of recalcitrant seeds, particularly those of crop species (Chin and Roberts, 1980). Two major unresolved issues emerged from that overview: there was no explanation of the basis of recalcitrant seed behaviour, and no successful storage regimes had been established. The present contribution deals with progress that has been made towards an understanding of the responses of post-harvest, recalcitrant seeds in terms of their cell biology.

Plants that produce recalcitrant seeds generally occur in habitats conducive to relatively rapid, if not immediate, seedling establishment, such as aquatic or marshy environments and humid forests, usually where there is no low temperature constraint (Roberts and King, 1980). In such environments there can be little selective advantage to maturation drying. The lack of maturation drying and the inability of recalcitrant seeds to tolerate much post-shedding desiccation, might be viewed as an evolutionary 'hangover', on the assumption that the acquision of these properties occurred subsequent to the development of the seed habit. In this regard it is interesting that many of the extant gymnosperms of tropical/sub-tropical distribution have seeds that are apparently recalcitrant (unpublished observations).

The storage behaviour of recalcitrant seeds is generally documented as a record of their short lifespan, even under fully (or almost fully) hydrated conditions, the moisture content at which damage occurs, and the lower limit of temperature tolerated (King and Roberts, 1980). However, recalcitrance is not an all-or-none phenomenon: there are varying degrees of recalcitrance. A study of the published data which are available has led to the proposal that

*Recent Advances in the Development and Germination of Seeds*
Edited by R.B. Taylorson
Plenum Press, New York

there is a continuum of recalcitrant seed types, which may be grouped as showing a minimum, moderate or high degree of recalcitrance (Farrant, Pammenter and Berjak, 1988a). According to that grouping, seeds in the first category will tolerate a fair amount of water loss and relatively low temperatures, the species concerned being indigenous to temperate and sub-tropical regions. Highly recalcitrant seeds produced by tropical forest and wetland species, will tolerate

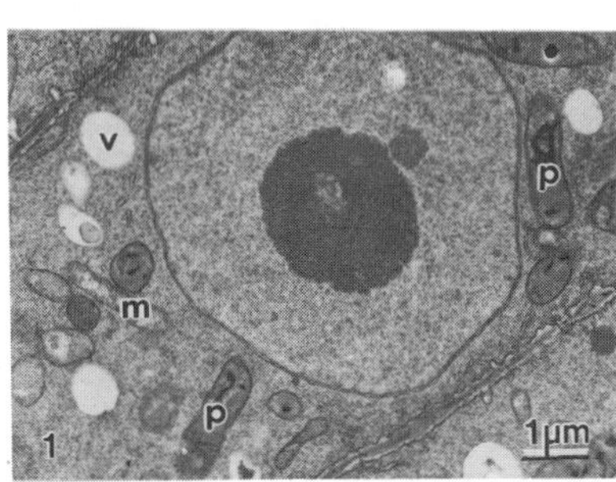


*A. marina;* newly-shed. v, vacuole, p, plastid, m, mitochondrion

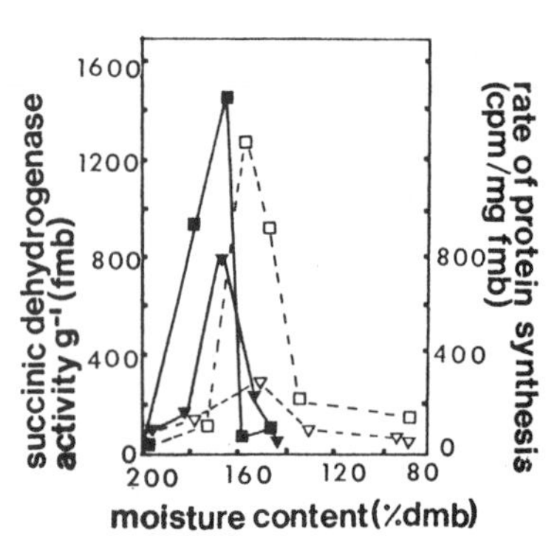


**Figure 1.** Succinic dehydrogenase activity (■,□) and protein synthesis rate (▼,▽) in relation to axis water content. *A. marina* seeds stored at 80% RH (——) and 10% RH (- - - -). (After Farrant et al., 1985).

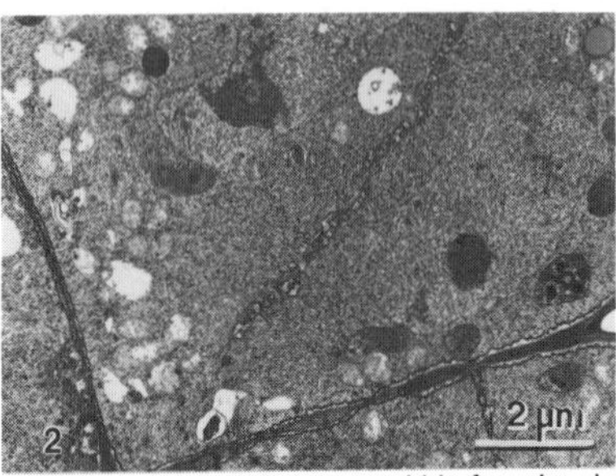


*A. marina;* cell division occurs within four days in wet storage

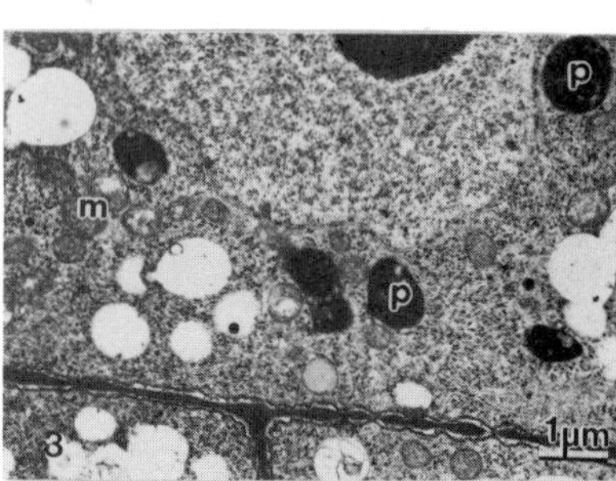


*A. marina;* 24 h after planting; p, plastid; m, mitochondrion

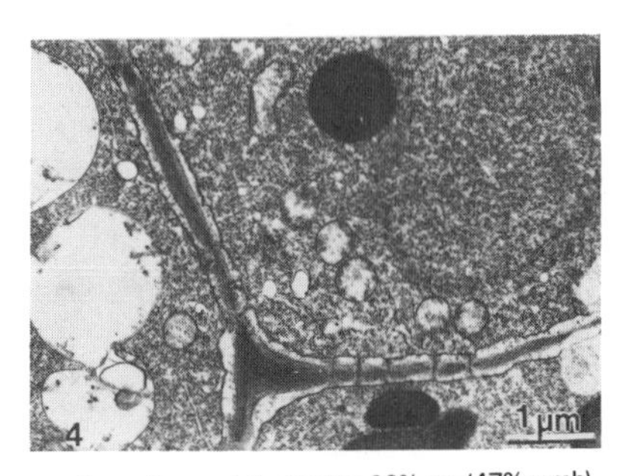


*A. marina;* rapidly dried to 90% mc (47% wmb)

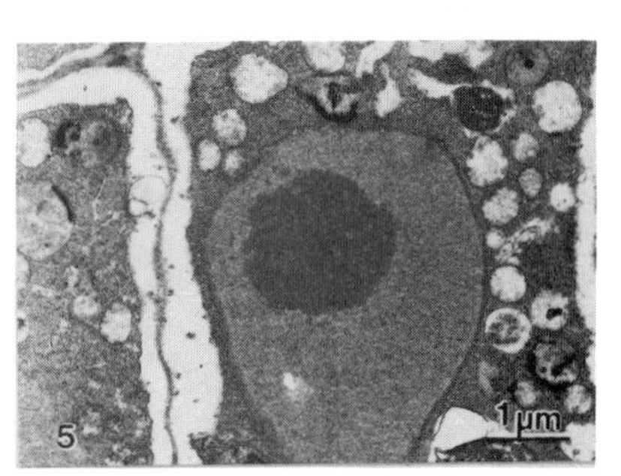


*A. marina;* slowly dried to 90% mc (47% wmb)

only a little water loss and are generally markedly temperature-sensitive. Seeds showing a moderate degree of recalcitrant behaviour are tropical in distribution and characterised by the intermediate degree of dehydration and mid-range temperatures tolerated.

Another general characteristic of recalcitrant seeds is the variation in moisture content at shedding, within any particular harvest of a single species. Results of a survey of 21 recalcitrant species, have shown that without exception there is a marked variation in moisture content on a seed-to-seed basis for both embryonic axes and storage tissues (Report of the International Seed Testing Association [ISTA] Seed Moisture Committee, 1989).

## SEEDS OF AVICENNIA MARINA - A CASE HISTORY

Avicennia marina is a mangrove species with highly recalcitrant seeds, that lose viability within 16 days. The embryonic axis, which has a moisture content (mc) of 160 - 190% dry mass basis [dmb] (62 - 67% wet/fresh mass basis [wmb/fmb]) has a well developed shoot apex and several defined root primordia when the seeds are shed.

Initial studies on the response of A. marina seeds to relatively slow desiccation (Berjak, Dini and Pammenter, 1984) showed an enhanced state of axis subcellular development despite 15% water loss in the short-term, compared with newly-shed material. Continued dehydration caused deleterious subcellular changes, culminating in seed death in 14 days, after the loss of 35% of the initial water content. Neither the enhanced development nor the deleterious changes resembled the disassembly accompanying maturation drying of orthodox seeds (Bain and Mercer, 1966; Klein and Pollock, 1968; Hallam, 1972). These results were interpreted as germination initiation of the hydrated seeds, which was curtailed by continued water loss (Berjak et al., 1984).

Subsequent studies on seeds maintained at their original water content (wet stored) showed that material stored for 3 - 4 days had a higher germination index than material stored for one day. If the index (Timson, 1965) was calculated from the start of storage rather than from the day of planting (after storage), the enhancement was eliminated. These results were interpreted as supporting the proposal of germination initiation in the hydrated seeds during storage (Pammenter, Farrant and Berjak, 1984).

Although A. marina seeds are hydrated, the ultrastructural situation upon shedding was shown to be one of relative quiescence (Farrant, Berjak and Pammenter, 1985). Root primordium cells had mitochondria characterised by electron transparent matrices and limited crista development, little polysome formation or endomembrane development, a lack of stored reserves in the plastids, and only a small degree of vacuolation (Plate 1). The levels of succinic dehydrogenase activity and protein synthesis were low (Fig. 1), in keeping with this appearance of relative quiescence. After a short period in storage, there was evidence of enhanced subcellular activity (Plate 2) and increases in both succinic dehydrogenase activity and protein synthesis occurred (Fig. 1). The subcellular events were similar to those characterising seeds planted out immediately upon shedding (Plate 3), although proceeding more slowly in storage (Farrant et al., 1985).

The progress of germination-associated events in wet-stored

recalcitrant seeds implies that they should become more desiccation sensitive with increasing storage time. Rapid dehydration to a constant moisture content using PEG 6000 solution, confirmed this (Farrant, Pammenter and Berjak, 1986). A reduction of 18% of the initial water content had no effect on newly-shed seeds, but caused considerable subcellular damage in seeds stored for 3 - 4 days and reduced the rate of subsequent germination (calculated according to Czabator, 1962). Fresh seeds that had been planted out for 24 hours were similarly adversely affected. For seeds stored for 10 days, the same degree of dehydration caused extensive ultrastructural derangement and reduced totality of germination as well. The rate of dehydration was also found to influence desiccation sensitivity (Plates 4 & 5), rapidly dried seeds surviving to lower water contents than those dried slowly (Farrant et al., 1985). The ultrastructure of root primordium cells showed that less germination-associated change had occurred in rapidly dried seeds than in those dried slowly to the same moisture content.

These data are consistent with the hypothesis that germination of recalcitrant seeds is initiated at, or around, shedding and as the germination-associated changes continue in storage the seeds become increasingly desiccation sensitive. Even under wet storage conditions the seeds ultimately lose viability, so to complete the germination process and establish seedlings, additional water is required. It has been suggested that this coincides with the stage of cell division and extensive vacuolation (Farrant, Pammenter and Berjak, 1988a).

Recalcitrant seeds should thus be viewed as developing seedlings rather than quiescent seeds, although the rate of development will vary considerably among species. A model based on the observations on A. marina (Fig. 2a) but suggested also to be applicable to other species, was developed to explain the behaviour of recalcitrant seeds in these terms (Farrant et al., 1986, 1988a). Although the abscissa is a time-dependent axis, it is not linearly so. The origin coincides with natural shedding and the rate of the germination process will vary widely among species.

## BROAD APPLICABILITY OF THE MODEL

Reference to the literature shows that there is wide variation among recalcitrant seed species in the moisture content at which viability is lost and the proportion of water loss that can be tolerated, and many species are also temperature sensitive (Chin and Roberts, 1980; Chin, Aziz, Ang and Samsidar, 1981; Corbineau and Côme, 1988). There is broad agreement that viability is best retained by storage of recalcitrant seeds in an hydrated state, at as low a temperature as is consistent with chilling sensitivity on a species basis. However, even under the most favourable conditions, viability will decline within a relatively short period (King and Roberts, 1980, 1982; Corbineau and Côme, 1988; Farrant, Pammenter and Berjak, 1989). Can the observations made on a wide variety of recalcitrant seed species be explained in terms of the proposed model?

Investigations have been carried out in our laboratory on seeds of several species showing differing degrees of recalcitrant behaviour, and all have responded in terms of the generalisations made in the model. An in-depth study was conducted on seeds of three unrelated species, Araucaria angustifolia, Scadoxus membranaceus and Landolphia kirkii (Farrant et al., 1989). There is considerable interspecific difference in initial moisture content and the time taken for newly-shed seeds to germinate.

The level of subcellular organisation increased in the meristematic tissue of the three species during short-term wet storage, differing only in the timing of the events and the pattern of reserve utilisation/deposition. In all cases mitochondrial differentiation, strong development of the endomembrane system, and assembly of polysomes occurred. Cell division - which is considered the ultimate marker event in germination initiation - occurred during wet storage in S. membranaceus and L. kirkii (Farrant et al., 1989) and has been subsequently observed for A. angustifolia. Shortly after this, although seed moisture content remained at the original level, viability declined. This is in accordance with the model, which suggests that additional water is required at this stage. Seeds of the three species all lost viability under desiccating storage conditions, but there was no correlation between the initial water content and proportional loss which could be tolerated. However, the rate at which water was lost under the same conditions varied and was correlated with the proportion of water loss tolerated. A. angustifolia seeds dried relatively slowly, but were the most desiccation-sensitive; seeds of L. kirkii dried the most rapidly, but tolerated the greatest water loss; those of S. membranaceus were intermediate for both parameters (Farrant et al., 1989).

The analysis of data about recalcitrant seeds is far from straightforward. In the first instance, it is important to stress that recalcitrant seeds may behave very differently depending on when in the season they are harvested, and on whether they have been naturally shed or the fruits picked. Because there is no clear-cut terminal event common to the development of recalcitrant seeds on the parent plant, it is often impossible to determine the absolute state of maturity on hand-harvesting. Thus such seeds may be pre-mature, at least physiologically, which could prolong the initial phase prior to the events of germination proper (Fig. 2). However, hand-harvesting might be the method of choice regarding collection of such seeds for experimental purposes, as it obviates the uncertainty arising about time of shedding of fallen seeds. We have also found considerable differences in both germination rate and wet storage lifespan among batches of seeds of the same species, hand-harvested early, midway or late in the season. Another point concerning data analysis is that, as the seeds of many of these species start to lose water from the time they are shed, 'newly-collected' and 'newly-harvested' often are not synonymous. Coupled with this is the fact that investigations may be carried out half-the-world away from the place of origin of the seeds. Under these circumstances, the history of the seeds from shedding may be difficult to ascertain precisely, and their packaging from collection to despatch may well allow water loss. Additionally, recalcitrant seeds being hydrated and metabolic, apparently cannot tolerate long-distance transport in an aircraft hold which is not adequately pressurised (Lins, pers. comm.). Further, if germination initiation on, or soon after, shedding is the rule for those recalcitrant seeds with well-differentiated (although not necessarily large) axes, then by the time they are available for storage manipulation, such seeds may well not be truly representative of the newly shed condition.

Investigations on seeds of Hevea brasiliensis illustrate these difficulties graphically (Berjak, 1989). Newly-shed seeds germinate within 10 days (Chin, 1980), although the embryonic axis is not fully differentiated on shedding (Berjak, Wesley-Smith and Mycock, 1988; Berjak, 1989). Thus they undergo the germination progression, per se,

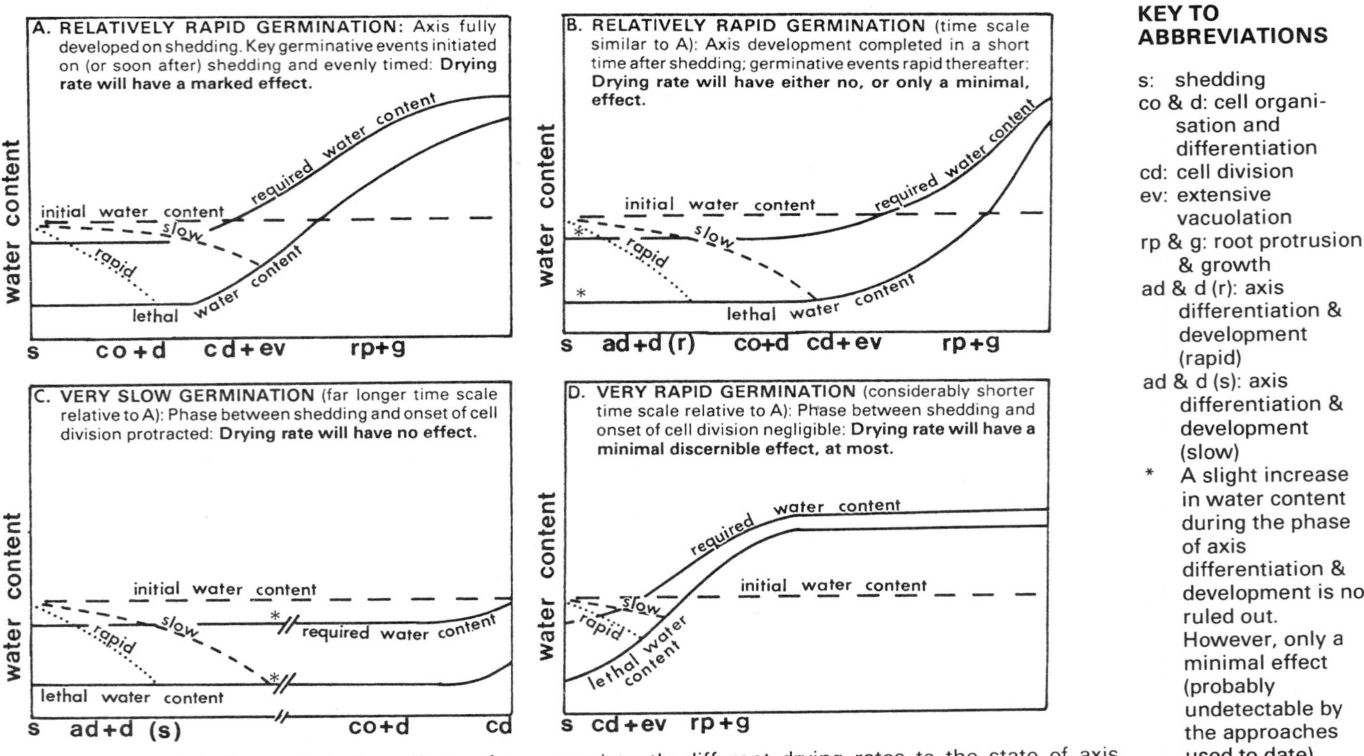


**Figure 2** relates dissimilar and similar effects of two consistently different drying rates to the state of axis development at shedding and the timing of the germination-related processes.

**KEY TO ABBREVIATIONS**

s: shedding
co & d: cell organisation and differentiation
cd: cell division
ev: extensive vacuolation
rp & g: root protrusion & growth
ad & d (r): axis differentiation & development (rapid)
ad & d (s): axis differentiation & development (slow)
* A slight increase in water content during the phase of axis differentiation & development is not ruled out. However, only a minimal effect (probably undetectable by the approaches used to date) would be expected.

rapidly (Fig. 2b). Comparison was made between seeds conveyed to the laboratory in closed polythene bags immediately they were shed, and those collected and transported over a number of days from some distance. The situation of axis cells from the latter showed that considerable subcellular deterioration had accompanied the modest water loss that had occurred, compared with the newly-shed condition (Plates 6 & 7). These seeds were able to repair sub-lethal intracellular damage when planted out, but germination was slow and totality was reduced. They were also able to effect a measure of subcellular repair during wet storage but showed a considerably curtailed lifespan, and lost viability at moisture contents significantly above those documented for this species by Chin et al. (1981).

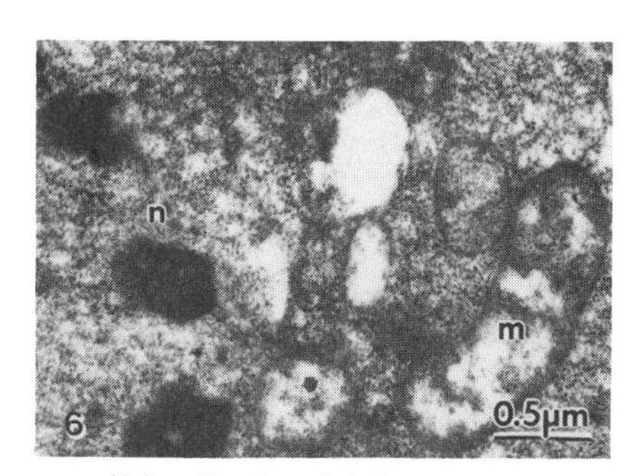


*H. brasiliensis;* c. 9 d after shedding, n, nucleus; m, mitochondrion

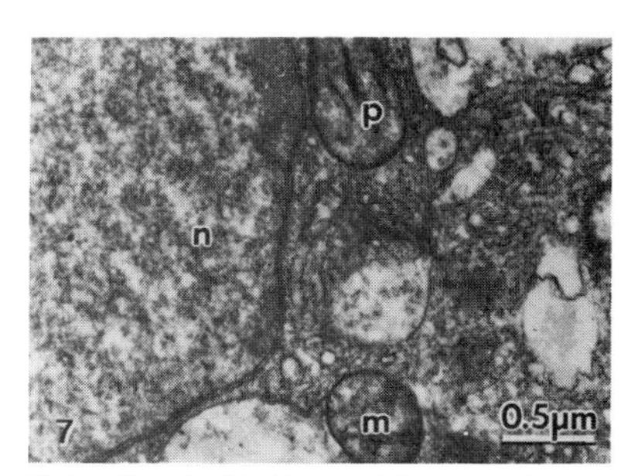


*H. brasiliensis;* 1 h after shedding; n, nucleus; m, mitochondrion; p, plastid

A further difficulty in the interpretation of results from experiments on recalcitrant seeds arises from the difference in moisture content between axis and storage tissues within individual seeds (Report of the ISTA Seed Moisture Committee, 1989). Thus one cannot compare data sets in which moisture content has been reported on a whole seed basis in one instance, and on the embryonic axes in another (e.g. for Araucaria spp.; c.f. Tompsett [1984] and Farrant et al. [1989]). This difficulty is exacerbated in cases where the embryonic axis constitutes only an insignificant fraction of the total seed volume, but is at a considerably higher moisture content than the storage tissues (Report of the ISTA Seed Moisture Committee, 1989). Assessment of the effects of moisture loss, and in particular, the effects of differential drying rates, can be complicated by the fact that some seeds may lose a certain proportion of water slowly, after which dehydration proceeds much more quickly under the same conditions (Prichard and Prendergast, 1986). This underlines the necessity for frequent sampling for moisture content determinations and viability assessments.

All 15 species of recalcitrant seeds investigated in our laboratory have shown enhanced germination rates following wet storage, interpreted as the consequence of germination initiation at, or shortly after, shedding. The eight species we have investigated ultrastructurally have all yielded evidence of the ongoing progression of germination under wet storage conditions. However, the time taken for visible germination to be manifested in wet storage (where this stage was reached), and the timing of the key events of the germination progression, varied among the species investigated.

These are considered to be critical issues when assessing the effects of differential drying rates. There are reports in the literature stating that differential drying rates apparently have no effect on the desiccation sensitivity of recalcitrant seeds (Tompsett, 1982, 1984, 1987; Probert and Longley, 1989). However, we are of the opinion that those results can be explained in terms of the rates of drying relative to the rates of germination in storage. A detailed discussion of Figure 2 might clarify the issue, noting that in all cases the origin of the abscissa corresponds to natural seed shedding.

Figure 2a illustrates the situation where the embryonic axis is fully differentiated on seed shedding. This is followed by a short period of organisation and differentiation at the cell level, a time scale of approximately 1 - 3 days being involved. During this period the tissue is at, or near, its relatively most desiccation tolerant. This phase is followed by that of cell division and extensive vacuolation, when desiccation sensitivity increases. Root protrusion and growth follow shortly thereafter. Differential drying rates normally achieved in the laboratory will have a marked effect. The seeds of *Avicennia marina* exemplify this behaviour.

However, many of the recalcitrant seed species are shed with relatively undifferentiated axes. Before cell division and subsequent growth can occur, axis differentiation must be completed. This phase, and subsequent germination, can be relatively rapid (Fig. 2b), as is the case for *Hevea brasiliensis*, where the germination of newly-shed seeds occurs within about 10 days (Chin, 1980). However, the period of relatively low desiccation sensitivity is extended compared with the case illustrated in Figure 2a. If the two drying rates are not sufficiently dissimilar, then a differential effect is unlikely to be apparent.

In some species the period of axis differentiation is quite extended and the subsequent germination process might also be slow (Fig. 2c). *Scadoxus membranaceus* exhibits this type of behaviour (Farrant et al., 1989). The period of low desiccation sensitivity can be of the order of months, and, unless drying rates are of this order, no differential effect will be apparent. Probert and Longley (1989) showed that for the recalcitrant seeds of the temperate aquatic grasses, *Zizania palustris* and *Spartina anglica*, enhanced rates and totality of germination (at 16°C) occurred after storage in the imbibed state at 2°C for several months. Although those data indicate a dormancy-breaking effect of low-temperature imbibed storage, this does not preclude some axis development and slow initiation of germination under those conditions.

Finally, recalcitrant seeds may germinate so rapidly that the relatively desiccation tolerant stage is practically obviated (Fig. 2d). In such cases, not even the most rapid drying rate is fast enough to prevent death of most of the seeds on dehydration. Thus there may be no practical difference between the effects of rapid and slower dehydration.

Very few studies on recalcitrant seeds provide data concerning rates of germination or information on changes occurring in storage. Thus it is difficult, if not impossible, to assess many published findings in terms of the model.

An additional problem is that it is often not simple to achieve differential drying rates of seeds. Some recalcitrant seeds are so large that it is not possible to dehydrate them rapidly, no matter how desiccating the conditions. For the genera Araucaria and Dipterocarpus (Tompsett, 1984 & 1987, resp.), it is the species with the largest seeds that dry the slowest and are the most desiccation sensitive. This is in agreement with results for other species (Farrant et al., 1989) and in accordance with the generalisations of the model.

Storage of recalcitrant seeds over saturated solutions of various salts is sometimes used to achieve differential drying rates. Although the seeds will come to different equilibrium moisture contents, the initial rate of desiccation (possibly to below the lethal moisture content) may be quite similar (see, e.g. Figs 1 & 5, Probert and Longley [1989]). This would be equivalent to the situations illustrated in Figures 2b and c. (The slow drying treatment used by Probert and Longley [1989] involved immersion of the seeds in a solution of PEG 6000 of a water potential of -10 MPa. It is not immediately apparent why those seeds took in excess of 90 days to approach equilibrium with this very concentrated solution).

Thus, in conducting experiments concerning the effects of differential drying rates on desiccation sensitivity, it is important that the drying rates and storage times used are commensurate with the rate of germination. An alternative approach is to test the desiccation sensitivity of the seeds (at the same rapid drying rate) after varying periods of storage under conditons that maintain their moisture content (Farrant et al., 1986). Maintenance of constant temperature during the drying treatments is also important, as is the stage of maturity at harvest or collection, if valid conclusions are to be drawn and comparisons made. Totality of germination alone following a particular treatment, reveals little about the vigour of the seeds. Use of a germination index, such as that of Czabator (1962) that considers rate of germination, is essential. For example, we have found that most seedlings produced by originally-debilitated seeds of Hevea brasiliensis, did not establish. Thus reporting on totality of germination assessed by root protrusion and growth, would have been entirely misleading. A final caveat: although the effects of micro-organisms are recognised, these are seldom accorded more than passing comment. However, their incidence and the changing patterns of species composition with changing seed moisture status (Berjak, Farrant, Mycock and Pammenter, 1989a) merits far closer examination, as the microflora might constitute a significant variable in the behaviour of recalcitrant seeds under different conditions (Plate 8).

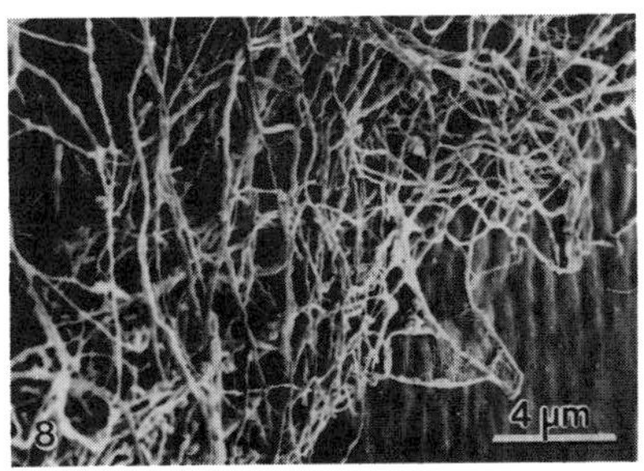

*A. marina;* internal fungal proliferation during wet storage

FLASH DRYING

Although Tompsett (1982) reported that drying rate had no significant effect on lethal water content in Araucaria hunsteinii, viability being completely lost around 14% moisture content (wmb), when Pritchard and Prendergast (1986) rapidly dried down excised embryos of A. hunsteinii to around 13% moisture content in an air-stream, 84% of the root meristems survived in culture. Chin et al. (1981) reported loss of viability of intact Hevea brasiliensis seeds at a moisture content of 20% (wmb). From our studies on differential drying between embryonic axes and storage tissue, we estimate axis moisture content when viability is lost to be about 40% (wmb) for this species. However, Normah, Chin and Hor (1986) have reported that isolated embryos survive to a water content of 16% when dehydration occurs in three hours. Those results support the contention that, provided dehydration occurs before the onset of cell division and extensive vacuolation, the more rapidly a seed can be dried, the more desiccation tolerant it is. We have obtained similar results with another four recalcitrant seed species, Castanospermum australe, Scadoxus membranaceus, Landolphia kirkii and Camellia sinensis: extremely rapid drying (approximately one hour) of excised axes in an air-stream permitted viability retention to much lower moisture contents than did drying of whole seeds. We have termed this extremely rapid dehydration of excised axes, flash drying (Berjak, Farrant, Mycock and Pammenter, 1989b). However, it is impossible to dry large, intact seeds to similar moisture contents in a matter of a few hours.

An investigation has been carried out on the responses of excised axes of the recalcitrant seeds of Landolphia kirkii to flash drying (Berjak et al., 1989b). Intact seeds showed the germination progression in wet storage that is typical of moderately to highly recalcitrant material (Farrant et al., 1989). Relatively slow drying over silica gel reduced viability from 95% to 50% in 15 days, concomitant with a axis water content decline from 220% to 120% (dmb), and after 20 days, when water content had further dropped to 49%, only 7% of the seeds remained viable. In contrast, flash drying allowed the reduction of water content to 13% (dmb) within 60 minutes, with almost full retention of viability. Flash drying was shown to cause a marked compaction of the entire cell content, with no ultrastructurally-visible deterioration of membranes, at least to a water content of 13%. On the other hand, considerable subcellular damage occurred in axes of seeds dried over silica gel such that, even at a water content of 120%, many cells were extensively deteriorated (Berjak et al., 1989b).

Investigations on the recalcitrant behaviour of individual seed species are generally curtailed in any one season due to its brevity and, to state the obvious, the fact that the seeds cannot be stored in their original condition. We are extremely fortunate in having been able to obtain large consignments of hand-harvested fruits of Camellia sinensis (tea) at weekly intervals. This species has a fruiting season lasting some three months locally, enabling detailed studies on the behaviour and responses of the seeds to be carried out. While previous records had suggested that these seeds might be recalcitrant, this had not been unequivocally established (King and Roberts, 1980). Thus our first task was to establish whether or not the seeds of C. sinensis are recalcitrant. It has since been ascertained that these seeds do exhibit all the characteristics that have come to be associated with recalcitrance (Devey, 1989).

Moisture content of newly-harvested seeds changed during the season from 169% in the first month to 280% three months later (63 - 74%, wmb). Early-harvested seeds germinated slowly and could be wet-stored for a longer period, while those harvested from midway through the season germinated far more rapidly and had a proportionally shorter period for which they could be successfully stored. This was a major consideration in the assessment and comparison of the characteristics of the material investigated. Total viability loss occurred within eight days of storage under desiccating conditions for early-harvested material (Devey, Pammenter and Berjak, 1986). Wet storage maintained seed water content constant, during which the germination-associated progression of subcellular events took place. These processes occurred more slowly in wet stored seeds, but were similar in all respects to those occurring in newly-harvested seeds which were immediately planted out (Devey, Pammenter and Berjak, 1987; Devey, 1989). Viability ultimately declined in seeds after long-term wet storage (Devey, 1989), as has been found for all other recalcitrant species.

Mid-season seeds of C. sinensis were used for an in-depth comparison of responses elicited by flash drying of excised axes with those of intact seeds subjected to slower drying over five days. Viability was assessed by growth of isolated axes in tissue culture in both cases. Figure 3 shows that flash-dried isolated axes maintained viability to considerably lower water contents than those excised from intact seeds that had been dehydrated over silica gel, correlating with differential electrolyte leakage.

In common with the embryos from several other recalcitrant seeds we have examined, the axis is very small and shows no radicle development nor discrete root meristem. Instead, the distal portion of the axis is characterised by a narrow band of potentially meristematic cells which is continuous with the cambium of the hypocotyl as a whole. This band of cells was chosen for ultrastructural investigation.

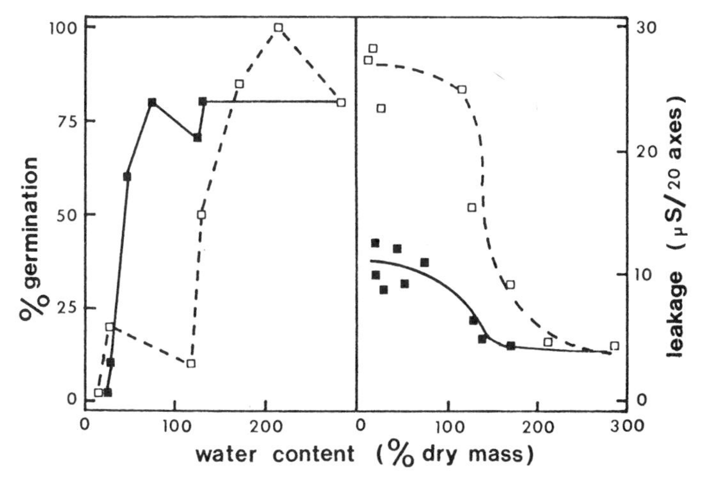


**Figure 3.** The relationship of viability (left) and leakage (right) with water content, for flash-dried (——) axes of *C. sinensis* and those excised from slowly-dried, intact seeds (- - - -).

Although these (and other) recalcitrant seeds are at their relatively most quiescent when newly-harvested, the subcellular organisation (Plates 9 & 10) indicated that they were metabolically active, which has been borne out by biochemical studies (Devey, 1989). The large nuclei, in which the nucleolus and patches of heterochromatin were well defined, dominated the cells, and fairly long rER profiles and polysomes were common. Golgi bodies occurred and well-defined, although relatively undifferentiated plastids largely devoid of storage product, were common. Mitochondria were small and spherical, the relatively well-developed cristae and dense matrix suggesting their activity. Protein vacuoles with variably-depleted contents, were a feature of these cells.

The seeds dried over silica gel showed a precipitous decline in viability between the third and fourth day concomitant with a drop in water content below 170% (63% [wmb]) (Fig. 3). The major subcellular change occurring in the cells of still-viable axes sampled on day 3, was a marked dilation and irregularity of the vacuoles, accompanied by an apparent thinning of the bounding membrane (Plate 11). Organelle matrices had become dense, as had the chromatin. Similar features characterised those cells of non-viable (day 5) material that were still essentially intact. However, even in these cells areas of localised lysis were frequent (Plate 12) and there were many regions where total cell collapse and wall fragmentation had occurred.

Flash drying elicited a very different spectrum of subcellular events. After 10 minutes, when water content had declined to an average of 143% (58% [wmb]), a marked compaction of mitochondria and plastids had occurred and polysomes persisted in the generally denser cytomatrix. The most striking event was the close association that had developed between the rER and many of the vacuoles (Plate 13). Water content remained essentially constant (143 - 132%) despite a further 20 minutes of flash drying, during which similar rER associations developed with the other organelles as well. Most commonly, more than one plastid along with other subcellular constituents appeared to become closely enwrapped by the rER (Plate 14). There was a marked drop in water content to 76% (43% [wmb]) between 30 and 40 minutes of flash drying, and it was only then that the plasmalemma showed significant inwards contraction (Plate 15). Despite the fact that association was apparently lost between the plasmalemma and some of the frequent plasmodesmata, plasmalemma rupture was not observed ultrastructurally. Leakage from these axes was substantially lower than that from material slowly dried down to similar water contents and viability had not declined (Fig. 3). Plate 15 shows that the nucleus had become extremely compacted, as had the cytomatrix, and consequently the ER-enwrapped structures were contrasted by means of a negative staining effect. After 60 minutes of flash drying, such extreme compaction of the cells had occurred, that subcellular detail was almost obscured (see Plate 18). However, even in these cells it could be seen that vacuolar integrity had apparently been retained, and there were no visible manifestations of plasmalemma rupture. The low levels of electrolyte leakage are in accordance with this observation. Although survival had declined somewhat by 60 minutes of flash drying, significant loss of viability occurred only after this, when axis water content was reduced below 48% (32% [wmb]).

Thus for both L. kirkii and C. sinensis, flash drying of excised axes not only permitted retention of viability to lower water contents than did drying of intact seeds over silica gel, but also elicited different ultrastructural responses, particularly subcellular

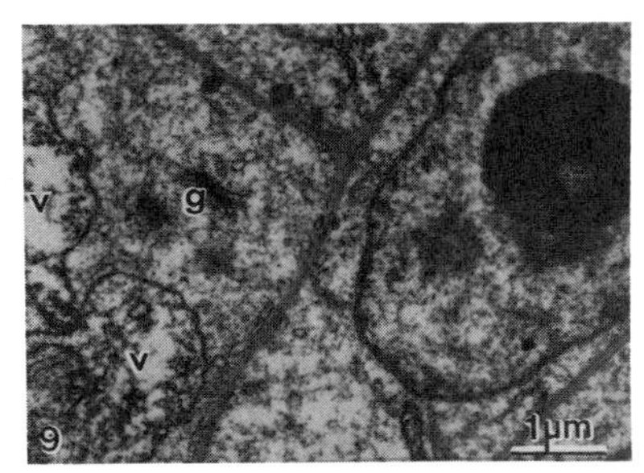

*C. sinensis;* newly-harvested; axis mc ±280% (±74% wmb); g, Golgi body; v. vacuole

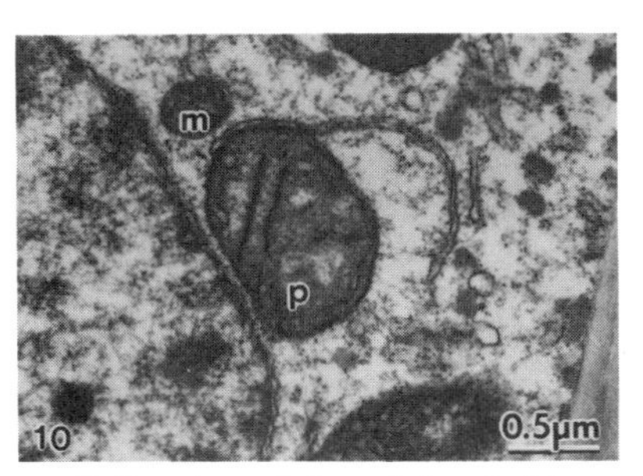

*C. sinensis;* newly-harvested; p, plastid; m, mitochondrion

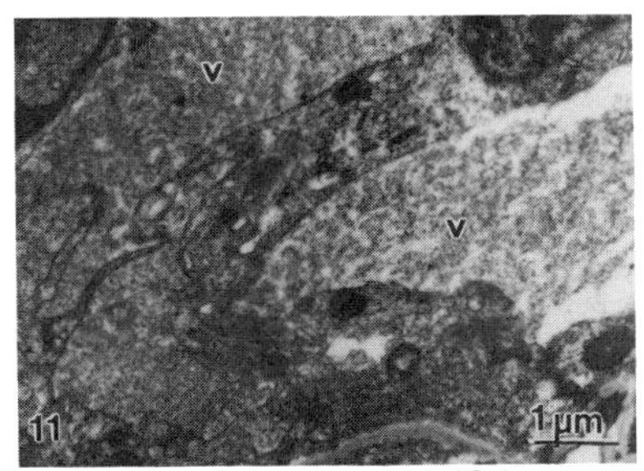

*C. sinensis;* intact seed dried 3 d; axis mc 170 - 130% (63-56%, wmb); v, vacuole

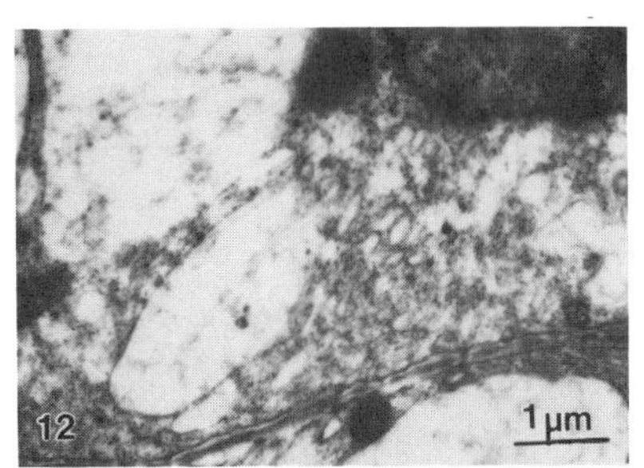

*C. sinensis;* intact seed dried 5 d; axis mc ±48% (±32% fmb); subcellular lysis

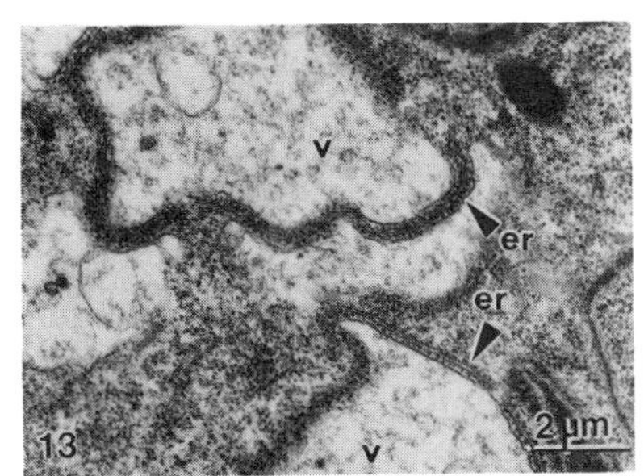

*C. sinensis;* 10 min. flash dried; mc ±143% (±58% fmb); note rER-vacuole association

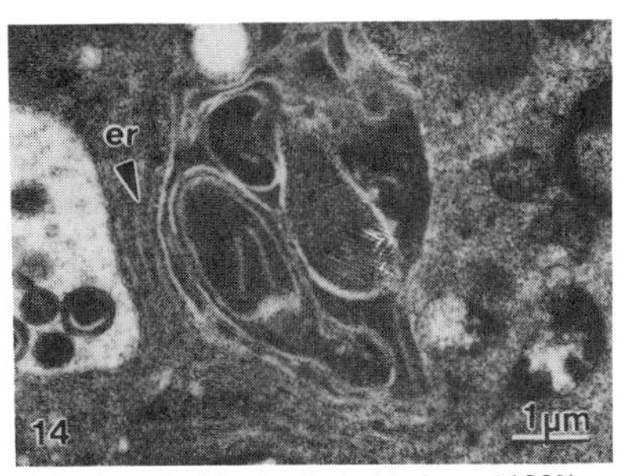

*C. sinensis;* 20 min. flash dried; mc ±133% (±57% fmb); organelle enwrapping by ER

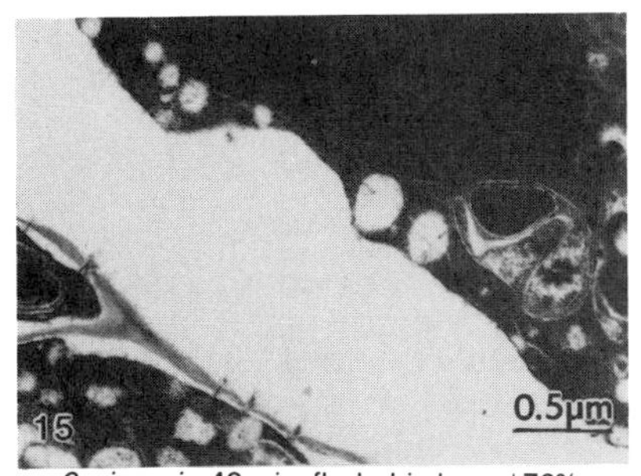

*C. sinensis;* 40 min. flash dried; mc ±76% (±43% wmb); plasmalemma contraction

compaction and maintenance of membrane integrity. It is not known whether these responses were the result of active processes or merely the passive consequence of rapid removal of water and the concentration of cellular contents.

The stabilisation of subcellular membranes in the dehydrated state is an important aspect of desiccation tolerance. In the hydrated state, water is considered to be firmly bound to the molecular components of the membrane surface. The involvement of polyols, and particularly oligosaccharides in membrane stabilisation upon dehydration (the 'water replacement hypothesis') is widely accepted (Crowe and Crowe, 1986a & b; Crowe, Crowe, Carpenter and Wistrom, 1987). While those authors describe trehalose as being primarily invoved in animal cells, sucrose in conjunction with raffinose and stachyose is suggested to be the most likely combination of oligosaccharides acting to stabilise membranes in dry orthodox seeds (Leopold and Vertucci, 1986; Caffrey, Fonseca and Leopold, 1988; Koster and Leopold, 1988). It is generally agreed that on dehydration of desiccation tolerant tissues, sugars may bind to phospholipid of the cell membranes, so replacing water and maintaining headgroup spacing.

Eleven percent of the dry mass of the embryonic axes excised from newly-shed seeds of C. sinensis is composed of a combination of small sugars, of which sucrose constitutes about 45%. These values are of the same order as those of some orthodox seeds investigated by Koster and Leopold (1988). However, the ratio of raffinose plus stachyose plus cellobiose to sucrose in C. sinensis seeds (0.1:1) is lower than for the orthodox seeds. This ratio appeared to increase on flash drying and during the initial stages of slow drying of intact seeds, but it is not known whether this is significant.

The Lens culinaris agglutinin (Sigma) having an affinity for α-D-mannosyl and α-D-glucosyl residues was used for lectin gold labelling. In fresh material, the gold label was located predominantly within the vacuoles (Plate 16). Following 10 minutes flash drying frequency of the label in the vacuoles had increased and gold labelling also occurred in the extraprotoplasmic space and plasmalemma invaginations. The label remained almost exclusively located within vacuoles, vesicles and extraprotoplasmic space (eps) following longer periods of flash drying, seemingly never becoming closely associated with the subcellular membranes (Plates 17 & 18). In cells of axes excised from intact seeds after slower drying, the disposition of the label appeared somewhat different. Not only was there a noticeable intra-organellar location of the gold particles, but the label was considerably more aggregated (as chains or clumps) within the vacuoles (Plate 19). This situation, which had already occurred after one day during which axis moisture content had declined from 283 - 213% (74 - 68% [wmb]), persisted while viability was retained. Like the situation in flash-dried material, the label did not become membrane associated.

Although giving interesting indications, the data from this lectin gold labelling cannot be unequivocally interpreted, a similar approach with lectins having different, but related affinities being necessary. It cannot be assumed that sugars are definitely not bound to the membranes, as such interaction might well interfere with the complementarity required for recognition by the lectin used. However, the surfaces of membrane preparations from these axes lyophilized in the presence of sucrose/raffinose/stachyose mixtures in the mass ratios used by Caffrey et al. (1988), retained a considerable number of gold particles. These preparations were processed for electron microscopy in exactly the same way as were the axes, thus there is nothing inherent in the techniques used that could have removed membrane-associated label.

It is possible that at a moisture content of 48 - 28% (32 - 22% [wmb]) there would be no tendency for water replacement by sugars. However, it has been suggested that sugars start to replace loosely-bound water on membranes in Phaseolus vulgaris seeds, at a water content of 33% [dmb] (Wolk, Dillon, Copeland and Dilley, 1989). In this regard, there are suggestions that an increase in the liquid-crystalline/gel phase transition temperature is initiated below a water content of 12 mol/mol phospholipid (Crowe and Crowe, 1986a, b). This represents a moisture content of 20 - 30% (dmb), assuming the cell water to be evenly distributed (Hoekstra and Van Roekel, 1988).

Wolk et al. (1989) make the point that, for the desiccation tolerant tissue with which they worked, the bound water fraction below 0.05g/g may not be replaceable, and Gaber, Chandrasekhar and Pattabiraman (1986) have indicated that some water remains bound to the phospholipid bilayer, no matter how severely desiccating the regime. Sorption isotherms for desiccation sensitive tissues have a hyperbolic form, as opposed to the reverse sigmoid form for desiccation tolerant tissues. This has been interpreted as indicating that strong water binding contributes only minimally, if at all, to the water sorption characteristics of these tissues (Vertucci and Leopold, 1987). However, the isotherms are such that water contents remain relatively high despite equilibration of the tissues at very low relative humidities (Leopold and Vertucci, 1986; Vertucci and Leopold, 1987). We have found that the sorption isotherm for embryonic axes of

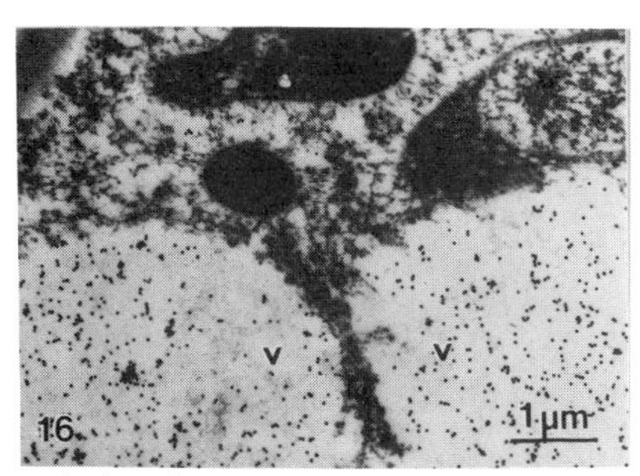


*C. sinensis;* newly-harvested; lectin gold label location principally intravacuolar

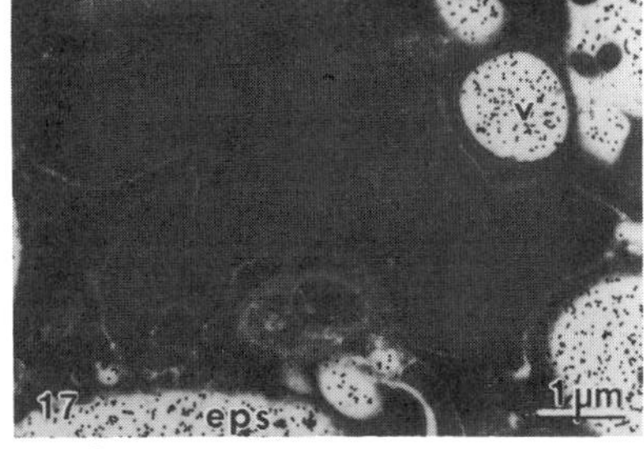


*C. sinensis;* typical disposition of the gold label after 20-40 min. flash drying

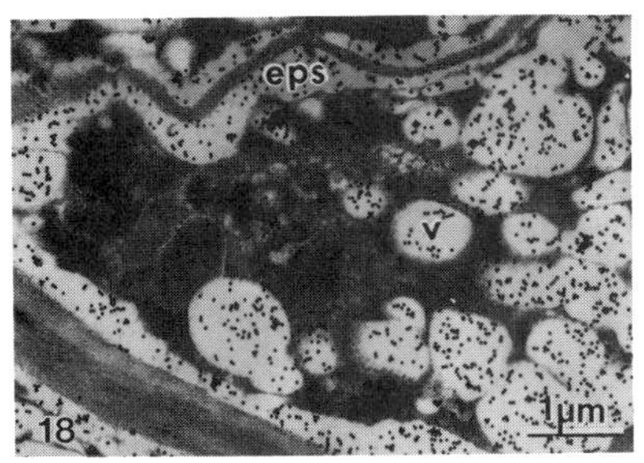


*C. sinensis;* 60 min. flash dried; gold labelling principally in vacuoles & eps

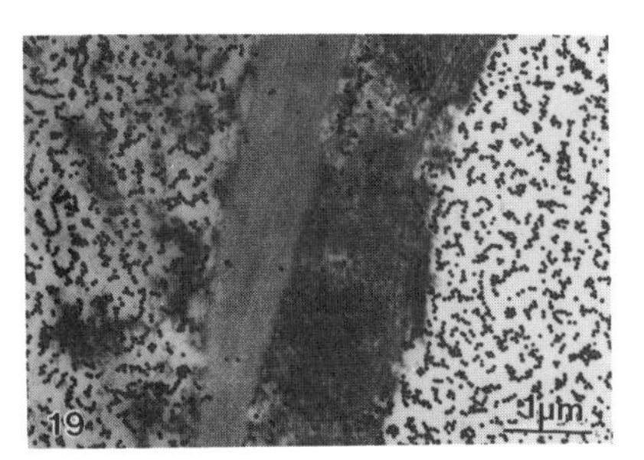


*C. sinensis;* intact seed dried 1 d; typical disposition of gold label

Avicennia marina is also typically hyperbolic and it is apparently impossible to remove all the water, despite equilibration at very low relative humidities. Although considerably more water can be removed from axes of C. sinensis, the isotherm also has a hyperbolic form.

Perhaps one could re-interpret the significance of the hyperbolic isotherms of desiccation sensitive tissue in terms of a significant fraction of water that is, and remains, so strongly bound that it cannot be removed, no matter how low the relative humidity of the equilibration system. It might not be replaced by sugars, because of the very tenacity of its binding. Although this interpretation is conjectural, it highlights the possibility that under normal circumstances, membrane integrity in recalcitrant seeds (and germinating orthodox seeds) demands a degree of bound water, considerably in excess of the tenacious fraction, that becomes perturbed below a minimal moisture content. Additionally, as the events of germination proceed, the demand for structured water might increase with increasing subcellular organisation, so increasing the lethal water content (Farrant et al., 1988a).

There is another interpretation of the rôle of sucrose and the other oligosaccharides in desiccation-stressed material. Timasheff (1982) has suggested that a protein remains stabilised in a protein-water-cosolvent (including sucrose) system by exclusion of unfavourable interactions between its surface and the cosolvent molecules, so giving rise to higher concentrations of water at the protein surface. Williams and Leopold (1989) have presented evidence that both the lipids and the non-lipid components in corn embryos exist in a highly viscous glassy state even at ambient temperature, below water contents of 0.12g/g. Those authors draw attention to the probable rôle of sucrose and raffinose, that together comprise 20% of the dry mass of the embryo, in vitrification of the non-lipid components. Glass formation is suggested to contribute to desiccation tolerance in terms of restricting molecular mobility and so imposing a stasis on biochemical activity (Williams and Leopold, 1989).

While the water content in the flash-dried axes of Camellia sinensis did not remotely approach 0.12g/g, there is no reason to assume that it was evenly distributed within individual cells, depending i.a., on the localisation of solutes. Viability was retained concomitant with the apparent exclusion of much of the sugar from the compacted cytomatrix (Plates 17 & 18 ). It is possible (although difficult to ascertain) that while predominantly strongly-bound water persists in the cytomatrix and nucleoplasm, viscous sugar solutions occur within vacuoles, vesicles and the extraprotoplasmic space. While, in C. sinensis, these are unlikely to assume the glassy state, such a situation might minimise perturbation and biochemical reactivity during rapidly-imposed desiccation stress. In flash-dried axes of Landolphia kirkii, water content could be successfully reduced to 0.13g/g (Berjak et al., 1989b): here perhaps, one could postulate the existence of glass, as suggested by Williams and Leopold (1989).

The ability of excised embryos to survive flash drying to remarkably low water contents poses the question as to whether this material is essentially orthodox in nature. Orthodoxy implies an inherent mechanism for desiccation tolerance, that might be expected to be expressed in conjunction with the appearance of novel proteins as a result of specific mRNA synthesis (Bewley, Kermode and Misra, 1989). Two-dimensional gel electrophoresis carried out on extracts of fully-viable fresh, flash-dried and slow-dried axes of Camellia sinensis have revealed that no new proteins appear as a result of either drying regime. The controlled subcellular de-differentiation associated with maturation drying of orthodox seeds (Bain and Mercer, 1966; Klein and Pollock, 1968; Hallam, 1972) did not occur on flash drying. Thus we consider as highly unlikely, the possibility of some inherent capacity that is not normally expressed. Rather, we are of the opinion that flash drying allows reduction of water content so rapidly that some vitrification might occur, permitting a measure of membrane stabilisation and imposing a stasis on metabolic activity. We have preliminary indications that flash-dried material may not be able to survive for any appreciable time at room temperature, and so may not be desiccation tolerant in the accepted sense. Presumably, if super-saturation or vitrification is a response to flash drying, crystallisation can subsequently occur. For desiccation tolerant organisms, the more rapid the dehydration, the more injurious its effects (Bewley and Krochko, 1982; Crowe and Crowe, 1986b). This is entirely contrary to the response of recalcitrant material to flash drying. However, if flash-dried recalcitrant material is not truly desiccation tolerant, this would resolve the apparent conflict.

There are several facets of recalcitrant seed behaviour that have not been considered in this review, including: 1. Developmental control that ensures an orthodox seed entering the phase of maturation drying, and which is absent or non-operative in the recalcitrant situation; 2. The reaction of the cotyledons and/or endosperm to desiccation and the interaction between the storage tissues and embryonic axis under various conditions; 3. General biochemical characterisation of recalcitrance; 4. The nature of the lesion(s) resulting from desiccation; 5. The issue of chilling sensitivity; and 6. The possibility of cryostorage of flash-dried material. Although there is some information and speculation about certain of these aspects, much remains to be ascertained.

## IN CONCLUSION - WHAT IS RECALCITRANCE?

A current working definition of recalcitrant seeds might be: seeds that are shed wet and cannot be dehydrated or stored. This working definition implies that there are two aspects to recalcitrance; intolerance to dehydration and inability to be stored. Both these aspects vary widely among recalcitrant seed species and there may be no numerical relationship between storage lifespan and desiccation sensitivity.

Lifespan in wet storage is related to the rate of axis development (if incomplete upon shedding) and germination (i.e. related to the time scale of the abscissa in Figure 2). Seeds that germinate rapidly when planted out will also undergo germination-associated events relatively rapidly in wet storage and will quickly reach the stage where additional water is required. Naturally slowly-germinating seeds will undergo these changes slowly in storage and so will have a longer lifespan.

Desiccation tolerance can be equated with the minimum water content to which tissue can be dried without viability loss. Sensitivity to dehydration of recalcitrant seeds increases with storage time and so inter-species comparisons should be made at the stage of shedding. It is our thesis that recalcitrant seeds cannot be dehydrated because they are well-differentiated at the subcellular level and either do not possess the genetic potential for de-differentiation (unlike orthodox seeds) or have it completely repressed. Flash-dried axes do show interspecific differences in the extent to which they can be dehydrated before marked viability loss (C. sinensis, 48%; L. kirkii, 13% [dmb]). If our hypothesis is correct, there should be a relationship between desiccation sensitivity (minimum lethal water content on almost instantaneous dehydration) and some measure of the degree of cellular differentiation, such as total membrane surface area.

As a result of all these factors it is extremely difficult to achieve a strict definition of recalcitrance, and perhaps it would be easier to reconsider the definition of orthodoxy. Orthodox seeds are those that are either shed in the dry state or can be dried and (this is the critical point) they can be successfully maintained in this state at moderate temperature for periods from months to years. A seed that fails to meet these criteria is not truly desiccation tolerant and is not orthodox.

In keeping with the terminology applied to plants in the vegetative phase, we propose the use of the term, poikilohydric, to describe seeds that can be maintained in equilibrium with ambient relative humidity for long periods (i.e. truly desiccation tolerant, or orthodox). Homoiohydric seeds, on the other hand, are those that cannot be so maintained (i.e. not truly desiccation tolerant, or recalcitrant).

ACKNOWLEDGEMENTS

We would like to thank our colleagues of the Plant Cell Biology Research Group for their enthusiastic assistance; and for financial support, the Foundation for Research Development, the Department of Agriculture & Water Supply, and the Anglo American/de Beers Chairman's Fund. Sapekoe (The Tea and Coffee Growers) are gratefully acknowledged for their unfailing weekly hand-harvesting of tea fruits, as are R.D. Heinsohn (L. kirkii fruit collection) and J. Coetzee [University of Pretoria] and J. Wesley-Smith (lectin/gold methodology).

REFERENCES

Bain, J. and Mercer, F.V. (1966), Subcellular organisation of the developing cotyledons of Pisum sativum L., Aust. J. Biol. Sci., 19:49.

Berjak, P. (1989), The basis of the storage behaviour of seeds of Hevea brasiliensis, (In Prep.).

Berjak, P., Dini, M. and Pammenter, N.W. (1984), Possible mechanisms underlying the differing dehydration responses in recalcitrant and orthodox seeds: desiccation-associated subcellular changes in propagules of Avicennia marina, Seed Sci. & Technol., 12:365.

Berjak, P., Wesley-Smith, J. and Mycock, D.J. (1988), Ultrastructural characteristics of stored seeds of Hevea brasiliensis, Proc. Electron Microsc. Soc. South. Afr., 18:101.

Berjak, P., Farrant, J.M., Mycock, D.J. and Pammenter, N.W. (1989a), Deterioration and pathology of recalcitrant (homoiohydrous) seeds, Proc. 22nd ISTA Congress, Edinburgh.

Berjak, P., Farrant, J.M., Mycock, D.J. and Pammenter, N.W. (1989b), Recalcitrant (homoiohydrous) seeds: the enigma of their desiccation-sensitivity, Seed Sci. & Technol. (In Press).

Bewley, J.D. and Krochko, J.E. (1982), Desiccation tolerance, in "Physiological Plant Ecology II. Encyclopedia of Plant Physiology", New Series, Volume 12B, O.L. Lange, P.S. Nobel, C.B. Osmond and H. Ziegler, eds, Springer, Berlin, Heidelberg, New York.

Bewley, J.D., Kermode, A.R. and Misra, S. (1989), Desiccation and minimal drying treatments of seeds of castor bean and Phaseolus vulgaris which terminate development and promote germination cause changes in protein and messenger RNA synthesis, Ann. Bot., 63:3.

Caffrey, M., Fonseca, V. and Leopold, A.C. (1988), Lipid-sugar interactions, Plant Physiol., 86:754.

Chin, H.F. (1980), Germination, in "Recalcitrant Crop Seeds", H.F. Chin and E.H. Roberts, eds, Tropical Press SDN.BDH , Kuala Lumpur.

Chin, H.F. and Roberts, E.H. (1980), "Recalcitrant Crop Seeds", Tropical Press SDN.BDH , Kuala Lumpur.

Chin, H.F., Aziz, M., Ang, B.B. and Samsidar, H. (1981), The effect of moisture and temperature on the ultrastructure and viability of seeds of Hevea brasiliensis, Seed Sci. & Technol., 9:411.

Crowe, J.H. and Crowe, L.M. (1986a), Stabilisation of membranes in anhydrobiotic organisms, in "Membranes, Metabolism and Dry Organisms", A.C. Leopold, ed., Comstock Publishing Associates, Ithaca and London.

Crowe, L.M. and Crowe, J.H. (1986b), Hydration-dependent phase transitions and permeability properties of biological membranes, in "Membranes, Metabolism and Dry Organisms", A.C. Leopold, ed., Comstock Publishing Associates, Ithaca and London.

Crowe, J.H., Crowe, L.M., Carpenter, J.F. and Wistrom, A. (1987), Stabilization of dry phospholipid bilayers and proteins by sugars, Biochem. J. 242,1.

Czabator, F.J. (1962), Germination value: An index combining speed and completeness of pine seed germination, For. Sci., 8:386.

Devey, R.M. (1989), Some studies on the storage behaviour of Camellia sinensis (tea) seeds, Unpublished M.Sc. Thesis, University of Natal, Durban.

Devey, R.M., Pammenter, N.W. and Berjak, P. (1986), Recalcitrance in Camellia sinensis, Proc. Electron Microsc. Soc. South. Afr. 16:97.

Devey, R.M., Pammenter, N.W. and Berjak, P. (1987), Comparison of the fate of cotyledonary reserves during storage and germination of recalcitrant tea seeds, Proc. Electron. Microsc. Soc. South. Afr., 17,103.

Farrant, J.M., Berjak, P. and Pammenter, N.W. (1985), The effect of drying rate on viability retention of recalcitrant propagules of Avicennia marina, S. Afr. J. Bot., 51:432.

Farrant, J.M., Pammenter, N.W. and Berjak, P. (1986), The increasing desiccation sensitivity of recalcitrant Avicennia marina seeds with storage time, Physiol. Plant., 67:291.

Farrant, J.M., Pammenter, N.W. and Berjak, P. (1988a), Recalcitrance - a current assessment, Seed Sci. & Technol., 16:155.

Farrant, J.M., Pammenter, N.W. and Berjak, P. (1988b), Development of the recalcitrant seeds of Avicennia marina, Proc. Electron Microsc. Soc. South. Afr., 18:109.

Farrant, J.M. Pammenter, N.W. and Berjak, P. (1989), Germination-associated events and the desiccation sensitivity of recalcitrant seeds . a study on three unrelated species, Planta, 178:189.

Gaber, B.P., Chandrasekhar, I. and Pattabiraman, N. (1986), The interaction of trehalose with the phospholipid bilayer: a molecular modeling study, in "Membranes, Metabolism and Dry Organisms", A.C. Leopold, ed., Comstock Publishing Associates, Ithaca and London.

Hallam, N.D. (1972), Embryogenesis and germination in rye (Secale cereale). I. Fine structure of the developing embryo, Planta, 104:157.

Hoekstra, F.A. and Van Roekel, T. (1988), Desiccation tolerance of Papaver dubium L. pollen during its development in the anther, Plant Physiol., 88:626.
King, M.W. and Roberts, E.H. (1980), Maintenance of recalcitrant crop seeds in storage, in "Recalcitrant Crop Seeds", H.F. Chin and E.H. Roberts, eds, Tropical Press SDN.BDH , Kuala Lumpur.
King, M.W. and Roberts, E.H. (1982), The imbibed storage of cocoa (Theobroma cacao) seeds, Seed Sci. & Technol., 10:535.
Klein, S. and Pollock, B.M. (1968), Cell fine structure of developing lima beans related to desiccation, Amer. J. Bot., 55,658.
Koster, K.L. and Leopold, A.C. (1988), Sugars and desiccation tolerance in seeds, Plant Physiol., 88:829.
Leopold, A.C. and Vertucci, C.W. (1986), Physical attributes of desiccated seeds, in "Membranes, Metabolism and Dry Organisms", A.C. Leopold, ed., Comstock Publishing Associates, Ithaca and London.
Normah, M.N., Chin, H.F. and Hor, Y.L. (1986), Desiccation and cryopreservation of embryonic axes of Hevea brasiliensis Muell. - Arg., Pertanika 9:299.
Pammenter, N.W., Farrant, J.M. and Berjak, P. (1984), Recalcitrant seeds: Short-term storage effects in Avicennia marina (Forsk.) Vierh. may be germination-associated, Ann. Bot., 54:843.
Pritchard, H.W. and Prendergast, F.G. (1986), Effects of desiccation and cryopreservation on the in vitro viability of embryos of the recalcitrant seed species Araucaria hunsteinii K. Shum., J. exp. Bot., 37:1388.
Probert, R.J. and Longley, P.L. (1989), Recalcitrant seed storage physiology in three aquatic grasses (Zizania palustris, Spartina anglica and Porteresia coarctata), Ann. Bot. 63:53.
Report of the ISTA Seed Moisture Committee (1989), Proc. 22nd ISTA Congress, Edinburgh.
Roberts, E.H. (1973), Predicting the storage life of seeds, Seed Sci. & Technol., 1:499.
Roberts, E.H. and King, M.W. (1980), The characteristics of recalcitrant seeds, in "Recalcitrant Crop Seeds", H.F. Chin and E.H. Roberts, eds, Tropical Press SDN.BDH , Kuala Lumpur.
Timasheff, S.N. (1982), Preferential interactions in protein-water-cosolvent systems, in "Biophysocs of Water", F. Franks and S. Mathias, eds, Wiley, Chischester.
Timson, J. (1965), A new method of recording germination data, Nature, 207:216.
Tompsett, P.B. (1982), The effect of desiccation on the longevity of seeds of Araucaria hunsteinii and A. cunninghamii, Ann. Bot., 50:693.
Tompsett, P.B. (1984), Desiccation studies in relation to the storage of Araucaria seed, Ann. appl. Biol., 105:581.
Tompsett, P.B. (1987), Desiccation and storage studies on Dipterocarpus seeds, Ann. appl. Biol., 110:371.
Vertucci, C.W. and Leopold, A.C. (1987), The relationship between water binding and desiccation tolerance in tissues, Plant Physiol., 85:232.
Williams, R.J. and Leopold, A.C. (1989), The glassy state in corn embryos, Plant Physiol., 89:977.

# PHYTIN SYNTHESIS AND DEPOSITION

John S. Greenwood,

Dept. of Botany, Univ. of Guelph
Guelph, Ont., N1G 2W1
Canada

## INTRODUCTION AND BACKGROUND

Reserves accumulated during seed development are not restricted to those of reduced carbon (starch and/or lipid) and reduced nitrogen and sulphur (storage, and other proteins). The remaining macronutrients - phosphorus, magnesium, potassium and calcium - are also sequestered within mature seeds, primarily as the organic phosphorus reserve, phytin.

This discussion is intended to deal most specifically with the possible biosynthetic pathways and mechanisms of intracellular deposition of phytin during seed development. However, some background information is required both to introduce the compound and stress the importance of the compound to the seed and growing seedling. Readers should consult the extensive lists of references contained within the reviews by Cosgrove (1966, 1980), Dalling and Bhalla (1984), Loewus (1974, 1983), Loewus and Loewus (1983), Lott (1980, 1984) Lott and Ockenden (1986) Nayini and Markakis (1986) and Reddy et al. (1982) for details.

### Chemical Nature and Occurrence

Strictly speaking, phytin is the mixed magnesium, potassium and calcium salts of _myo_-inositol hexa_kis_phosphoric acid ($mIP_6$, phytic acid, Fig. 1) that can be isolated from mature seeds; phytic acid and phytate refer to the free acid, and any salt of $mIP_6$, respectively (Cosgrove, 1966). Phytin is of widespread occurrence in higher plants but most important to this discussion is the fact that phytin appears to be ubiquitous in mature seeds of the numerous genera studied to date. It typically represents the major phosphorus store within these seeds (Pfeffer, 1872; Cosgrove, 1966; see partial listings in Lott, 1980, 1984; Lott et al., 1985 and Reddy et al., 1982).

### Location within the Mature Seed

Phytin is the primary component of the globoid inclusions of protein bodies in mature seeds (Fig. 2) although in many of the leguminous species, where globoid inclusions are small or absent, phytin may be bound to the storage protein component of the organelles. Direct chemical analyses of isolated protein bodies and globoids and indirect evidence by electron

*Recent Advances in the Development and Germination of Seeds*
Edited by R.B. Taylorson
Plenum Press, New York

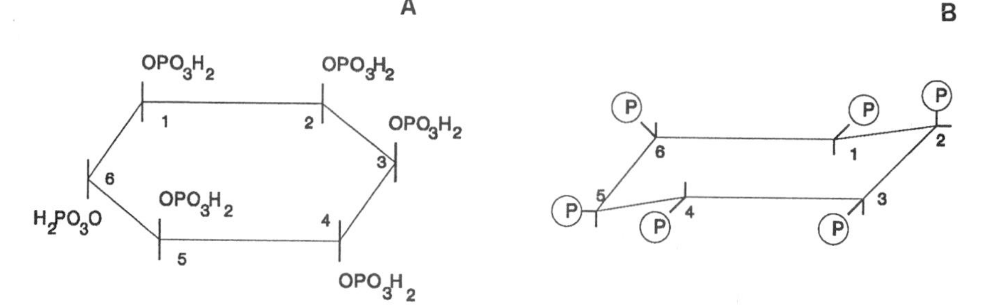


Figure 1. Structure (A) and chair conformation (B) of myo-inositol hexakis-phosphoric acid (phytic acid).

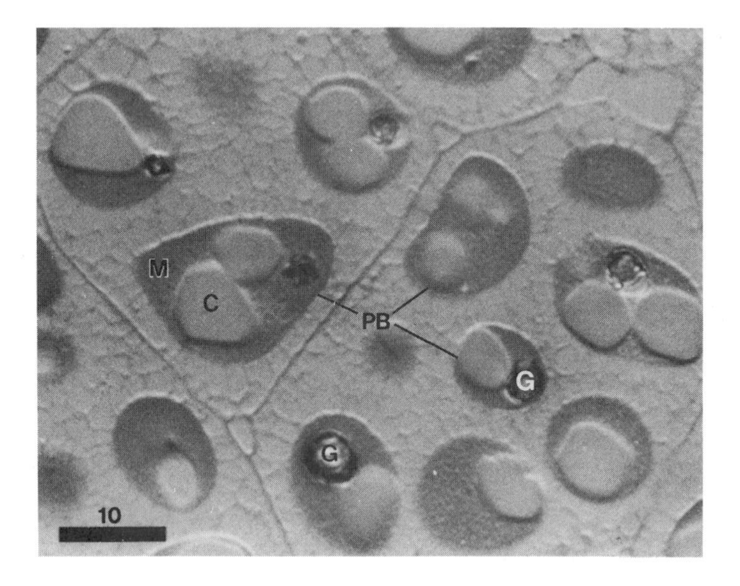


Figure 2. Protein bodies in mature castor bean seed endosperm. Three components of the protein bodies (PB) are indicated, the proteinaceous crystalloid (C), the proteinaceous matrix (M) and phytin-containing globoids (G). Bar units in um.

microscopic energy dispersive X-ray spectroscopy (EDS) of a variety of mature seed tissues have demonstrated that phytin is restricted to the protein bodies (Lott, 1980, 1984; Lott et al., 1985; Reddy et al., 1982). In mature seeds of dicotyledonous species phytin is generally located in the protein bodies throughout those tissues that are involved in protein storage: cotyledon, embryonic axis or endosperm. For example, in lettuce phytin and reserve proteins and lipids are stored primarily in the cotyledons, whereas the endosperm is the site of carbohydrate (galactomannan) storage (Halmer et al., 1978). In castor bean, reserve lipids, protein and phytin are found primarily in the endosperm, and to a lesser extent in the embryonic tissues (Greenwood et al., 1984). In most cereal grains the majority of phytin is restricted to protein bodies in cells of the living aleurone layer of the endosperm; small amounts are found in the embryonic tissues. Maize is an exception, where approximately 90 percent of the phytin in the kernel is present in the protein bodies of the scutellar tissues (see Lott, 1984). In seeds where the cotyledons are the major storage organs the distribution of phytin, or of the cations associated with phytic acid, may also vary between tissue type (provascular and epidermal vs. palisade and spongy mesophyll) and organ (embryo axis vs. cotyledon) (Lott et al., 1979).

Physiological Roles of Phytin

Phytin as a Phosphorus, myo-Inositol and Cation Reserve. Although it comprises only one- to eight percent of the dry weight, phytic acid represents the major phosphorus store, accounting for fifty- to ninety percent of the total phosphorus present in mature seeds (see listings and references in Lott, 1984; Reddy et al., 1982). Phytin is enzymatically catabolized by the enzyme, phytase, within the storage tissues of seeds following germination. Released phosphate and myo-inositol are transported to and utilized by the growing seedling (reviewed by Dalling and Bhalla, 1984, and Loewus, 1983). Thus there is little doubt that the compound acts as the major reserve of both phosphorus and myo-inositol in seeds.

The role of phytin as a cation reserve in mature seeds is similarly beyond question. The covalently-linked phosphate moieties of phytic acid carry strong negative charges; phytic acid is able to chelate metallic cations of $Ca^{2+}$, $Mg^{2+}$ and $K^{+}$ (and others including $Fe^{2+}$, $Ba^{2+}$, $Mn^{2+}$, and $Zn^{2+}$)(Lott, 1984). The association of these cations with protein body globoids is easily demonstrated using EDS analysis and by non-aqueous fractionation techniques followed by chemical analysis (see Lott, 1980, 1984; Lott et al., 1985; Reddy et al., 1982). Cations are released during the enzymatic hydrolysis of phytin following germination and become freely available to the growing seedling (Cosgrove, 1980; Lott, 1980, 1984). It has not been clearly determined, however, if phytin acts as the principal store for metallic cations in seeds. Support for this view comes from the studies of the accumulation of phytic acid and P, $Mg^{2+}$, and $K^{+}$ in the developing endosperm of rice by Ogawa et al. (1979a). Results from EDS analyses of unfixed freeze-dried sections of tissue demonstrated that the elements were mobilized from the starchy endosperm to the aleurone layer, the site of phytin accumulation, with development. In combination with chemical analyses, which demonstrated that the available cations of $Mg^{2+}$ and $K^{+}$ balanced the negative charge of phytic acid phosphorus, the data strongly suggest that phytin is indeed the principal storage form for $Mg^{2+}$ and $K^{+}$ in the rice grain. Similar and more extensive studies should be performed using seeds of other genera.

Phytic Acid and Inorganic Phosphate Homeostasis. Matheson and Strother (1969) demonstrated that the inorganic phosphate content per unit fresh weight of germinated wheat seedlings remained constant over a fourteen day study, the hydrolysis of phytic acid being the source of inorganic phosphate as the fresh weight of the seedlings increased (Fig. 3A). Strother (1980) later demonstrated the same phenomenon using seedlings of barley, broad bean, pea and lettuce and suggested that the catabolism of phytic acid following germination may be rigidly controlled to maintain inorganic phosphate levels at a constant level. A similar argument was presented by Samotus (1965) from studies of inorganic phosphate homeostasis and phytic acid metabolism in potato tubers. Phytase activity in vivo may be modulated by inorganic phosphate, as has been demonstrated in in vitro assays (see Dalling and Bhalla, 1984).

Controlled synthesis of phytic acid may provide a similar method of maintenance of inorganic phosphate levels during seed development and desiccation. The concentration of inorganic phosphate within the endosperm of developing castor bean is maintained at a relatively constant level throughout the period of rapid reserve accumulation (Fig.3B). Water stress during development hastens the formation of phytic acid in wheat (Williams, 1970), as does increasing the supply of inorganic phosphorus to developing seeds (Asada et al., 1962; Raboy and Dickinson, 1987). Phytic acid synthesis may be regulated, at least in part, by the level of inorganic phosphorus within or available to the developing seed.

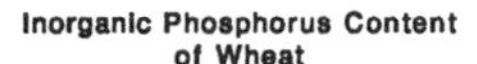


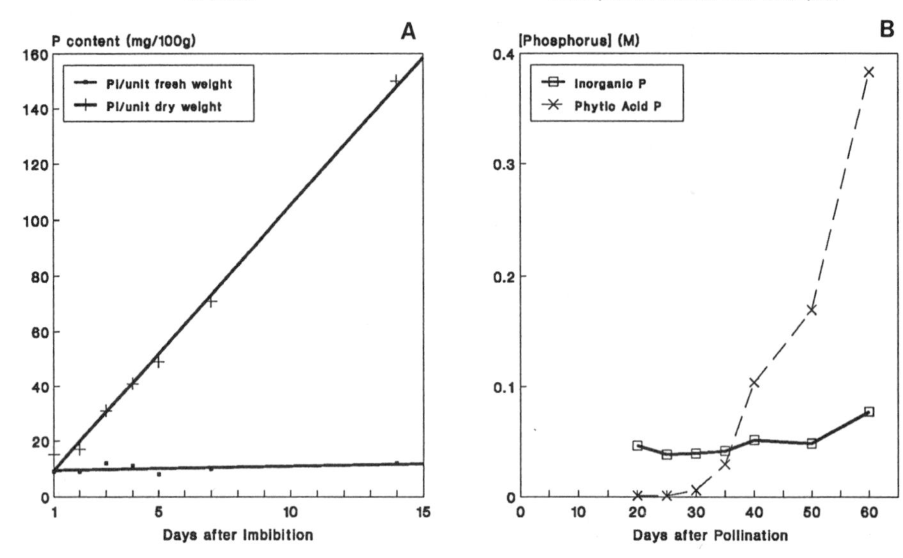


Figure 3. Inorganic phosphorus homeostasis following germination and during seed development. A) Inorganic phosphorus content following imbibition in wheat. Note that, on a per g fresh weight basis, inorganic phosphorus levels remain constant. (Redrawn from Matheson and Strother, 1969) B) Inorganic- and phytic acid phosphorus concentration during development of castor bean endosperm. Note, again, that inorganic phosphorus levels remain fairly constant throughout development. (Calculated from data supplied by Greenwood and Bewley, 1982, and Greenwood et al., 1985)

Suggested Roles in Quiescence and as an Energy Store. The synthesis of phytic acid via the complete phosphorylation of myo-inositol is ATP dependent. Sobolev and Rodionova (1966) suggested that during the latter stages of seed development the demand for ATP required for rapid synthesis of phytic acid would reduce ATP levels in the cells, thereby inducing dormancy or quiescence. This role is in doubt since Williams (1970) demonstrated that the major accumulation of phytic acid in the aleurone of non-stressed wheat grains, and in the embryo and scutellum of water-stressed grains, coincided with an increase in ATP levels. A decline in ATP concentration did coincide with the period of rapid phytic acid accumulation in the aleurone of stressed grains, but levels of ATP were restored immediately afterwards, well prior to quiescence.

Williams (1970) and Sharma and Dieckert (1975) proposed that the chelation of cations by phytic acid during seed development may aid in regulating cellular metabolism, and possibly induce quiescence, by reducing ionic cofactors necessary for enzyme activity. As yet, there is no experimental evidence supporting this view. Indeed, the studies of Ogawa et al. (1979a,b) tend to negate this role. The movement of $K^+$ from the starchy endosperm to the aleurone layer of rice is delayed until the "dough" stage of development, well after the majority of phytic acid accumulated in the aleurone. Association of $K^+$ with globoids in the aleurone was similarly delayed compared to that of $Mg^{2+}$ and $Ca^{2+}$. Potassium acts as a cofactor for starch synthesis in rice; movement to the aleurone and association with globoids follows the major increases in starch within the grain. The results of Ogawa et al. (1979a,b) suggest that ions become associated with phytic acid as their involvement in other metabolic pathways declines.

Due to its high level of phosphorylation, several researchers have suggested that phytic acid may act in energy generation in germinating seeds (Biswas and Biswas, 1965; Biswas et al., 1975, 1978b; Tanaka et al., 1976a), phosphate groups being transferred enzymatically to ADP or GDP. Biswas and Biswas (1965) and Biswas et al. (1975, 1978b) isolated enzymes from germinated and dry mung bean seeds that were able to transfer phosphate from $^{32}P$-labelled phytic acid to GDP and ADP, resulting in the formation of labelled GTP and ATP. Only the high-energy axial phosphate of phytic acid (C2-position, Fig. 1 B) was able to be transferred to ADP (Biswas et al., 1978b); the reaction was found to be reversible. However, the phytic acid:ribonucleoside diphosphate phosphotransferase reaction is unlikely to be important in seed germination. Bieleski (1973) calculated that, if all the phosphate contained in phytic acid could be transferred to ADP over the 6-10 day period of utilization by the growing seedling, the ATP generated per g fresh weight would be equivalent to that produced by respiratory metabolism within 10 minutes. Of course the value would be one-sixth of this if only the C2-phosphate of phytic acid is involved in the transfer.

## ACCUMULATION AND MECHANISM OF DEPOSITION DURING DEVELOPMENT

### Accumulation during Seed Development

Rapid accumulation of phytin occurs during the maturation phase of seed development, the period of rapid cell expansion and reserve accumulation, paralleling that of starch or storage lipid and total storage protein. In most cases accumulation continues at a reduced rate during the maturation drying phase.

Correlative studies of the accumulation of reserves, including phytin, during seed development have been rare, but the relative timing of the accumulation of reserves in castor bean seed has been investigated (Greenwood et al., 1984). The period of rapid accumulation of phytin lags behind that of total protein and lipid accumulation both in the major reserve tissue, the endosperm, and in the embryonic tissues (Fig. 4 A and B). Microscopical observations of protein body formation, although more qualitative, indicate that phytin and storage protein accumulate simultaneously in protein bodies of individual cells (Greenwood and Bewley, 1985).

### Sites of Accumulation and Mechanisms of Deposition

Protein bodies of vacuolar origin are the final sites of accumulation of phytin in storage parenchyma cells of developing seeds. It has been suggested that the protein bodies are the sites of synthesis of phytic acid during seed development. If this is indeed the case, then the mechanism of deposition of phytin during seed development would be self-evident.

Sobolev and Rodionova (1966) and Tanaka et al. (1976b) attempted to demonstrate that protein bodies were capable of synthesizing phytic acid independently. Both investigations used protein bodies, isolated from developing seeds of sunflower and aleurone layers of mature rice, respectively, in a cell-free biosynthetic assay following the incorporation of $^{32}$Pi, [$^{32}$P]-ATP or [$^{3}$H]-myo-inositol into acid-soluble myo-inositol phosphate fractions. Sobolev and Rodionova (1966) obtained labelled phytic acid from their reaction mixtures. However, the role of the protein bodies in the synthesis of phytic acid was ambiguous. The protein bodies used were isolated by differential centrifugation; organelle purity was not established. Reaction mixtures included mitochondria, isolated from the same developing seeds, as ATP generators for the duration of the reactions. Unidentified organelles, co-sedimenting with either the protein bodies or the mitochondria, would have been included in the reaction mixtures. Additionally, the authors did not determine if phytic acid could be synthesized in their reaction mixture in the absence of protein bodies. Similarly, the non-

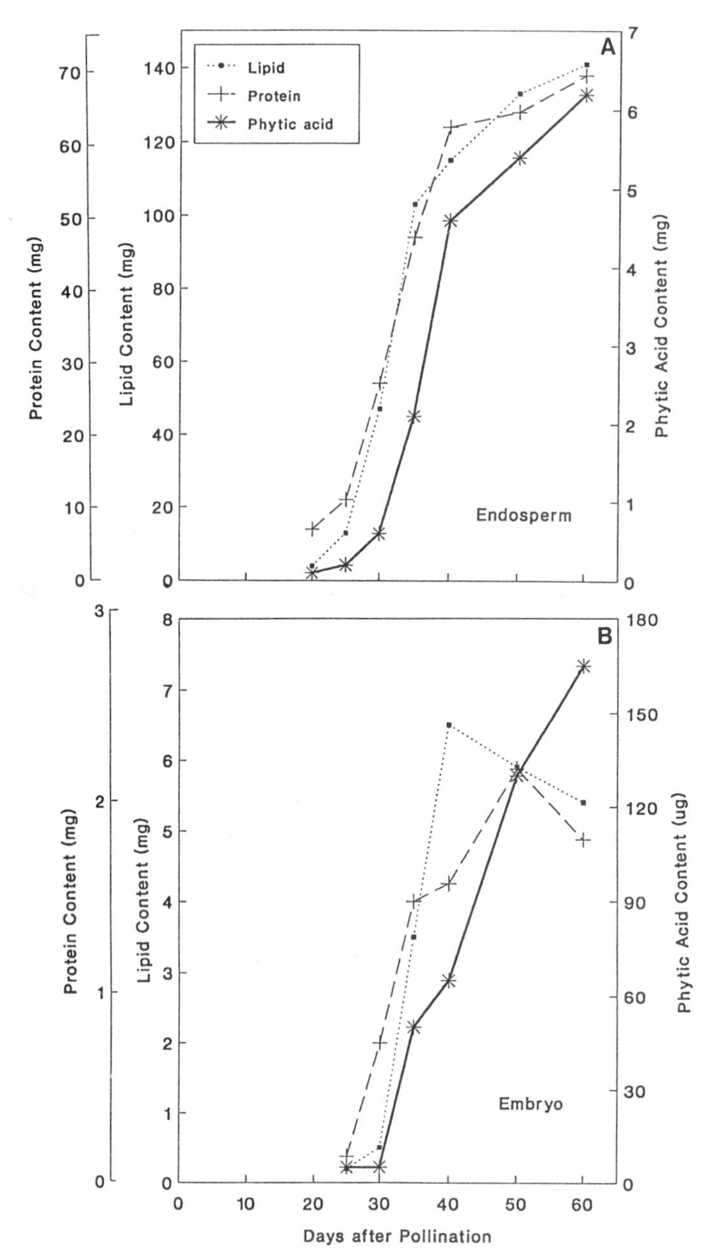


Figure 4. Reserve accumulation in developing endosperm (A) and embryo (B) of castor bean.

aqueous isolation procedures used by Tanaka et al. (1976b) would not have excluded other organelles or cytoplasmic proteins from the reaction mixtures. Furthermore, Tanaka et al. did not obtain labelled phytic acid from the reactions; only labelled mI-1-P was obtained. Thus the conclusion reached in both studies, that the protein bodies are the sites of phosphorylation of myo-inositol to form phytic acid, is not fully justified.

A second hypothesis regarding the intracellular site of phytic acid biosynthesis has been proposed based on results of light and electron microscopic studies of developing castor bean seed endosperm (Greenwood and Bewley,

1984). Phytin does not accumulate as globoid inclusions in the protein bodies during the initial phases of reserve synthesis. Instead, particles having the same characteristics as phytin-rich globoids collect in the cytoplasm surrounding the protein bodies of the developing storage parenchyma cells (Fig. 5 A-D.). The number of cytoplasmic particles decreases with continued development, as globoids within the protein bodies increase in size (Fig. 5 A-D).

The particles are also seen at the ultrastructural level (Fig. 5 E), often in close association with the cisternal rough endoplasmic reticulum (C.E.R., Fig. 5 F). The particles also become associated with the membrane of the developing protein body (Fig. 6 B), raising the possibility that deposition into the protein body may involve vesicular fusion, although the dynamics of the process are impossible to demonstrate using conventional microscopy. The inherent electron density of the particles makes it difficult to determine if they are membrane bound. In unstained sections, and occasionally in stained sections, dense particles can be seen lying within the R.E.R. cisternae, and some particles free in the cytoplasm are surrounded by a membrane (Fig. 6 A, B, E). At later stages of development, phytin and storage protein appear to be deposited together within the protein body (Fig. 6 C).

The cytoplasmic particles and globoids are equally soluble in dilute acetic acid and susceptible to digestion by non-specific alkaline phosphatase, both observations being consistent with the assumption that the particles and globoids are primarily composed of phytin. EDS analyses of the cytoplasmic and C.E.R. particles and globoids give identical elemental spectra, providing further evidence that the three are of similar composition (data not shown, see Greenwood and Bewley, 1984). These observations led us to suggest that phytin is initially synthesized in the cytoplasm, in or in close association with the C.E.R., then transported to and deposited within the developing protein bodies via C.E.R.-derived vesicles (outlined in Fig. 6 E). The proposed mechanism is similar to that of storage protein synthesis and deposition in other dicot seeds, with the exception that the Golgi complex does not appear to be involved (see Chrispeels, 1983; Greenwood and Chrispeels, 1985). Possible involvement of the Golgi cannot be dismissed, however. Few Golgi profiles were evident in the cells of developing castor bean endosperm, probably as a result of the short term fixation and embedding procedures used for maximum retention of phytin. Under better fixation conditions, Golgi complexes are indeed prevalent in cells of the developing castor bean endosperm, and are involved in the transport of storage proteins to the protein bodies (M.J. Brown, J.S. Greenwood, in preparation). The involvement of the C.E.R. in the synthesis and intracellular transport of phytin awaits verification, both by demonstrating that the C.E.R. is involved in at least a portion of the synthetic pathway and by confirming the observations above using developing seeds of other genera. The mechanism of targetting of phytin to the protein bodies awaits elucidation. However, cytoplasmic phytin particles have been found in developing cotyledonary cells of Pisum (Lott et al., 1985) and in the cotyledonary cells of germinated, phosphate-supplemented, castor bean (Organ, 1988).

## Pathways of Phytic Acid Biosynthesis

Attempts to ascertain the pathway of phytic acid biosynthesis were initially conducted using intact developing seeds, either by determining relative amounts of the various phosphorylated esters of myo-inositol during development or by following the incorporation of $^{3}H$- or $^{14}C$-labelled myo-inositol and inorganic-$^{32}P$ into myo-inositol phosphates (Asada and Kasai, 1962; Asada et al., 1968, 1969; Ogawa et al., 1979b; Saio, 1964; Sobolev, 1962). Two different mechanisms of phytic acid biosynthesis have been proposed based on these, and other, studies.

Asada and coworkers suggested that at least a portion of the phosphorylation process occurs via bound or complexed myo-inositol phosphate intermediates. Studies were conducted using developing rice and wheat (Asada and Kasai, 1962; Asada et al., 1968, 1969). The most consistent finding of

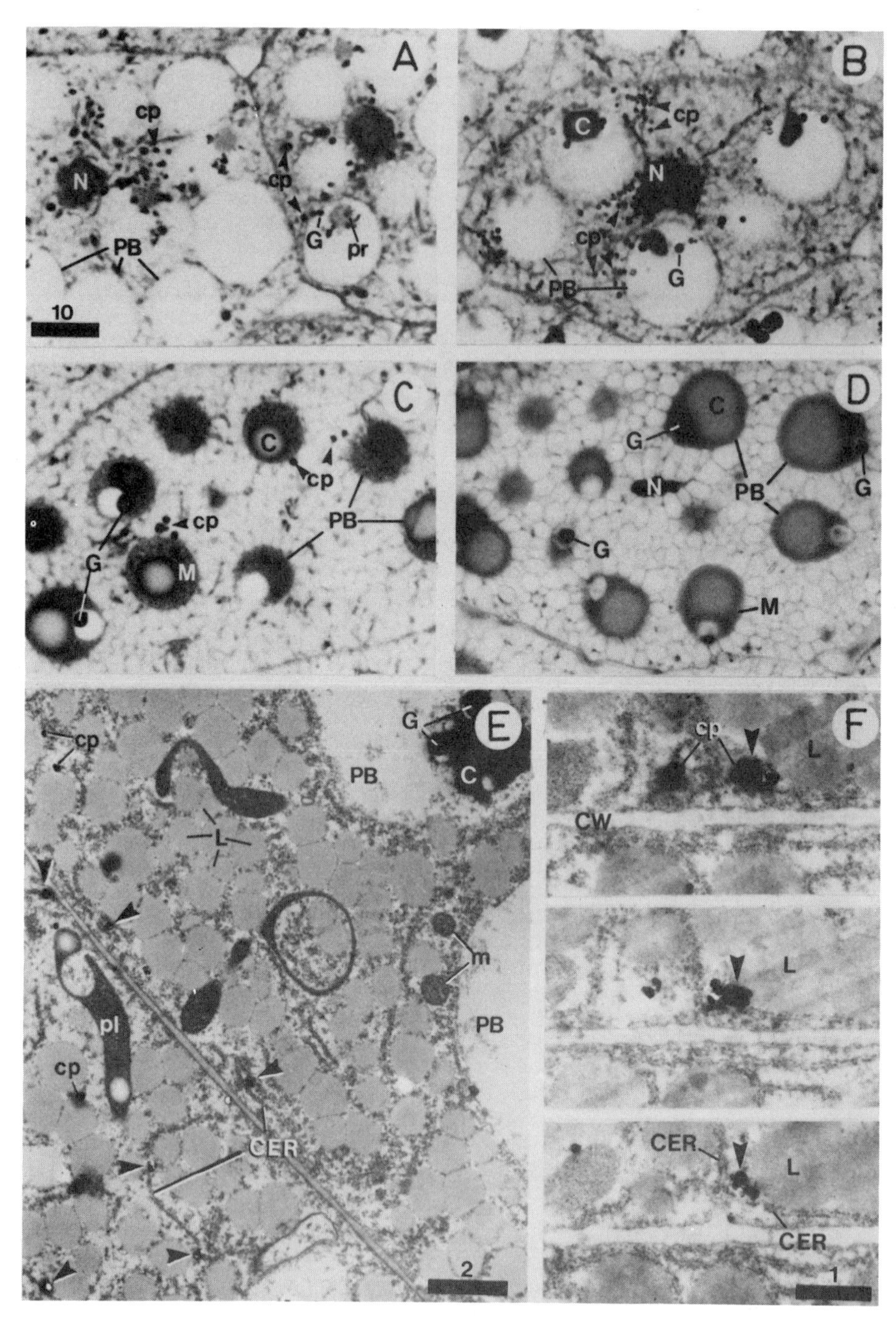


Figure 5 A-F Occurrence of phytin in the cytoplasm and its association with the cisternal rough endoplasmic reticulum during endosperm cell development in castor bean seed. A to D. Particles with the same staining characteristics as protein body globoids are present in the cytoplasm during early and mid stages of development (A,B,C) but are absent at maturity (D). E. Electron micrograph illustrating phytin particles (arrows) in the cytoplasm of developing castor bean

these studies was that, of the numerous phosphorylated esters of myo-inositol, only myo-inositol monophosphate (most probably mI-1-P) and $mIP_6$ were labelled when seeds were fed [$U^3H$]- or [$U^{14}C$]-myo-inositol or $^{32}Pi$ during the ripening process. Phosphorylated intermediates, myo-inositol bis- to pentakisphosphates ($mIP_2$ to $mIP_5$), were not found. If phytic acid were to be synthesized via a stepwise phosphorylation of myo-inositol, then the intermediates should be detectable, unless the kinetics of the reaction so favour production of the hexakisphosphate ester that intermediates would not accumulate. Based on their evidence, Asada and coworkers suggested that the synthesis of phytic acid involves an initial phosphorylation of myo-inositol to mI-1-P. The monophosphate ester is then bound to some compound "X", phosphorylated to completion in an ATP-dependent manner, and then released as phytic acid (outlined in Figure 7, scheme A). Compound "X" was detected (Asada et al., 1969) but neither it nor the enzyme(s) involved in the phosphorylation of the bound intermediate(s) have been identified.

In the second proposed mechanism, a direct step-wise phosphorylation of myo-inositol or mI-1-P to finally yield $mIP_6$ is implied. Early studies indicated that, in addition to $mIP_6$, lower phosphorylated esters of myo-inositol, $mIP_3$ and $mIP_4$, were present in extracts from a variety of developing seeds and suggested, albeit somewhat equivocally, that the synthesis of phytic acid proceeded via a stepwise phosphorylation of myo-inositol (Sobolev, 1962). Enzymatic digestion of phytic acid during extraction could explain the presence of these lower phosphorylated intermediates (Cosgrove, 1980). The mechanism has gained favour, however, since enzymes capable of phosphorylating myo-inositol and higher myo-inositol phosphates have been isolated and characterized (see below).

## Enzymes Involved in the Step-wise Phosphorylation of myo-Inositol

Most of our knowledge of the enzymes involved in phytic acid biosynthesis has been obtained from studies of quiescent and germinated mung bean seeds, and resting-stage turions of *Lemna gibba*. Developing seeds have been rarely used as enzyme sources, but, then again, very few studies of the enzymatic pathway of phytic acid biosynthesis have been conducted.

The Initial Phosphorylation of myo-Inositol. One feature common to the proposed pathways of phytic acid biosynthesis, outlined in Figure 7, is that myo-inositol monophosphate (mIP) is the initial substrate for the remaining ATP-dependent phosphorylation reactions. Synthesis of mIP occurs in two ways, via the cyclization of glucose-6-P to 1L-mI-1-P by 1L-mI-1-P synthase and via an ATP-dependent phosphorylation of free myo-inositol to 1L-mI-1-P by myo-inositol kinase (see Fig. 7).

The $NAD^+$-dependent and $NH_4^+$-activated 1L-mI-1P synthase (EC 5.5.1.4) is important in the synthesis of phytic acid since it is probably the sole enzyme that links carbohydrate metabolism to myo-inositol formation (Loewus and Loewus, 1983). Free myo-inositol is derived from the product of the synthase reaction via the action of $Mg^{2+}$-dependent 1L-mI-1-phosphatase (EC 3.1.3.25).

---

Fig. 5 (cont'd.)
endosperm cells. The particles are often found in association with cisternal (rough) endoplasmic reticulum. F. Serial sections illustrating that phytin particles apparently free in the cytoplasm may be associated with cisternal endoplasmic reticulum above or below the plane of section. Bar units in um. (from Greenwood and Bewley, 1984 and Greenwood, 1983).

C, crystalloid protein; CER, cisternal endoplasmic reticulum; cp, cytoplasmic phytin particles; G, globoid; L, lipid body; M, matrix protein; m, mitochondrion; N, nucleus; PB protein body; pl, plastid, pr, storage protein.

myo-Inositol is then capable of entering into a variety of reactions and pathways including the production of non-cellulosic cell wall polysaccharides, the synthesis of phosphoinositides and myo-inositol : phosphoinositide exchange, and production of bound auxins. It is possible that free myo-inositol can be scavenged for phytic acid biosynthesis through a conversion back to 1L-mI-1-P by the action of ATP-and $Mg^{2+}$-dependent myo-inositol kinase (EC 2.7.1.64) (Loewus and Loewus, 1983).

1L-mI-1-P synthase has been isolated from a number of developing seeds and appears to act as the sole enzyme involved in the production of 1L-mI-1-P during maturation of mung bean seed (Majumder and Biswas, 1973). myo-Inositol kinase has been isolated and partially purified from germinating mung bean (Dietz and Albersheim, 1965; English et al., 1966) and has been implicated in the synthesis of phytic acid during the germination process (Majumder et al., 1972) but was not detected in extracts from developing mung bean seed (Majumder and Biswas, 1973); its possible role in the formation of phytic acid during seed maturation is in doubt (Biswas et al., 1984).

Subsequent Phosphorylation to myo-Inositol Hexakisphosphate. Enzymes responsible for the ATP-dependent phosphorylation of mIP to $mIP_6$ have been identified and isolated from turions of *Lemna gibba* entering quiescence (Bollman et al., 1980) and from germinating seeds of mung bean (reviewed by Biwas et al., 1984). The proposed pathways are outlined in Figure 7, schemes B and C.

*Lemna gibba* (duckweed) synthesizes phytic acid under conditions that promote quiescence. Bollman et al. (1980), using affinity chromatography, isolated three separate enzymes or enzyme complexes from *Lemna* that, if considered in combination, were capable of phosphorylating myo-inositol to $mIP_6$ (Fig. 7, scheme B). The first enzyme involved was myo-inositol kinase (described above); 1L-mI-1-P synthase activity was not detected in this vegetative system. A second enzyme, or enzyme complex, catalyzed the ATP-dependent phosphorylation of mI-1-P to $mIP_3$ *in vitro*. The exact structure of the product was not determined. The third unnamed ATP-dependent enzyme or enzyme complex phosphorylated $mIP_3$ to $mIP_6$. Intermediates of either reaction ($mIP_2$ in the first, $mIP_4$ and $mIP_5$ in the second) were not detected. No details concerning the amount of protein used in the assays were given and specific activities of the enzymes and kinetics of the reactions were not determined. A

---

Figure 6. Association of the cytoplasmic phytin particles with the endoplasmic reticulum, protein body membrane and storage protein in developing castor bean endosperm cells. A. A phytin particle closely associated with the endoplasmic reticulum, surrounded by membrane. B. Phytin particles in close association with the endoplasmic reticulum and protein body membrane. A few particles appear to be membrane bound. C and D. Phytin is occasionally found embedded in proteinaceous material associated with the endoplasmic reticulum (C) and appears to be co-deposited into the protein body with storage protein (D). E. Electron micrograph illustrating the mechanism of phytic acid deposition proposed by Greenwood and Bewley, 1984. Phytin is synthesized in or in association with the cisternal ER (I A and B) and packaged into ER-derived vesicles (II A and B, note membrane surrounding particle in A, II). The phytin particles are then transported through the cytoplasm to the protein body membrane (III and IV, A and B) where the fusion of protein body and vesicular membranes releases phytin into the protein body (V A and B). The particles then condense to form the globoid (VI B). Bar units in um.

CER, cisternal endoplasmic reticulum; cp, cytoplasmic phytin particles; CW, cell wall; G, globoid; L, lipid body; M, matrix; m, mitochondrion; p, phytin; PB, protein body; PBM, protein body membrane; pl, plastid; pr, protein; Vs, vesicle; VsM, vesicular membrane.

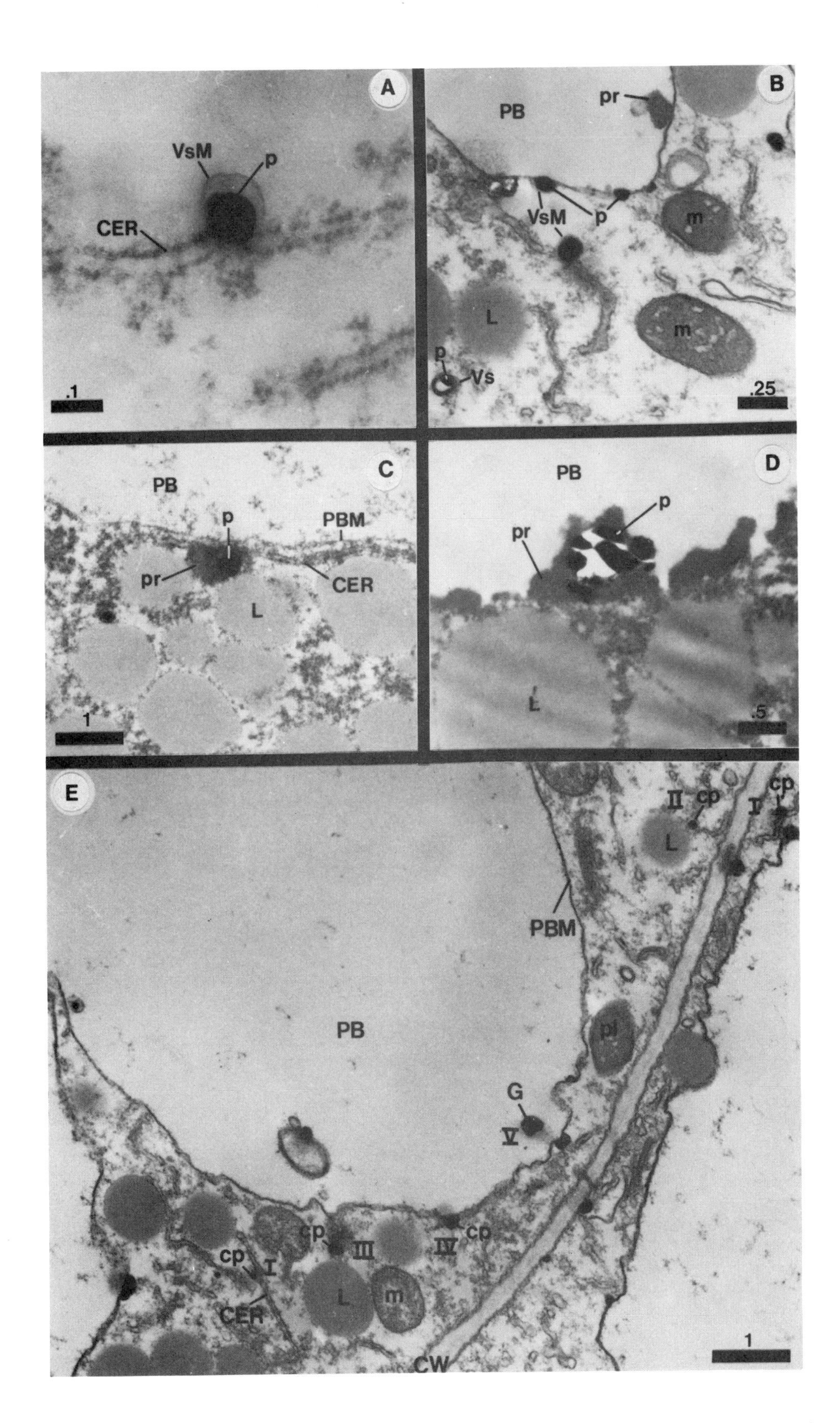
A
VsM
p
CER
.1
B
PB
pr
VsM
p
m
L
p
Vs
.25
C
PB
p
PBM
pr
CER
L
1
D
PB
p
pr
L
.5
E
II
cp
I
cp
L
PBM
pl
PB
G
V
cp
III
IV
cp
I
CER
L
m
CW
1

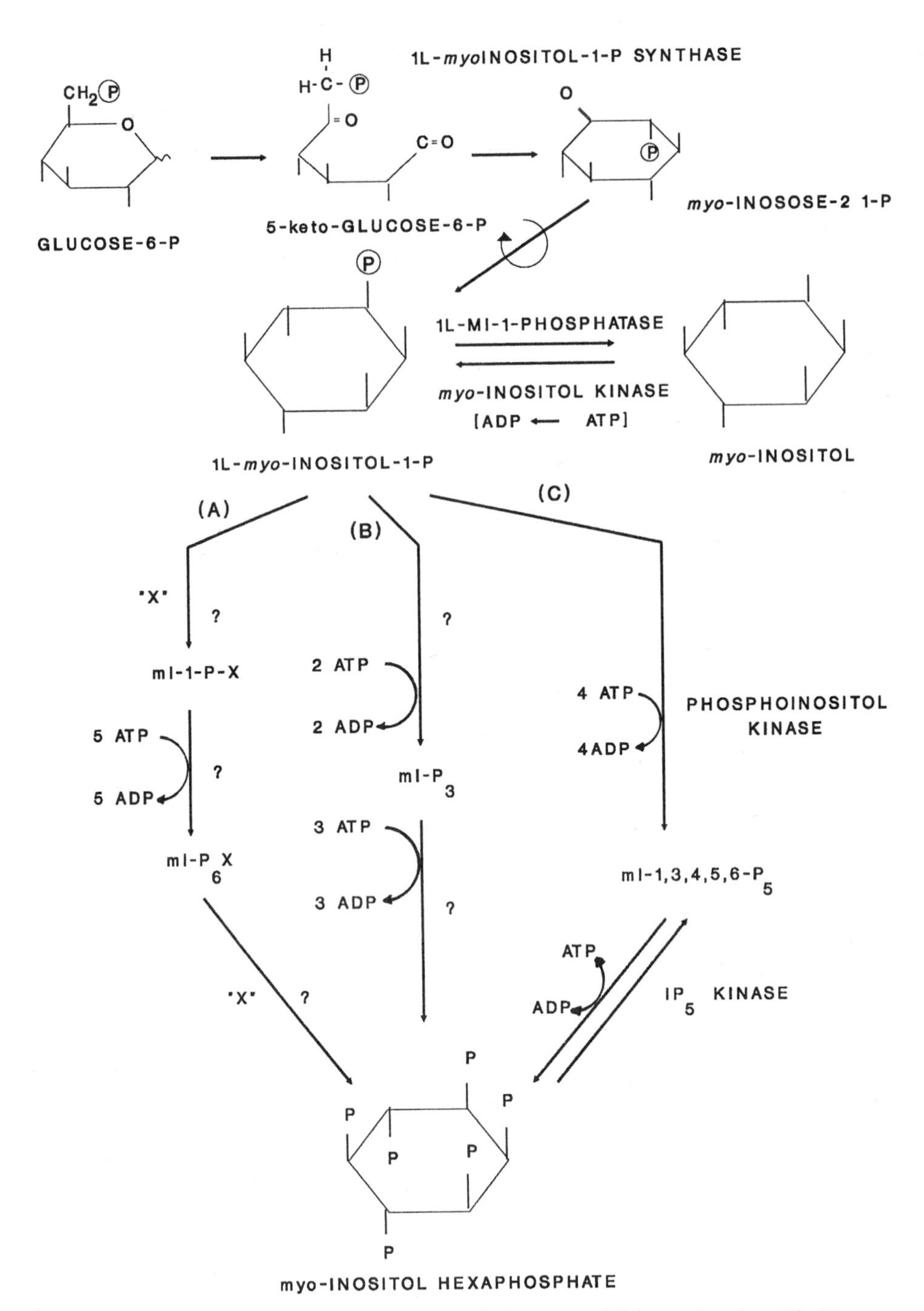


Figure 7. Outline of proposed pathways of phytic acid biosynthesis. Phytic acid biosynthesis begins with the cyclization of glucose-6-P to 1L-mI-1P by the action of 1L-mI-1-P synthase (top, circular arrow denotes rotation of the molecule). Free myo-inositol and 1L-mI-1-P can be interconverted by the action of 1L-mI-1-phosphatase and myo-inositol kinase. Three mechanisms have been proposed for the ATP-dependent phosphorylation of 1L-mI-1-P to phytic acid. Schemes outlined are from Asada and co-workers (A), Bollman et al., 1980 (B) and Biswas and co-workers (C), described in text.

relatively low rate conversion of $mIP_3$ to $mIP_6$ was noted. This may be explained by the fact that a mixture of twenty possible trisphosphate isomers was used as substrate. It is likely that the enzyme is capable of phosphorylating only one of the $mIP_3$ isomers.

The most intensive studies of the enzymes involved in the subsequent phosphorylation of 1L-mI-1P to phytic acid have been conducted by Biswas and co-workers (reviewed by Biswas et al., 1984). Most were performed using germinating mung bean seeds as the enzyme source, following the discovery that the seeds both catabolized and synthesized phytic acid during the first 36 h following imbibition (Mandal and Biswas, 1970). According to the studies, the phosphorylation of 1L-mI-1-P to $mIP_6$ is catalyzed by two enzymes, phosphoinositol kinase and $mIP_6$:ADP phosphotransferase (otherwise known as $mIP_5$ kinase).

Phosphoinositol kinase, isolated both from germinating and developing seeds (Chakrabarti and Biswas, 1981; Majumder and Biswas, 1973; Majumder et al., 1972) phosphorylates 1L-mI-1-P to mI-1,3,4,5,6-$P_5$ in a step-wise manner (Fig. 7, scheme C). The Km for mIP, using a mixture of monophosphate isomers, was found to be $1.5 \times 10^{-5}$M. The kinase is not able to phosphorylate the C-2 position. The enzyme is capable of phosphorylating mI-2-P and higher phosphorylated esters of myo-inositol, providing that C-2 is phosphorylated, to form phytic acid. More recent evidence indicates that two forms of the enzyme exist in germinating mung bean seeds. In both cases the phosphorylation reaction is $Mg^{2+}$-or $Mn^{2+}$-dependent, $Mn^{2+}$ being most effective. ATP is the most effective phosphate donor.

According to Biswas and co-workers, phosphorylation of mI-1,3,4,5,6-$P_5$ to $mIP_6$ is catalyzed by the reversed $mIP_6$:ADP phosphotransferase reaction (Biswas et al., 1978a,b, 1984). The enzyme has been isolated from germinating mung bean seeds (Biswas et al., 1978b) and from developing seeds (Biswas et al., 1978a), although no details concerning the latter isolation were given. In the $mIP_5$ kinase reaction, the enzyme is specific for mI-1,3,4,5,6-$P_5$ and is ATP-dependent. Both $Mg^{2+}$ and $Mn^{2+}$ stimulate the reaction, $Mn^{2+}$ being more effective (Biswas et al., 1978b). Details concerning the kinetics of the kinase reaction were not provided.

There has been no additional evidence published in support of any of the phytic acid biosynthetic pathways proposed by Asada and co-workers, Bollman et al., (1980) or Biswas and co-workers, outlined in Figure 7. In defense of the enzymatic pathway proposed by Biswas et al., a $mIP_5$ kinase has recently been isolated from quiescent soybean seeds and purified. Preliminary results indicate that the enzyme has specificity for the 1,3,4,5,6-P ester of myo-inositol and has other properties similar to those of the $mIP_6$:ADP phosphotransferase isolated from mung bean by Biswas and co-workers (B.Q. Phillippy, USDA-ARS, New Orleans, Louisiana, personal communication). However, the applicability of either of the proposed enzymatic pathways to the situation occurring in developing seeds is unknown, since details concerning the pathways were obtained either from a fully vegetative or (primarily) from a germinating seed system with little reference back to developing seeds. As with the proposal concerning the site of phytin biosynthesis, similar studies on a number of genera are required in order to confirm, or refute, either or both of the suggested pathways.

In light of the last statement above, some comment should be made with regard to future attempts at determining the phytic acid synthetic pathway. With the current interest in phosphatidylinositol and phosphatidylinositol-phosphates and their role in signal transduction in animal and in plant cells (the literature in this area is extensive, see recent articles by Irvine et al., 1989; Morse et al., 1987, Rincon and Boss, 1987 and the review by Hokin, 1985), it is reasonable to speculate that these membrane lipids may be involved in the phytic acid biosynthesis. Products of the hydrolysis of phosphatidylinositol and phosphatidylinositol-phosphates include specific myo-inositol mono- to tri-phosphates, all of which could act as initial substrates or intermediates of phytic acid biosynthesis, but the possible involvement of inositol-containing lipids in the phytic acid biosynthetic pathway has been virtually ignored.

Future attempts to determine the synthetic pathway of $mIP_6$ in vivo must involve monitoring all possible sources of inositol phosphates, including the inositol-containing lipids.

CONCLUDING STATEMENT

Despite the fact that we have known that phytin is a major phosphorus reserve in seeds for well over seventy years, we are not at all certain of the mechanisms involved in phytin biosynthesis or deposition. Our current understanding is based on a very limited number of studies from a very few laboratories. These studies must be repeated using seeds from a variety of genera. Further, attempts at elucidating the pathway of phytic acid biosynthesis, using germinating seed or non-seed systems, must be related back to the developing seed. Fortunately, there are a number of analytical and preparative procedures now available that should aid in this research. Cryo-preservation and cryo-electron microscopy provide alternatives to the standard methods of tissue preparation, preventing the solubilization and movement of phytin within tissues. This should provide for more accurate localization and identification of phytin by EDS analysis and assist in determining the mechanism of deposition. HPLC procedures have been developed for separation of various inositol phosphates and the isomers of individual inositol phosphates (Phillippy and Bland, 1988), providing very specific substrates for enzyme characterization. The commercial availability of highly purified phosphatidylinositol and specific phosphatidylinositol-polyphosphates should assist in determining if these membrane lipids are involved in phytic acid biosynthesis. Relatively recent advances in protein separation and purification procedures should also provide for better separation and possible purification of the enzymes involved in phytic acid biosynthesis.

REFERENCES

Asada, K., Kasai, Z., 1962, Formation of myo-inositol and phytin in ripening rice grains, Plant Cell Physiol., 3: 397

Asada, K., Tanaka, K., Kasai, Z., 1968, Phosphorylation of myo-inositol in ripening grains of rice and wheat. Incorporation of phosphate-$^{32}P$ and myo-inositol-$^{3}H$ into myo-inositol phosphates, Plant Cell Physiol., 9: 185

Asada, K., Tanaka, K., Kasai, Z., 1969, Formation of phytic acid in cereal grains, Ann. N.Y. Acad. Sci., 165: 801

Bieleski, R.L., 1973, Phosphate pools, phosphate transport and phosphate availability Ann. Rev. Plant Physiol., 24: 225

Biswas, B.B., Ghosh, B., Majumder, A.L., 1984, myo-Inositol polyphosphates and their role in cellular metabolism. A proposed cycle involving glucose-6-phosphate and myo-inositol phosphates, Subcell. Biochem., 10: 237

Biswas, B.B., Biswas, S., Chakrabarti, S., De, B.P., 1978, A novel metabolic cycle involving myo-inositol phosphates during formation and germination of seeds, In: Wells, W.W., Eisenberg, F. Jr. (eds.), Cyclitols and Phosphoinositides, Academic Press, New York

Biswas, S., Biswas, B.B., 1965, Enzymatic synthesis of guanosine triphosphate from phytin and guanosine diphosphate, Biochim. Biophys. Acta, 108: 710

Biswas, S., Burman, S., Biswas, B.B., 1975, Inositol hexaphosphate guanosine diphosphate phosphotransferase from Phaseolus aureus, Phytochem., 14: 373

Biswas, S., Maity, I.B., Chakrabarti, S., Biswas, B.B., 1978, Purification and characterization of myo-inositol hexaphosphate : adenosine diphosphate phosphotransferase from Phaseolus aureus, Arch. Biochem. Biophys., 185: 557

Bollman, O., Strother, S., Hoffmann-Ostenhof, O., 1980, The enzymes involved in the synthesis of phytic acid in Lemna gibba. (Studies on the biosynthesis of cyclitols. XL), Mol. Cell. Biochem., 30: 171

Chakrabarti, S., Biswas, B.B., 1981, Two forms of phosphoinositol kinase from germinating mung bean seeds, Phytochem., 20: 1815

Chrispeels, M.J., 1983, The Golgi apparatus mediates the transport of phytohemagglutinin to the protein bodies in bean cotyledons, Planta, 158: 140

Cosgrove, D.J., 1966, The chemistry and biochemistry of inositol polyphosphates, Pure Appl. Chem., 16: 209

Cosgrove, D.J., 1980, Inositol phosphates, their chemistry, biochemistry and physiology, Elsevier Scientific Pub. Co., Amsterdam

Dalling, M.J., Bhalla, P.L., 1984, Mobilization of nitrogen and phosphorus from endosperm, In Murray, D.R. (ed.), Seed physiology, volume 2. Germination and reserve mobilization, Academic Press, Sidney

Dietz, M., Albersheim, P., 1965, The enzymatic phosphorylation of myo-inositol, Biochem. Biophys. Res. Comm., 19: 598

English, P.D., Dietz, M., Albersheim, P., 1966, myo-Inositol kinase: partial purification and identification of product, Science, 151: 198

Greenwood, J.S., Bewley, J.D., 1982, Seed development in Ricinus communis L. cv. Hale. I. Descriptive morphology, Can. J. Bot., 60: 1751

Greenwood J.S., Bewley, J.D., 1984, Subcellular distribution of phytin in the endosperm of developing castor bean: a possibility for its synthesis in the cytoplasm prior to deposition within protein bodies, Planta, 160: 113

Greenwood, J.S., Bewley, J.D., 1985, Seed development in Ricinus communis cv. Hale (castor bean). III. Pattern of storage protein and phytin accumulation in the endosperm, Can. J. Bot., 63: 2121

Greenwood, J.S., Chrispeels, M.J., 1985, Immunocytochemical localization of phaseolin and phytohemagglutinin in the endoplasmic reticulum and Golgi complex of developing bean cotyledons, Planta, 164: 295

Greenwood, J.S., Gifford, D.J., Bewley, J.D., 1984, Seed development in Ricinus communis cv. Hale (castor bean). II. Accumulation of phytic acid in the developing endosperm and embryo in relation to the deposition of lipid, protein and phosphorus, Can. J. Bot., 62: 255

Halmer, P., Bewley, J.D., Thorpe, T.A., 1978, Degradation of the endosperm cell walls of Lactuca sativa L., cv. Grand Rapids, Planta, 139: 1

Hokin, L.E., 1985, Receptors and phosphoinositide-generated second messengers, Ann. Rev. Biochem., 54: 205

Irvine, R.F., Letcher, A.J., Lander, D.J., Drobak, B.K., Dawson, A.P., Musgrave, A., 1989, Phosphatidylinositol(4,5)bisphosphate and phosphatidylinositol(4)phosphate in plant tissues, Plant Physiol., 89: 888

Loewus, F.A., 1974, The biochemistry of myo-inositol in plants. Rec. Adv. Phytochem., 8: 179

Loewus, F.A., 1983, Phytate metabolism with special reference to its myo-inositol component, In Nozzolillo, C., Lea, P.J., Loewus, F.A. (eds.), Mobilization of reserves in germination, Plenum Press, New York

Loewus, F.A., Loewus, M.W., 1983, Myo-inositol: its biosynthesis and metabolism, Ann. Rev. Plant Physiol., 34: 137

Lott, J.N.A., 1980, Protein Bodies. In Tolbert, N. (ed.), The biochemistry of plants, a comprehensive treatise, volume 1. Cells and organelles, Academic Press, New York

Lott, J.N.A., 1984, Accumulation of seed reserves of phosphorus and other minerals. In Murray, D.R.,(ed.), Seed physiology, volume 1. Development, Academic Press, Sidney

Lott, J.N.A., Ockenden, I., 1986, The fine structure of phytate-rich particles in plants. In Graf, E, (ed.), Phytic acid: Chemistry and applications, Pilatus Press, Minneapolis

Lott, J.N.A., Spitzer, E., Vollmer, C.M., 1979, Calcium distribution in globoid crystals of Cucurbita cotyledons, Plant Physiol., 63: 847

Lott, J.N.A., Randall, P.J., Goodchild, D.J., Craig, S., 1985, Occurrence of globoid crystals in cotyledonary protein bodies of Pisum sativum as influenced by experimentally induced changes in Mg, Ca and K contents of seeds, Austral. J. Plant Physiol., 12: 341

Majumder, A.L., Biswas, B.B., 1973, Metabolism of inositol phosphates. Part V. Biosynthesis of inositol phosphates during ripening of mung bean (Phaseolus aureus) seeds, Indian J. Exp. Biol., 11: 120

Majumder, A.L., Mandel, N.C., Biswas, B.B., 1972, Phosphoinositol kinase from germinating mung bean seeds, Phytochem., 11: 503

Mandal, N.C., Biswas, B.B., 1970, Metabolism of inositol phosphates: part II - Biosynthesis of inositol phosphates in germinating seeds of Phaseolus aureus, Indian J. Biochem., 7: 63

Matheson, N.K., Strother, S., 1969, The utilization of phytate by germinating wheat, Phytochem., 8: 1349

Morse, M.J., Crain, R.C., Satter, R.L., 1987, Phosphatidylinositol cycle metabolites in Samanea saman pulvini, Plant Physiol., 83: 640

Nayini, N.R., Markakis, P., 1986, Phytases. In Graf, E., (ed.), Phytic acid: Chemistry and applications, Pilatus Press, Minneapolis

Ogawa, M., Tanaka, K., Kasai, Z., 1979a, Accumulation of phosphorus, magnesium and potassium in developing rice grains: Followed by electron microprobe X-ray analysis focusing on the aleurone layer, Plant Cell Physiol., 20: 19

Ogawa, M., Tanaka, K., Kasai, Z., 1979b, Phytic acid formation in dissected ripening rice grains, Agr. Biol. Chem., 43: 2211

Organ, M.G., 1988, Phytin synthesis in germinated castor bean, MSc thesis, University of Guelph

Pfeffer, W., 1872, Untersuchungen uber die protein-korper und die bedeutung des asparagins beim keimen der samen. Jb. Wiss. Bot., 8: 529

Phillippy, B.Q., Bland, J.M., 1988, Gradient ion chromatography of inositol phosphates, Analyt. Biochem., 175: 162

Raboy, V., Dickinson, D.B., 1987, The timing and rate of phytic acid accumulation in developing soybean seeds, Plant Physiol., 85: 841

Reddy, N.R., Sathe, S.K., Salunkhe, D.K., 1982, Phytates in legumes and cereals, Adv. Food Res., 28: 1

Ricon, M., Boss, W.F., 1987, myo-Inositol triphosphate mobilizes calcium from fusogenic carrot (Daucus carota L.) protoplasts, Plant Physiol., 83: 395

Saio, K., 1964, The change in inositol phosphates during the ripening of rice grains, Plant Cell Physiol., 5: 393

Samotus, B., 1965, Role of phytic acid in potato tuber, Nature 206: 1372

Sharma, C.B., Dieckert, J.W., 1975, Isolation and partial characterization of globoids from aleurone grains of Arachis hypogaea seed, Physiol. Plant., 33: 1

Sobolev, A.M., 1962, Occurrence, formation and utilization of phytin in higher plants, Usp. Biol. Khim., 4: 248

Sobolev, A.M., Rodionova, M.A., 1966, Phytin synthesis by aleurone grains in maturing sunflower seeds, Soviet Plant Physiol., 13: 958

Strother, S., 1980, Homeostasis in germinating seeds, Ann. Bot., 45: 217

Tanaka, K., Nishitomi, T., Ogawa, M., Yoshida, T., Kasai, Z., 1976a, Formation of adenosine triphosphate in isolated aleurone particles of rice grains, Agr. Biol. Chem., 40: 1313

Tanaka, K., Yoshida, T., Kasai, Z., 1976b, Phosphorylation of myo-inositol by isolated aleurone particles of rice, Agr. Biol. Chem., 40: 1319

Williams, S.G., 1970, The role of phytic acid in the wheat grain, Plant Physiol., 41: 376

# CALCIUM-REGULATED METABOLISM IN SEED GERMINATION

Stanley J. Roux

Department of Botany
The University of Texas at Austin
Austin, TX 78713

## INTRODUCTION

The past decade has seen a dramatic growth of interest in $Ca^{2+}$ as a stimulus-response coupler in plant cells. Key discoveries that ignited this interest were that plant cells, like animal cells, maintain their cytosolic free $Ca^{2+}$ concentration ($[Ca^{2+}]_{cyt}$) near $10^{-7}$ M (Gilroy et al., 1986; Bush and Jones, 1987; Brownlee et al., 1987); plant cells contain a variety of $Ca^{2+}$-binding proteins that can become metabolic regulators at $[Ca^{2+}]_{cyt}$ of $10^{-6}$ or above (Biro et al., 1984; Harmon et al., 1987; Boustead et al., 1989), and stimulus-response coupling in many different plant cells can be blocked by $Ca^{2+}$ chelation, $Ca^{2+}$ channel blockers, and antagonists of $Ca^{2+}$-binding proteins (Daye et al., 1984; Serlin and Roux, 1984; Saunders and Hepler, 1983). There is now a major and rapidly expanding literature implicating $Ca^{2+}$ in the control of many aspects of plant metabolism, growth and development, including gravitropism (Roux and Serlin, 1987), photomorphogenesis (Roux et al., 1986), photosynthesis (Homann, 1987), mitosis (Lambert and Vantard, 1986) and stress responses (Kauss, 1987), to name only a few of the best studied.

How do these discoveries relate to an understanding of seed germination? Although there is an extensive literature documenting a nutritive role for $Ca^{2+}$ in seed germination (see below), there are, in fact, few reports that test whether $Ca^{2+}$ plays a major role as a regulator coupling environmental and hormonal stimuli to germination. Interest in this topic derives mainly from reports that intracellular $Ca^{2+}$ controls a number of important processes and enzyme activities associated with seed germination and that agents which critically influence the germination of many different seeds (e.g., light and hormones) exert their regulatory effects in many instances through the mediation of $Ca^{2+}$. Indirectly, interest follows also from the fact that $Ca^{2+}$ helps to control the transition from dormancy to germination in fern spores and in pollen grains.

Although investigations linking $Ca^{2+}$-controlled events to seed germination have not been a major theme in recent journal



*Recent Advances in the Development and Germination of Seeds*
Edited by R.B. Taylorson
Plenum Press, New York

publications, the data that are available make it plausible to propose that there may be such a link. Especially in the context of a workshop which encourages thinking about and critical evaluation of relatively untested ideas, it will be useful to selectively review what is known about $Ca^{2+}$ as an intracellular regulatory agent in plants, with particular emphasis on its role in those metabolic contexts likely to be important for seed germination. The chapter will conclude with a discussion of key experiments that could clarify how extensively $Ca^{2+}$ is employed as a signal amplifier for agents that stimulate or inhibit seed germination.

## Calcium as a Nutrient for Germination

There is a fairly extensive literature documenting a nutritive role for $Ca^{2+}$ in seed germination. Over the last several decades dozens of reports have described that under specific conditions (temperature, salinity, etc.), seeds germinate better if mM levels of $Ca^{2+}$ are available in the surrounding soil or medium (Koller, 1972). In this context, $Ca^{2+}$ is probably playing mainly a nutritive role, sustaining certain basic functions, such as membrane stability, needed for the growth events of germination. Clarkson and Hanson (1981) discuss several, mainly extracellular, nutritive functions of $Ca^{2+}$ that would require relatively high concentrations of the ion. Undoubtedly, some of the "nutritive" $Ca^{2+}$ is used for intracellular regulatory roles, too, so the distinction is somewhat arbitrary. Most of the emphasis below will be on those functions of intracellular $Ca^{2+}$ that can be initiated by micromolar concentration changes and that would be likely to have important consequences for germination metabolism.

## Agents That Regulate Seed Germination Can Utilize $Ca^{2+}$ as a Transducing Signal

Light and hormones are two kinds of agents well known to regulate seed germination (Mayer and Poljakoff-Mayber, 1989). Among the photoreceptors controlling light-induced seed germination, phytochrome is the most studied (Frankland and Taylorson, 1983). The principal hormones that have been implicated as inducers or inhibitors of seed germination are the gibberellins, ABA (Fincher, 1989), the cytokinins and ethylene (Kuan and Huang, 1988).

Reports on the involvement of $Ca^{2+}$ in regulating the responses of plants to phytochrome and to various hormones have been accumulating in the literature for about a decade now (Roux and Serlin, 1987). However, to rigorously show that $Ca^{2+}$ helps to transduce an environmental or hormonal stimulus into a physiological response, one would ultimately have to identify the specific $Ca^{2+}$-dependent step in the sequence of biochemical changes leading to the response. To date this has not been done for any stimulus-response pathway in plants. The indirect evidence for the critical involvement of $Ca^{2+}$ as a transducing signal for phytochrome and for some hormones in plants is strong and consists of the following three kinds of results: The stimulus (e.g., red light or hormone application) induces changes in calcium transport in responding tissue (Hale and Roux, 1980; dela Fuente, 1984); antagonists that block calcium

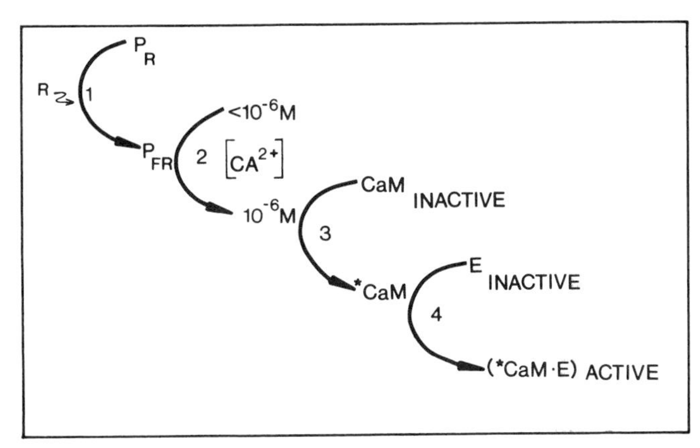


Fig. 1. A model describing a hypothetical cascade of events initiated by light-activated phytochrome. 1, Phytochrome activated by red light (R); 2, Pfr-mediated increase in $[Ca^{2+}]_{cyt}$; 3, activation of cytosolic CaM; 4, binding of activated CaM (*CaM) to *CaM-dependent enzymes (E), thus activating them.

uptake or that block the action of $Ca^{2+}$-binding proteins block the conversion of the stimulus into the usual response (Serlin and Roux, 1984; Elliott, 1986); agonists that promote calcium uptake can substitute for the stimulus in inducing the response (Perdue et al., 1988; Saunders and Hepler, 1982).

The model of a $Ca^{2+}$-based stimulus transduction predicted from such data is a familiar one (Fig. 1). Applying this model to seed germination, one would predict that an environmental or chemical stimulus that could induce germination would initiate the process by stimulating an increase of the $[Ca^{2+}]_{cyt}$ in cells in some tissue critical for seed germination; e.g., the aleurone layer, endosperm cells, or the enlarging radicle of cereal grains. This would result in the activation of one or more $Ca^{2+}$-binding regulatory proteins, which would, in turn, activate some enzyme(s) critical for germination metabolism. As a first test of this model, it would be important to identify $Ca^{2+}$-binding and $Ca^{2+}$-activated proteins in seeds.

## Potential Targets of $Ca^{2+}$ Action in Seeds

When a change in $[Ca^{2+}]_{cyt}$ serves to couple a stimulus to a response, the rise in $[Ca^{2+}]_{cyt}$ is often transitory and is usually over a relatively narrow range of $[Ca^{2+}]_{cyt}$ values below $10^{-5}$ M (Williamson and Ashley, 1982; Alkon and Rasmussen, 1988). For this reason the "coupling" is usually through the mediation of proteins that bind $Ca^{2+}$ with a high affinity. These proteins may either be enzymes or regulatory proteins that change the activity of enzymes by binding to them.

The best known $Ca^{2+}$-binding protein in plants is calmodulin. Among the first characterized calmodulins in plants was peanut seed calmodulin (Anderson et al., 1980). Not only

Table 1. Change in Calmodulin Levels in Radish Seeds During Germination

| Incubation Conditions | Calmodulin Level (ug/g fr wt)$^{-1}$ |
|---|---|
| 0 h | 25.5 |
| 24 h, $H_2O$ | 52.7 |
| 24 h, + ABA | 29.3 |
| 24 h, + Fusicoccin (FC) | 62.6 |
| 24 h, + ABA & FC | 50.2 |

Taken from Cocucci and Negrini, Plant Physiol. (1988).

is calmodulin present in seeds, but its content in some seeds can be modulated by chemicals that promote or inhibit germination. Cocucci and Negrini (1988) have found that in the early phases of radish seed germination, the content of calmodulin rises and the level of a calmodulin-binding protein, which was thought to be a calmodulin inhibitor, decreases. The increase of calmodulin is promoted correspondingly with germination promotion by fusicoccin; and the increase is blocked by ABA (Table 1), which simultaneously also blocks germination. The authors concluded that calmodulin could play an important role in the regulation of germination of radish seeds.

To confirm this hypothesis it will be necessary to identify a calmodulin-regulated enzyme in seeds that can affect germination. The list of enzymes that are reported to be calmodulin-regulated in plants is not very long (Piazza, 1988), and even this list contains only three or four enzymes that have been purified and demonstrated to be under calmodulin control in a defined *in vitro* assay. Among the enzymes postulated to be controlled by calmodulin, several have been found in pre-germinated or newly germinated tissues (Table 2). These include two wheat germ protein kinases (Polya and Micucci, 1984), a phosphatidic acid phosphatase in bean cotyledons (Paliyath and Thompson, 1987), and a DNA-binding NTPase in pea buds (Chen et al., 1987). There is a neutral lipase activity in castor bean lipid bodies that is stimulated by micromolar $Ca^{2+}$ (Hills and Beevers, 1987), but whether this stimulation is through the mediation of calmodulin or some other $Ca^{2+}$-binding protein (see below) is not clear.

Table 2. Some Calmodulin-Stimulated Enzymes in Plants [a]

| |
|---|
| NAD kinase (Anderson et al., 1980) |
| Microsomal $Ca^{2+}$ ATPase (Dieter and Marme, 1981) |
| Phosphatidic acid phosphatase (Paliyath and Thompson, 1981) |
| Protein kinases from wheat germ (Polya and Micucci, 1984) |
| Chromatin-associated NTPase (Chen et al., 1987) |

[a] All enzyme preparations tested were partially purified.

Calmodulin is not the only protein in plants that binds $Ca^{2+}$ with a high affinity. Others that have been described include protein kinase C (PKC) (Schafer et al., 1985; Elliott and Kokke, 1987), annexins (Clark et al., 1988; Boustead et al., 1989), a phospholipid-independent kinase (Harmon et al., 1987), and a calpain-like protein (Clark et al., 1989). Although the function of none of these proteins is known in plants, the functions of PKC, annexins, and calpain are at least partially understood in animals, and one could speculate that they may have similar functions in plants.

PKC is known to control the activity of a score of different enzymes in animal cells (Nishizuka, 1986); evidence that it may function also in plant cells comes mainly from inhibitor studies (Dharmawardhane et al., 1989). In animal cells annexins are thought to initiate membrane fusion in exocytosis and to help link cytoskeletal elements to the plasma membrane (Burgoyne and Geisow). The localization of plant annexin-like proteins along the plasma membrane (Clark et al., 1988) would suggest they may play similar roles. Calpains are $Ca^{2+}$-dependent proteases that serve to regulate the controlled turnover of specific proteins. Evidence that they occur in plants is thus far strictly immunological: by Western blot analysis, there are antigens in plants that cross-react with purified antibodies to animal calpains, and the plant antigens have molecular weights similar to the animal calpains (Clark, Goll and Roux, unpublished). Determining whether PKC, annexins, and calpain are present in seeds will greatly aid an assessment of $Ca^{2+}$ function in germination.

The $Ca^{2+}$-binding proteins discussed above would all be activated by micromolar changes in $[Ca^{2+}]_{cyt}$. There are also some seed enzymes that show significant stimulation by $Ca^{2+}$ only at much higher concentrations when they are assayed _in vitro_. Examples of these would be a lipoxygenase from soybean seeds (Dressen et al., 1982) and a phospholipase D from rice bran (Lee, 1989a). Although in principle, these enzymes could also be targets of $Ca^{2+}$ action in seeds (Lee, 1989b), it is not clear what microenvironment of the seed would respond to changes in the concentration of $Ca^{2+}$ in the mM range. More likely, either the _in vitro_ effects of $Ca^{2+}$ on these enzymes are relatively non-specific or some aspect of the extraction or the handling of these enzymes radically altered their affinity for $Ca^{2+}$ or their regulatory properties.

## Calcium-Regulated Secretion in Seeds

There is an extensive literature implicating $Ca^{2+}$ in the regulation of secretion in plants (see references in Cunninghame and Hall, 1986). In seeds, studies on $Ca^{2+}$-stimulated secretion concern mainly the secretion of amylase from aleurone cells in cereal grains (Chrispeels and Varner, 1967; Jones and Jacobsen, 1983; Mitsui et al., 1984). Although in virtually all cases, the release of $\alpha$-amylase from aleurone cells is a post-germinative event, it is a good example of a secretion function in seeds, and certainly some secretion events in seeds are important for germination. All authors studying this phenomenon report that relatively high $Ca^{2+}$ concentrations, in the mM range, are required to maximize the secretion rates. As yet there is no information on which

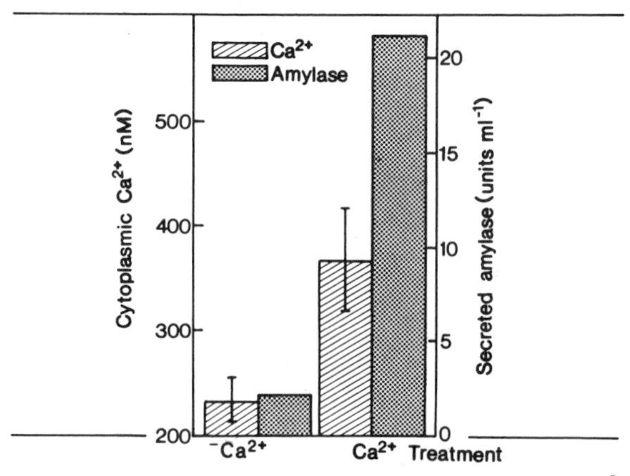


Fig. 2. The effect of extracellular $Ca^{2+}$ on the $[Ca^{2+}]_{cyt}$ and on α-amylase secretion from barley aleurone protoplasts. Taken from Bush & Jones (1988).

molecular target $Ca^{2+}$ is modifying to promote the stimulation. Recently, however, Bush and Jones (1988) reported important findings that implicate an intracellular target, and one that can respond to micromolar changes in $[Ca^{2+}]_{cyt}$.

Their study utilized barley aleurone protoplasts, which show characteristics of $Ca^{2+}$-stimulated secretion in the presence of $GA_3$ similar to those in isolated aleurone layers (Bush et al., 1986). The protoplasts were loaded with the $Ca^{2+}$-responsive dye, indo-1 so that their $[Ca^{2+}]_{cyt}$ could be monitored when the protoplasts were stimulated to secrete α-amylase by $Ca^{2+}$. They found that when the $Ca^{2+}$ concentration of the protoplast incubation medium was changed from 10 uM to 20 mM, the $[Ca^{2+}]_{cyt}$ of the protoplasts was elevated from 200 nm to ca. 350 nM and secretion was stimulated (Fig. 2). It was possible to induce an increase in $[Ca^{2+}]_{cyt}$ without adding $Ca^{2+}$ to the medium by raising the pH from 4.5 to 6.5. This pH change induces an increase in $[Ca^{2+}]_{cyt}$ from 235 nM to over 500 nM. Protoplasts incubated at pH 6.5 with no added $Ca^{2+}$ produce and secrete as much α-amylase as cells incubated at pH 4.5 with 20 mM $Ca^{2+}$ added.

These results strongly suggest that $[Ca^{2+}]_{cyt}$ changes in the submicromolar range help to regulate secretion. Since the $Ca^{2+}$-binding proteins most associated with secretion regulation in animals are the annexins (Burgoyne and Geisow, 1989), an investigation into the possible occurrence and function of annexins in aleurone cells seems warranted.

Although the secretion of α-amylase from aleurone cells is one of the best studied model systems for elucidating the role of $Ca^{2+}$ in secretion, other enzymes whose secretion is stimulated by $Ca^{2+}$ may also be important for the germination process. In addition to hydrolytic enzymes that help to mobilize food reserves in seeds, wall peroxidases, which have been implicated in growth control (Fry, 1986), also are

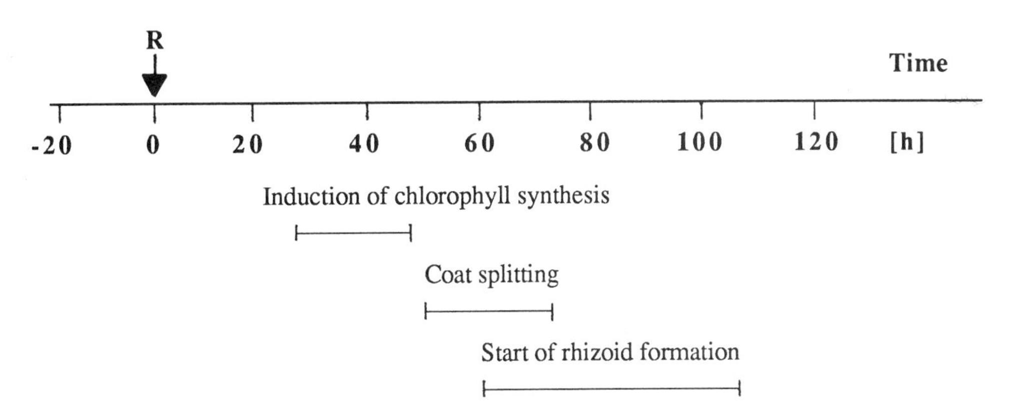


Fig. 3. Time course of germination of *Dryopteris* spores.

secreted by a mechanism that is stimulated by $Ca^{2+}$ (Sticher et al., 1981). The growth and protrusion of the radicle is the definitive sign that germination has occurred, so control of the rate of secretion of peroxidases to the wall could play an important role in initiating the growth required for germination.

## Fern Spore Germination: a Model System for Studying the Effects of $Ca^{2+}$ on Germination

The germination of the spores of *Onoclea* and *Dryopteris* is greatly stimulated by phytochrome, and this response requires $Ca^{2+}$ in the medium (Wayne and Hepler, 1984; Scheuerlein et al., 1989). Although seed germination is a very different process from spore germination, the simplicity of the spore system recommends it as a model for understanding how $Ca^{2+}$ could participate in a transduction chain leading from light stimulation to an emergence from dormancy.

The initiation of this transduction chain can be achieved with a brief red light (R) irradiation, which converts phytochrome to its physiologically active Pfr form. Scheuerlein et al. (1989) recently completed a study of how soon after an R treatment $Ca^{2+}$ is needed for germination to proceed. It takes about three days for the R-treated spore to progress from dormancy to germination (= rhizoid emergence), but this developmental progression is marked by several easily visualized morphological and biochemical changes that occur before germination (Fig. 3). Interestingly, spores irradiated in the absence of $Ca^{2+}$ remain capable of responding to $Ca^{2+}$ as late as 40 h after R (Fig. 4), when the first of the visible hallmarks of the germination progression can be visualized. Thus, although spores need external $Ca^{2+}$ to germinate, the metabolic step which requires this $Ca^{2+}$ does not occur until about 40 h after the germination process is initiated. Furthermore, although some product of Pfr action is needed for $Ca^{2+}$ to induce germination, Pfr *itself* does not have to be present at the time when $Ca^{2+}$ exerts its influence.

Since the level of external $Ca^{2+}$ needed to promote germination is micromolar, it is likely that $Ca^{2+}$ is having

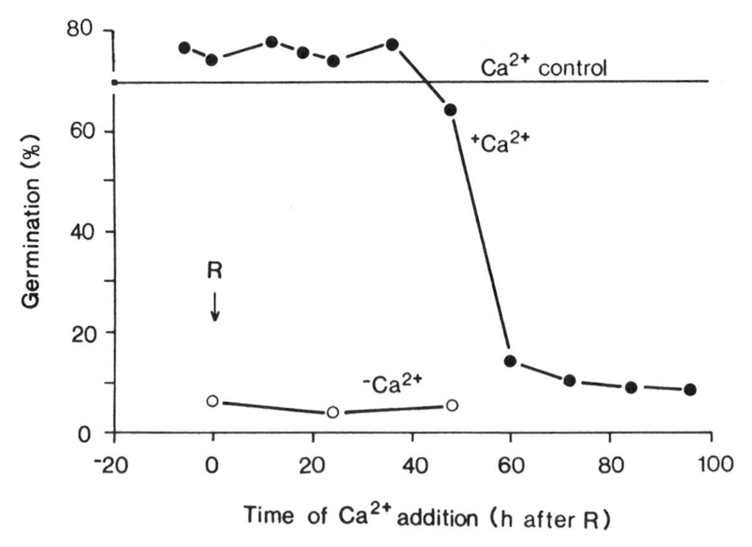


Fig. 4. Effects on fern spore germination of delaying the addition of $Ca^{2+}$ to culture medium (Scheuerlein et al., 1989).

its effects through the mediation of some intracellular $Ca^{2+}$-binding protein. This raises the question of how soon after R does the $[Ca^{2+}]_{cyt}$ change. To answer this question, our laboratory, in collaboration with M. Poenie, has succeeded in loading the fluorescent $Ca^{2+}$-indicator dye, fura-2 into fern spores without disrupting germination (Scheuerlein, Roux and Poenie, unpublished). Initial indications are that this technique will allow us to measure dynamic R-induced changes in $[Ca^{2+}]_{cyt}$ at various critical stages of the germination process. Eventually we hope to identify the key metabolic events that are regulated by $Ca^{2+}$ to promote germination.

CONCLUSION

Several different enzymes and processes in seeds that are regulated by $Ca^{2+}$ and important for germination have been identified. In most of the cases described $Ca^{2+}$ exerted its effects at micromolar levels. Given the large amounts of bound but exchangeable $Ca^{2+}$ in seeds and in the soils in which they would usually germinate, the $Ca^{2+}$ needed for regulatory roles would almost never be limiting in the natural environment.

If it is almost never limiting, how do we evaluate the role of $Ca^{2+}$ in seed germination? In simple systems (e.g., fern spore germination) the use of $Ca^{2+}$ chelators, $Ca^{2+}$-channel blockers, $Ca^{2+}$ agonists, etc. have provided valuable information. Given that seed germination is a complex process involving the interaction of several different tissues and of many distinct biochemical and biophysical processes in these tissues, trying to evaluate the effects of agonists and antagonists on intact seeds may not be so useful an exercise. Rather, more credible insights on the role of $Ca^{2+}$ in seed germination will come from identifying targets of $Ca^{2+}$ action in seeds, specifying their locale, and determining which functions they regulate. Now that we know, for example, that sub-micromolar changes in the $[Ca^{2+}]_{cyt}$ of aleurone cells can trigger α-amylase release, there is a strong rationale for studying which $Ca^{2+}$-binding protein or enzyme transduces the $Ca^{2+}$ signal into changes in the rate of enzyme production and release. In this context, it will be useful to investigate whether annexins, which help to regulate secretion in animal cells, are present and function in aleurone cells.

Spore germination may seem far removed from seed germination, but it can serve as a simple model system for discovering how $Ca^{2+}$ can promote the transition from dormancy to germination. There are some intriguing $Ca^{2+}$-related parallels in the germination of fern spores and seeds. For example, just as in radish seeds, the content of calmodulin begins low and rises during the germination process in fern spores (Scheuerlein and Roux, unpublished). Of course, this example may turn out to be only a curious coincidence. Nonetheless, if and when the targets of $Ca^{2+}$ action in Pfr-induced spore germination are discovered, their potential relevance to Pfr-induced seed germination should be evaluated.

In summary, the biochemical dissection of enzyme activities and other processes in seeds has revealed that a number of these are regulated by $Ca^{2+}$. Given that most of these discoveries are very recent and that the study of $Ca^{2+}$-binding proteins in plants is in its infancy, many more instances of $Ca^{2+}$ involvement in seed germination metabolism will probably be found in future years. Valuable insights on the mechanisms of germination control will surely result.

## ACKNOWLEDGEMENTS

The work of the author referenced in this review was supported by grants from NASA (NAGW 1519) and NSF (DCB-8716572).

## REFERENCES

Alkon, D.L. and Rasmussen, H. 1988. A spatial-temporal model of cell activation. Science 239:998.

Anderson, J. M., Charbonneau, H., Jones, H. P., McCann, R. O., and Cormier, M. J., 1980, Characterization of the plant nicotinamide adenine dinucleotide kinase activator protein and its identification as calmodulin, Biochemistry, 19:3113.

Biro, R. L., Daye, S., Serlin, B. S., Terry, M. E., Datta, N., Sopory, S. K., and Roux, S. J., 1984, Characterization of oat calmodulin and radioimmunoassay of its subcellular distribution, Plant Physiol., 75:382.

Boustead, C. M., Smallwood, M., Small, H. Bowles, D. J., and Walker, J. H., 1989, Identification of calcium-dependent phospholipid-binding proteins in higher plant cells, FEBS Lett., 244:456.

Brownlee, C., Wood, J. W., and Briton, D., 1987, Cytoplasmic free calcium in single cells of centric diatoms. The use of fura-2, Protoplasma, 140:118.

Burgoyne, R. D., and Geisow, M. J., 1989, The annexin family of calcium-binding proteins, Cell Calcium, 10:1.

Bush, D. S., and Jones, R. L., 1987, Measurement of cytoplasmic

calcium in aleurone protoplasts using indo-1 and fura-2, Cell Calcium, 8:445.
Bush, D. S., and Jones, R. L., 1988, Cytoplasmic calcium and α-amylase secretion from barley aleurone protoplasts, Euro. J. Cell Biol., 46:466.
Chen, Y. R., Datta, N., and Roux, S. J., 1987, Purification and partial characterization of a calmodulin-stimulated nucleoside triphosphatase from pea nuclei, J. Biol. Chem., 262:10689.
Chrispeels, M. J., Varner, J. E., 1967, Gibberellic acid enhanced synthesis and release of α-amylase and ribonuclease by isolated barley aleurone layers, Plant Physiol., 42:398.
Clark, G., Dauwalder, M., and Roux, S. J., 1988, Partial purification and characterization of a calcimedin-like protein in peas, Plant Physiol., 86 (Suppl.):714.
Clark, G., Dauwalder, M., and Roux, S., 1989, Partial purification and characterization of calcium-phospholipid binding proteins in peas, J. Cell Biol., 109:in press.
Clarkson, D. T., and Hanson, J. B., 1980, The mineral nutrition of higher plants, Ann. Rev. Plant Physiol., 31:239.
Cocucci, M., and Negrini, N., 1988, Changes in the levels of calmodulin and of a calmodulin inhibitor in the early phases of radish (*Raphanus sativus* L.) seed germination, Plant Physiol., 88:910.
Cunninghame, M. E., and Hall, J. L., 1986, The effect of calcium antagonists and inhibitors of secretory processes on auxin-induced elongation and fine structure of *Pisum sativum* stem segments, Protoplasma, 133:149.
Daye, S., Biro, R. L., and Roux, S. J., 1984, Inhibition of gravitropism in oat coleoptiles by the calcium chelator, ethyleneglycol-bis-(B-aminoethyl ether)-N, N'-tetraacetic acid, Physiol. Plant., 61:449.
Dela Fuente, R. K., 1984, Role of calcium in the polar secretion of indoleacetic acid, Plant Physiol., 76:342.
Dharmawardhane, S., Rubinstein, B., and Stern, A. I., 1989, Regulation of transplasmalemma electron transport in oat mesophyll cells by sphingoid bases and blue light, Plant Physiol., 89:1345.
Dreesen, T. D., Dickens, M., and Koch, R. B., Partial purification and characterization of a calcium stimulated lipoxygenase from soybean seeds, Lipids, 17:964.
Elliott, D. C., 1986, Calcium involvement in plant hormone action, in: "Molecular and Cellular Aspects of Calcium in Plant Development", A. J. Trewavas, ed., Plenum Press, New York.
Fincher, G. B., 1989, Molecular and cellular biology associated with endosperm mobilization in germinating cereal grains, Ann. Rev. Plant Physiol. Plant Mol. Biol., 40:305.
Frankland, B., and Taylorson, R., 1983, Light control of seed germination, in: "Photomorphogenesis", W. Shropshire, Jr., and H. Mohr, eds., Springer-Verlag, Berlin.
Fry, S. C., 1986, Cross-linking of matrix polymers in the growing cell wall of angiosperms, Ann. Rev. Plant Physiol., 37:165.
Gilroy, S., Hughes, W. A., and Trewavas, A.J., 1986, The measurement of intracellular calcium levels in protoplasts from higher plants, FEBS, 199:217.
Hale II, C. C. and Roux, S. J., 1980, Photoreversible calcium fluxes induced by phytochrome in oat coleoptile cells, Plant Physiol., 65:658.

Harmon, A. C., Putnam-Evans, C., and Cormier, M. J., 1987, A calcium-dependent but calmodulin-independent protein kinase from soybean, Plant Physiol., 83:830.
Hills, M. J., and Beevers, H., 1987, $Ca^{2+}$ stimulated neutral lipase activity in castor bean lipid bodies, Plant Physiol. 84:272.
Homann, P. H., 1987, The relations between the chloride, calcium, and polypeptide requirements of photosynthetic water oxidation, J. Bioener. B., 19:105.
Jones, R. L., Jacobsen, J. V., 1983, Calcium regulation of the secretion of α-amylase isozymes and other proteins from barley aleurone layers, Planta, 158:1.
Kauss, H., 1987, Some aspects of calcium-dependent regulation in plant metabolism, Ann. Rev. Plant Physiol., 38:47.
Khan, A. A., and Huang, X.-L., 1988, Synergistic enhancement of ethylene production and germination with kinetin and 1-aminocyclopropane-1-carboxylic acid in lettuce seeds exposed to salinity stress, Plant Physiol., 87:847.
Koller, D., 1972, Environmental control of seed germination, in: "Seed Biology", T. T. Kozlowski, ed., Academic Press, New York.
Lambert, A.-M., and Vantard, M., 1986, Calcium and calmodulin as regulators of chromosome movement during mitosis in higher plants, in: "Molecular and Cellular Aspects of Plant Development", A. J. Trewavas, ed., Plenum Press, New York.
Lee, M. H., 1989, Phospholipase D of rice bean. II. The effects of the enzyme inhibitors and activators on the germination and growth of root and seedlings of rice, Plant Science, 59:35.
Lee, M. H., 1989, Phospholipase D of rice bran. I. Purification and characterization, Plant Science, 59:25.
Mayer, A. M., Poljakoff-Mayber, A., 1989, "The Germination of Seeds", 4th ed., Pergamon Press, New York.
Mitsui, T., Christeller, J. T., Hara-Nishimura, I., and Akazawa, T., 1984, Possible roles of calcium and calmodulin in the biosynthesis and secretion of α-amylase in rice seed scutellar epithelium, Plant Physiol., 75:21.
Nishizuka, Y., 1986, Studies and perspectives of protein kinase C, Science, 233:305.
Paliyath, G., and Thompson, J. E., 1987, Calcium- and Calmodulin-regulated breakdown of phospholipid by microsomal membranes from bean cotyledons, Plant Physiol., 83:63.
Perdue, D. O., LaFavre, A. K., and Leopold, A. C., 1988, Calcium in the regulation of gravitropism by light, Plant Physiol., 86:1276.
Piazza, G. J., 1988, Calmodulin in plants, in: "Calcium-Binding Proteins", M. P. Thompson, ed., CRC Press, Florida.
Polya, G. M., and Micucci, V., 1984, Partial purification and characterization of a second calmodulin-activated $Ca^{2+}$-dependent protein kinase from wheat germ, Biochim. Biophys. Acta, 785:68.
Roux, S. J., and Serlin, B. S., 1987, Cellular mechanisms controlling light-stimulated gravitropism: role of calcium, CRC Critical Reviews in Plant Sciences, 5:205.
Roux, S. J., Wayne, R. O., and Datta, N., 1986, Role of calcium ions in phytochrome responses: an update, Physiol. Plant., 66:344.
Saunders, M. J., and Hepler, P. K., 1982, $Ca^{2+}$ ionophore A23187 stimulates cytokinin-like mitosis in Funaria, Science, 217:943.

Saunders, M. J., and Hepler, P. K., 1983, Calcium antagonists and calmodulin inhibitors block cytokinin-induced bud formation in Funaria, Develop. Biol., 99:41.
Schafer, A., Bygrave, F., Matzenauer, S., and Marme, D., 1985, Identification of a calcium- and phospholipid-dependent protein kinase in plant tissue, FEBS Lett., 187:25.
Scheuerlein, R., Wayne, R., and Roux, S. J., 1989, Calcium requirement of phytochrome-mediated fern spore germination: No direct phytochrome-calcium interaction in the phytochrome-initiated transduction chain, Planta 178:25.
Serlin, B. S., and Roux, S. J., 1984, Modulation of chloroplast movement in the green alga Mougeotia by the $Ca^{2+}$ ionophore A23187 and by calmodulin antagonists, Proc. Natl. Acad. Sci. USA, 81:6368.
Wayne, R. O., and Hepler, P. K., 1974, The role of calcium ions in phytochrome-mediated germination of spores of Onoclea sensibilis L., Planta 160:12.
Williamson, R. E., and Ashley, C. C., 1982, Free $Ca^{2+}$ and cytoplasmic streaming in the alga Chara, Nature, 296:647.

# TRANSPORT AND TARGETING OF PROTEINS TO PROTEIN STORAGE VACUOLES (PROTEINS BODIES) IN DEVELOPING SEEDS

**Maarten J. Chrispeels and Brian W. Tague**

**Department of Biology**
**University of California, San Diego**
**La Jolla, CA 92093-0116**

SUMMARY

The vacuoles of plant cells contain a variety of proteins including acid hydrolases, storage proteins and plant defense proteins. During seed development, the central vacuoles of the storage parenchyma cells accumulate large amounts of all three classes of these proteins. We have studied the biosynthesis, transport, posttranslational modifications and accumulation in developing legume cotyledons of acid hydrolases (e.g. α-mannosidase), storage proteins (e.g. phaseolin), and plant defense proteins (e.g. phytohemagglutinin and α-amylase inhibitor). Transport of proteins to vacuoles is mediated by the secretory system (endoplasmic reticulum and Golgi apparatus) and correct targeting of protein to vacuoles requires positive sorting information. This information is contained within the polypeptide domain of the vacuolar glycoprotein phytohemagglutinin (PHA). When the gene for PHA is introduced into yeast (*Saccharomyces cerevisiae*) cells, the resulting protein is targeted to yeast vacuoles. By expressing in yeast, chimeric genes consisting of the signal peptide and various portions of the PHA coding region with the gene for yeast invertase, we were able to show that the vacuolar targeting domain of PHA is in an amino-proximal region between amino acids 14 and 43 of the mature protein. Experiments are now under way to determine whether the same domain of PHA can target yeast invertase to plant vacuoles (protein bodies in tobacco seeds).

## INTRODUCTION

The vacuole of the plant cell is at once the largest and the least understood of all cellular organelles. All living plant cells have vacuoles and all vacuoles have small amounts of protein. Certain cell-types have vacuoles that contain large amounts of storage proteins and plant defense proteins. Such protein storage vacuoles are found primarily in seeds, and humanity derives most of its protein nutrition from the vacuolar proteins of seeds. Our ability to manipulate genes and transform important crop plants is opening up the prospect of improving the nutritional quality as well as the digestibility of these proteins. In addition, we can begin to think of vacuoles as repositories for the accumulation (and easy harvest for subsequent processing) of peptides and proteins that do not normally occur in plants. This will require that we understand how proteins are targeted to vacuoles and the properties that determine their accumulation. The recent progress in this field is the subject of this review.



*Recent Advances in the Development and Germination of Seeds*
Edited by R.B. Taylorson
Plenum Press, New York

## VACUOLES IN PLANT CELLS

Unlike animal cells, which are filled entirely with cytoplasm and organelles, in a typical plant cell much of the space is occupied by a large central vacuole. For example, in a leaf mesophyll cell, a thin layer of peripheral cytoplasm is sandwiched between the plasma membrane and the vacuolar membrane or tonoplast. Such vacuoles, containing dilute solutions of organic acids, mineral ions and sugars, allow plant cells to enormously increase in size without the need to synthesize an equivalent volume of protein-rich cytoplasm. Vacuoles often contain half-digested organelles, and the demonstration that all vacuoles contain acid hydrolases gave rise to the proposal, now widely accepted, that vacuoles constitute the lysosomal compartment of the plant cells (Matile, 1975).

Not all plant cells have the characteristic large central vacuole described above. Young, dividing cells have many small vacuoles scattered in the cytoplasm. Cell expansion is accompanied by the fusion of these small vacuoles and the formation of a central vacuole. Seed storage tissues, such as cotyledons or endosperm, have numerous small protein-filled vacuoles (up to several thousand per cell) often referred to as protein bodies. Proteins are deposited in such vacuoles during seed formation and catabolized during and after germination, providing the growing seedling with amino acids. In addition to storage proteins, such protein storage vacuoles contain lectins, enzyme inhibitors, and acid hydrolases.

Plant vacuoles have many properties in common with fungal vacuoles (e.g. *Neurospora crassa* or *Saccharomyces cerevisiae*) which are also acidic compartments that function as temporary storage depots and sites of macromolecular hydrolysis. The presence of acid hydrolases and of an evolutionarily-conserved membrane-bound proton ATPase in the tonoplasts of plant vacuoles and the membranes surrounding the vacuoles of fungi and the lysosomes of mammalian cells (Nelson, 1988) further emphasizes the structural and functional relatedness of the acidic compartments of all eucaryotic cells.

## PROTEINS IN VACUOLES

All vacuoles examined so far, including the protein storage vacuoles of seeds, contain α-mannosidase, and this enzyme is often used as a vacuolar marker enzyme because most of the α-mannosidase activity present in a tissue-extract is located within the vacuoles (Boller and Kende, 1979). Vacuoles also contain other acid hydrolases such as β–N-acetylglucosaminidase, phosphatase, pryophosphatase, carboxypeptidase C, phospholipase D, ribonuclease, and proteinase (Matile, 1978). All these enzymes have pH optima of 4-6. However, all cell types may not contain all these enzymes within their vacuoles. Of particular interest is acid proteinase which is abundantly present in the vacuoles of leaves and suspension-cultured cells, but is nearly absent from the protein storage vacuoles of developing and mature seeds. The mobilization during seedling growth of the storage proteins that are deposited in these vacuoles is the result of the biosynthesis of acid proteinase(s) that is transported into the vacuoles as the seedlings grow (Baumgartner et al., 1978). We presume, although there is as yet no evidence, that protein storage vacuoles in vegetative tissues (see below) are also characterized by low levels of proteinase activity. Hydrolases with anti-fungal properties that can be characterized as plant defense proteins, such as chitinase, lysosyme and β-glucanase are also present in vacuoles (Boller and Vögeli, 1984; Mauch and Staehelin, 1989). It should be noted, however, that plant tissues may also contain extracellular forms of these and other acid hydrolases.

Typical seed storage proteins are found in the protein storage vacuoles present in embryos and associated reserve tissues such as the endosperm. These storage proteins fall into three large families: the 12 S globulins (legumin-family), the 7 S globulins (vicilin-family), and the 2 S albumins. The cDNAs for many of these proteins have been cloned and there is extensive sequence conservation in widely separated plant families. For example, the 12 S and 7 S globulins, which were first thought to be characteristic of the legumes, have now also been found in the cereals. It should be noted here that the prolamin-filled protein bodies in the endosperm of corn (and probably sorghum) are probably derived from the endoplasmic reticulum, and do not fit the description of protein storage vacuoles (Khoo and Wolf, 1970; Larkins and Hurkman, 1978).

Recent research shows that storage proteins also accumulate in the vacuoles of parenchyma cells in vegetative tissues (leaves, bark, tubers, roots). For example, a vegetative storage protein (VSP) accumulates in the paraveinal mesophyll of soybean leaves as the plants get ready to flower and is catabolized during seed-fill. This protein accumulates to much higher levels in leaves and stems when soybean plants are continuously depodded. The cDNA for this vacuolar glycoprotein has now been cloned and its nucleotide sequence determined (Staswick, 1988, 1989). It has recently been shown that the abundant potato tuber protein patatin is a vacuolar glycoprotein (Sonnewald et al., 1989). It is not known for certain, however, that patatin is a storage protein. Because it has an enzymatic activity (lipolytic acyl hydrolase), it could be a vacuolar plant defense protein (see below).

Vacuoles contain two types of antifeedants that are considered to be plant defense proteins: lectins and enzyme inhibitors. The well-known lectins (such as concanavalin A, soybean agglutinin, and phytohemagglutinin) all occur in the protein storage vacuoles of seeds. Some of these lectins have been shown to have antifeedant properties towards insects and to be toxic to mammals. Homologous proteins such as arcelin (Osborn et al., 1988), which also belong to the lectin family of proteins, although they may not have any carbohydrate-binding properties, also act as antifeedants towards insects. Protein storage vacuoles of seeds also contain a variety of enzyme inhibitors such as the Kunitz trypsin inhibitor, the Bowman-Birk trypsin inhibitor, and an inhibitor of α-amylase. The inhibitors do not inhibit the catalytic activities of endogenous plant enzymes but are active against mammalian or insect enzymes. Enzyme inhibitors that deter insect feeding also accumulate in the vacuoles of leaves when plants are wounded. For example, wounding of tomato and potato leaves causes the accumulation of protease inhibitors I and II in mesophyll vacuoles. Inhibitor I inhibits trypsin only, and inhibitor II inhibits both trypsin and chymotrypsin (Walker-Simmons and Ryan, 1977).

## TRANSPORT, BULKFLOW AND TARGETING

Transport of proteins to the plant vacuole is mediated by the secretory system: the endoplasmic reticulum (ER), the Golgi apparatus, and their associated elements (e.g. transition elements and trans-Golgi network). This system is responsible for protein secretion and for transport of proteins to the vacuole. Arrival of a protein at a specific location (vacuole or extracellular space) may depend on the interaction of a specific targeting domain of the transported protein with a transport receptor or it may simply be by bulkflow (Fig. 1). Experiments with mammalian cells and yeast cells show that transport to the cell surface of secretory proteins and plasma membrane-anchored proteins is by bulkflow, while transport to lysosomes requires positive sorting information involving the interaction of a targeting domain with a targeting receptor (Kelly, 1985). The best-understood case of this type of targeting mechanism concerns the mannose-6-phosphate receptors that bind to modified glycans with mannose-6-phosphate groups on lysosomal hydrolases for their delivery to lysosomes. The receptors cycle between the trans-Golgi network and a pre-lysosomal (endosomal) compartment. For the yeast vacuolar enzyme carboxypeptidase Y, the targeting information is contained in a polypeptide domain at the amino-terminus of the protein. This amino terminal domain is lost after the protein arrives in the vacuole. As with mammalian lysosomal hydrolases, a posttranslational modification appears to be intimately associated with targeting. The absence of vacuolar targeting information in a transported protein, the malfunction of the targeting machinery (e.g. a mutation in the receptor), or improper physiological conditions (incorrect pH in the trans-Golgi cisternae) may result in secretion of a protein via the default or bulkflow pathway.

## COTRANSLATIONAL AND POSTTRANSCRIPTIONAL MODIFICATIONS OF VACUOLAR PROTEINS

Vacuolar proteins are extensively modified along the transport pathway and a summary of these modifications is shown in Table 1. The entry of proteins into the secretory system depends on the presence of a signal peptide, and all derived amino acid sequences of vacuolar proteins so far obtained have typical signal sequences. Passage into the lumen of the ER is accompanied by the removal of the signal sequence (presumably cotranslationally) and the attachment of glycans to specific asparagine residues of some proteins. Although many vacuolar proteins are glycoproteins with asparagine-linked glycans either in their mature form or as transport intermediates, other vacuolar proteins have no covalently attached glycans (e.g. the 12 S globulins). The glycans which are transferred have the typical $Glc_3Man_9$ $(GlcNAc)_2$ composition and structure also found on nascent mammalian and yeast glycoproteins. These glycans are trimmed to yield the typical high-mannose glycan: $Man_9$ $(GlcNAc)_2$. Release into the ER is accompanied by the formation of disulfide bonds on some vacuolar proteins (e.g. the 12 S globulins) and the oligomerization of subunits. Most vacuolar proteins examined so far appear to be oligomers. Some oligomers have biological activity at this point (e.g. phytohemagglutinin and wheat germ agglutinin), whereas others do not (e.g. concanavalin A and bean α-amylase inhibitor). For the latter, further processing is necessary to bring about a conformational change that imparts biological activity. However, further processing may also take place with the former.

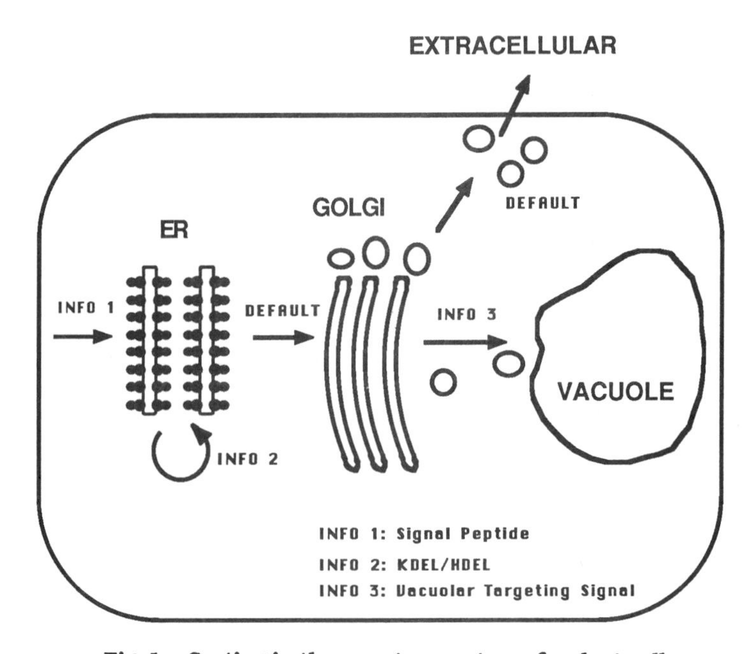


**Fig. 1. Sorting in the secretory system of a plant cell.**

**Table 1. Co- and Posttranscriptional Modifications of Vacuolar Proteins**

In the endoplasmic reticulum:
- Removal of the signal peptide
- Addition of high-mannose glycans
- Removal of glucose residues from the glycans
- Formation of disulfide bonds
- Incorporation of metal ions
- Formation of oligomers

In the Golgi apparatus:
- Removal of sugar residues from some glycans
- Addition of sugar residues to the same glycans
- Proteolytic processing

In the protein storage vacuoles:
- Removal of sugar residues
- Proteolytic processing
- Additional formation of oligomers

Some of the asparagine-linked glycans, those that are accessible to modifying enzymes, are acted upon by glycosidases and glycosyltransferases in the Golgi apparatus to bring about the formation of complex glycans. A complex series of reactions results in the formation of glycans that have a $Man_3 (GlcNAc)_2$ core with a $\beta$,1->2 xylose residue and an $\alpha$,1->3 fucose residue. Such small complex glycans have been found on a number of plant glycoproteins (Faye et al., 1989). The reactions that lead to the formation of these glycans occur primarily in the Golgi apparatus, although the final removal of terminal GlcNAc residues takes place after the glycoproteins have reached the vacuoles.

Many vacuolar proteins undergo posttranslational proteolytic processing usually after they arrive in the vacuole. In some cases, the polypeptide is simply clipped in two or three smaller polypeptides, all of which remain together in the oligomer. In other cases, a domain is lost either at the N-terminus, at the C-terminus, or in the middle of the polypeptide. For some polypeptides, processing comes to completion (all polypeptides are processed), while for others processing is only partial (some polypeptides remain unprocessed). Processing may also depend on the environment. For example, phytohemagglutinin is not processed in bean cotyledons, but is processed (more than 50% of all polypeptides) when the gene is expressed in tobacco seeds.

## TRANSPORT TO THE PLANT VACUOLE IS NOT BY BULKFLOW

In mammalian cells, mistargeting of lysosomal enzymes occurs when the pH of the transport pathway is raised by treatment of the cells with weak bases such as chloroquine (Gonzalez-Noriega et al.,1980) or the pH of the trans-Golgi complex is raised by ionophores such as monensin (see Tartakoff, 1983). This disruption of the normal recognition process between ligand and receptor results in the secretion of lysosomal enzymes (transport by bulk flow). When cotyledons of peas (Craig and Goodchild, 1984) and jackbeans (Bowles et al., 1986) are treated with monensin, storage proteins and lectins that normally accumulate in protein storage vacuoles were mistargeted to the plasma membrane and were found to accumulate on the cell wall-plasma membrane interface. This finding indicates that a receptor-mediated and pH-modulated vacuolar transport process probably also operates in plant cells. Additional indirect evidence for a receptor-mediated process comes from the observation that transport rates for different proteins can be vastly different in the same cell (Higgins et al.,1983). To test the hypothesis that transport to the vacuole requires positive sorting information, and is not by bulk flow, we studied the subcellular location of two proteins, the genes for which were introduced into tobacco via *Agrobacterium*-mediated transformation. The two genes we used were the dlec2 gene which encodes the PHA-L polypeptide of phytohemagglutinin, the seed lectin of the bean (*Phaseolus vulgaris*), with its own 3' and 5' sequences (Sturm et al., 1988), and a chimeric gene consisting of the 5' and signal peptide sequences of PHA-L and the coding sequence and 3' sequence of a cytosolic protein, the PA2 albumin of pea (Dorel et al., 1989). The objective of this experiment was to introduce these two proteins into the secretory system, and compare their subsequent transport, posttranslational modification and targeting. One protein is a known vacuolar protein (phytohemagglutinin), while the other one (PA2) is a known cytosolic protein that we presume to be devoid of vacuolar targeting information. Both proteins accumulated to relatively high levels (0.5-1% of total protein) in tobacco seeds, entered the endoplasmic reticulum where they were cotranslationally glycosylated, proceeded to the Golgi apparatus, where some of the glycans were posttranslationally modified (Fig. 2), and passed through the Golgi apparatus. After passing through the Golgi complex, phytohemagglutinin was correctly targeted to the protein storage vacuoles of tobacco seeds (Sturm et al., 1988), while PA2 did not accumulate there (Dorel et al., 1989). The subcellular location of PA2 is not known with certainty, but the most likely explanation of our results is that PA2 is in the periplasmic/extracellular space. Confirmation of this postulate must await the expression of PA2 in different organs and the application of standard subcellular fractionation techniques.

**Table 2. Proteolytic Processing of Vacuolar Proteins**

**Five different types of processing are consistent with the targeting of proteins to vacuoles.**

**NO PROTEOLYTIC PROCESSING (except signal peptide)**

Phytohemagglutinin
Soybean agglutinin
β subunit of soybean β–conglycinin
Tomato proteinase inhibitor II
Pea vicilin
Phaseolin
Patatin

**C-TERMINAL DOMAIN LOST**

Soybean glycinin
Wheat germ agglutinin
Barley lectin

**CLEAVAGE AND RELIGATION**

Concanavalin A

**N-TERMINAL DOMAIN LOST**

Napin
Sulfur-Rich protein of Brazil Nut
Tomato proteinase inhibitor I
Potato proteinase inhibitor I
Sweet potato sporamin
α and α' subunits of soybean β-conglycinin

**INTERNAL CLEAVAGE WITH OR WITHOUT LOSS OF A DOMAIN**

12 S Globulins
Sulfur-Rich protein of Brazil Nut
Napin
Castor bean ricin
Castor bean *(Ricinus communis)* agglutinin
Pea lectin
Pea vicilin

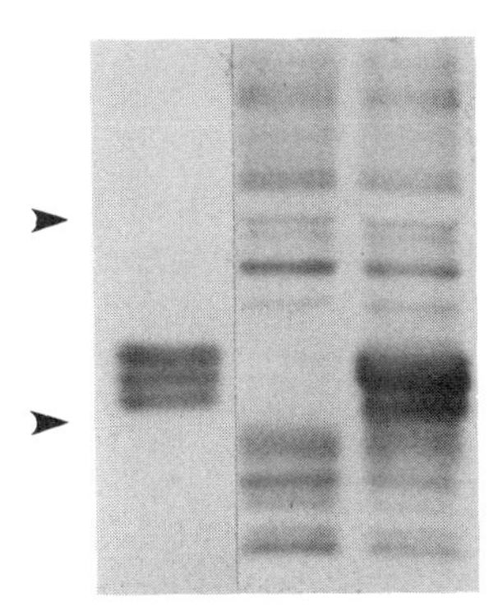

**Fig. 2. The chimeric protein *phalb* has complex glycans indicating that it entered the secretory system and passed through the Golgi complex. Extracts of control seeds (lane 2) or transformed seeds (lanes 1, 3) were fractionated by SDS-PAGE and immunoblotted with antiserum to albumin (lane 1) or antiserum against complex glycans (lanes 2, 3). The reaction of the phalb protein in lane 3 shows that it has complex glycans indicating that it passed through the Golgi apparatus (see Dorel et al., 1989).**

## GLYCANS ARE NOT NECESSARY FOR TARGETING TO VACUOLES

Many vacuolar and lysosomal proteins are glycoproteins, and glycans have been shown to play an important role in the targeting of lysosomal enzymes in mammalian cells. In mammalian cells, certain high-mannose glycans on the lysosomal enzymes are modified in the Golgi apparatus by enzymes whose action results in the formation of mannose-6-phosphate groups. The targeting of these enzymes to the lysosomes is mediated by membrane-bound receptors that cycle between the trans-Golgi network and a pre-lysosomal compartment (Brown et al.,1986). The association/dissociation of the enzyme-receptor complex depends on a lowering of the pH along the transport pathway. Could a similar mechanism operate in plant cells? To investigate this question, we eliminated the glycan-attachment sites of PHA by site-directed mutagenesis (Voelker et al., 1989). The high-mannose glycan of PHA is attached to $Asn^{12}$-$Glu^{13}$-$Thr^{14}$ , while the second complex glycan is attached to $Asn^{60}$-$Thr^{61}$ $Thr^{62}$. In one mutant, $Thr^{14}$ was changed to $Ala^{14}$ , in a second mutant, $Asn^{60}$ was changed to $Ser^{60}$, and a third mutant incorporated both changes. The resulting mutant genes were expressed in tobacco using the β-phaseolin promoter, and the fate of the proteins was examined.

The gene for PHA-L was mutated to remove the two glycan attachment sites ($Thr^{14}$ ->$Ala^{14}$ and $Asn^{60}$->$Ser^{60}$) and the mutated gene introduced into tobacco. PHA-L was located with affinity-purified antibodies to PHA-L followed by goat anti-rabbit IgG coupled to colloidal gold. (Courtesy of Eliot Herman, USDA, Beltsville, MD.; see also Voelker et al., 1989)

Analysis of the tobacco seeds showed that PHA with only one glycan or without any glycans was correctly targeted to the protein storage vacuoles (Voelker et al., 1989). The subcellular localization of PHA without glycans in transgenic tobacco seeds is shown in Fig. 3. We also found that the absence of either the complex glycan or the high-mannose glycan did not alter the processing of the other glycan. These results indicate that the targeting signal of this vacuolar protein should be contained in its polypeptide domain. Our results confirm earlier work by Bollini et al. (1985) who showed that tunicamycin, an inhibitor of asparagine-linked glycosylation, does not inhibit the transport of unglycosylated PHA to the protein storage vacuoles of bean cotyledons. Furthermore, a number of vacuolar proteins, such as the 11 S globulins, do not have covalently attached glycans, indicating again that the targeting signal must be in the polypeptide domain.

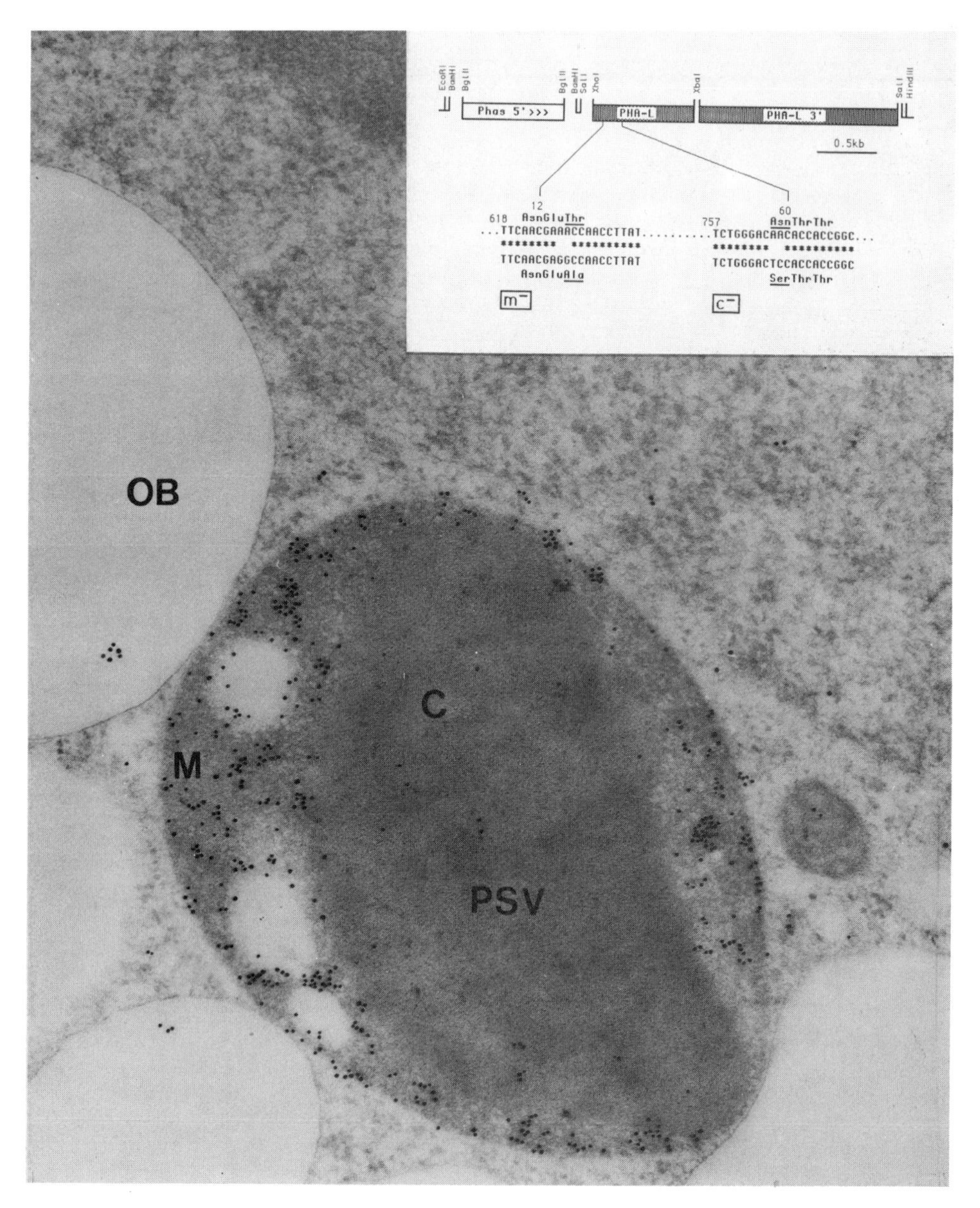


**Fig. 3. Construction of hybrid genes, site-directed mutagenesis to remove glycosylation sites, and immunological localization of phytohemagglutinin without glycans in the protein storage vacuoles of transgenic tobacco seeds (for details, see Voelker et al., 1989).**

As a first step in studying the targeting signal of a plant vacuolar protein at the molecular level, we wished to determine whether yeast (*Saccharomyces cerevisiae*) cells would recognize this signal and target PHA to the yeast vacuole. The gene for PHA-L was cloned into the yeast expression vector, pYE7, under control of the acid phosphatase (*PHO5*) promoter. Deletion of phosphate from the culture medium resulted in the expression of the PHA-L gene and the accumulation of PHA-L in the cells up to 0.1% of total yeast protein. PHA-L in yeast was glycosylated, as expected for a protein that enters the secretory system, and all the glycans were of the high-mannose type as expected for a yeast glycoprotein. Cell fractionation studies showed that nearly all the PHA-L accumulated in vacuoles (Tague and Chrispeels, 1987). The next series of experiments was aimed at determining if the vacuolar targeting signal of PHA is contained within a linear domain of the polypeptide. We carried out unidirectional deletions of the PHA-L gene starting from the carboxy-terminus of the protein and fused these deletions to a truncated yeast invertase gene that starts at the third amino acid of the mature protein. After transformation of yeast with these chimeric constructs, we selected translational fusions on the basis of their invertase activity and determined the subcellular distribution of invertase as well as its glycosylation status. A similar approach was used by Johnson et al. (1987) to define the vacuolar targeting domain of a yeast carboxypeptidase Y. Our data indicate that an amino terminal portion of PHA comprised of a 20 amino acid signal sequence and 43 amino acids of the mature protein is sufficient to target invertase to the yeast vacuole. The vacuolar targeting domain appears to be located between amino acids 14 and 43 of the mature protein. To arrive at this conclusion, we made and tested a large number of constructs, three of which are shown in Fig. 4. Experiments to test whether the same portion of PHA is sufficient to target yeast invertase to the protein bodies of tobacco seeds are presently in progress. Our preliminary results show that the yeast invertase gene is expressed in tobacco seeds and that enzyme activity is present in the dry seeds. Whether invertase synthesized in tobacco seeds of plants that have been transformed with any of these constructs is transported to the protein storage vacuoles, or is secreted, remains to be determined.

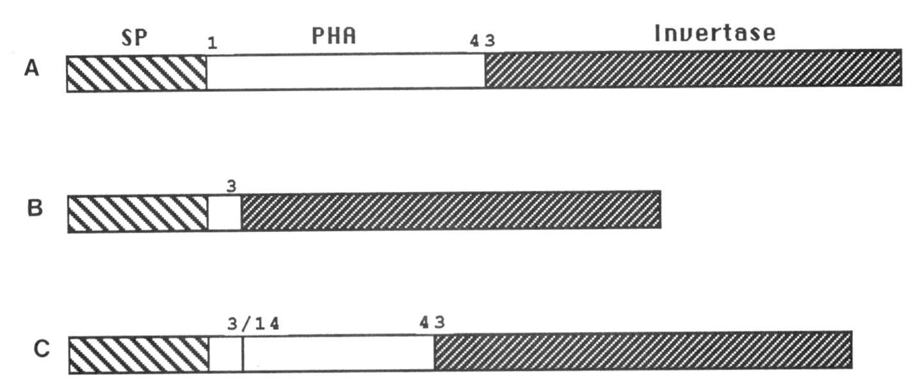


**Fig. 4. Defining the vacuolar targeting domain of PHA in yeast. The chimeric genes consist of portions of the PHA gene (signal peptide and N-terminal domain of the mature protein) and the yeast invertase gene. In yeast transformed with construct A, invertase goes to the vacuole, with construct B invertase is secreted, and with construct C, it is also vacuolar. This defines the vacuolar targeting domain as being between amino acids 14 and 44 of the mature protein.**

## LEGUME VACUOLAR PROTEINS CONTAIN A CONSERVED SEQUENCE THAT IS ALSO FOUND IN YEAST CARBOXYPEPTIDASE Y

A comparison of the amino acid sequences of legume storage proteins and lectins, and of yeast carboxypeptidase Y shows a conserved motif, which in PHA has the sequence $L^{18}$QRD (Fig. 5). The glutamine residue is completely conserved in all legume lectins sequenced to date. The functionality of this sequence has been tested for both carboxypeptidase Y (CPY) and PHA. A single amino acid change from $L^{3}$QRP in normal carboxypeptidase Y (targeted to the yeast vacuole) to $L^{3}$KRP resulted in the complete mistargeting (secretion) of this vacuolar enzyme in yeast cells (Valls et al., 1987). A three amino acid change from LQRD to LEGN in a portion of PHA that is 43 AA long (in addition to the signal peptide) and used in a chimeric construction with yeast invertase resulted in the secretion of invertase from the cells rather than its transport to the vacuoles. Thus, the sequence conservation that we observe is in a region that is functionally important for proper targeting in yeast cells.

| | | | | |
|---|---|---|---|---|
| | | 1 (mature) | | |
| PHA-L | | SNDIYFNFQRF--NETNLI | **LQR**DASV-SSSGQ | LRLTNLNGNGEPRVG-SLGR... |
| PHA-E | | ASQTSFSFQRF--NETNLI | **LQR**DATV-SSKGQ | LRLTNVNDNGEPTLS-SLGR... |
| SBA | | AETVSFSWNKFVPKQPNMI | **LQG**DAIVTSSG-K | LQLNKVDENGTPKPS-SLGR... |
| ConA | STHE | TNALHFMFNQFSKDQKDLI | **LQG**DA-TTGTEGN | LRLTRVSSNGSPQGS-SVGR... |
| Pea Lectin | | TETTSFLITKFSPDQQNLI | **FQG**DGYTT-KE-K | LTLTKAVKN-------TVGR... |
| Favin | T | DEITSFSIPKFRPDQPNLI | **FQG**GGYTT-KE-K | LTLTKAVKN-------TVGR... |
| | | (N-proximal) | | |
| Phaseolin | | ...SDNSWNTLFKNQYGHIRV | **LQR**FDQQSKRLQNL... | |
| | | (C-proximal) | | |
| Pea Legumin | | ...TSSVINDLPLDVVAATFK | **LQR**DEARQLKSNN... | |
| | | (C-proximal) | | |
| Ricin | | ...LSTAIQESNQGARASPIQ | **LQR**DGSKFSVYDV... | |
| | | 1 (pro) | | |
| CPY (yeast) | | IS | **LQR**PLGLDKDVLL... | |

**Fig. 5. Amino acid sequence comparisons of lectins, storage proteins, and yeast carboxypeptidase Y showing a conserved sequence in the vacuolar targeting domain of PHA and carboxypeptidase Y.**

The LQRD motif which is present in yeast carboxypeptidase Y and PHA is also found in a number of other plant vacuolar proteins (Fig. 5). However, it is absent from other plant vacuolar proteins (for example, wheat germ agglutinin, proteinase inhibitor(s), napin and other 2 S storage proteins). So far, we have shown that the LQRD region is sufficient for targeting invertase to the yeast vacuole. However, is it also necessary, or will another segment of PHA towards the C-terminal end also target invertase to the yeast vacuole?

## FUTURE EXPERIMENTS

Future experiments will be directed at answering the following questions. Is the domain which targets PHA to yeast vacuoles also necessary to target it to plant vacuoles? We have recently introduced the same chimeric constructs into tobacco and are now studying the subcellular localization of yeast invertase in tobacco seeds. Is secretion the bulk-flow or default pathway? So far, we were not able to show that phalb was secreted in the storage parenchyma cells of the cotyledons. By using a different promoter, we will express phalb in leaves and study its secretion by protoplasts. What is the identity of the sorting receptor? Using iodinated PHA, we are attempting the identity with which protein(s) PHA interacts. Can we obtain vacuolar protein sorting (vps) mutants? These are mutants that secrete their vacuolar enzymes, because there is a defect in the sorting apparatus or its operation. We will attempt to obtain mutants of *Arabidopsis thaliana* that secrete their vacuolar acid hydrolases.

## REFERENCES


Baumgartner, B., Tokuyasu, K. T., and Chrispeels, M. J., 1978, Localization of vicilin peptidohydrolase in the cotyledons of mungbean seedlings by immunofluorescence microscopy, J. Cell Biol., 79:11.

Boller, T., and Kende, H., 1979, Hydrolytic enzymes in the central vacuole of plant cells, Plant Physiol., 63:1123.

Boller, T., and Vögeli, U., 1984, Vacuolar localization of ethylene-induced chitinase in bean leaves, Plant Physiol., 74:442.

Bollini, R., Ceriotti, A., Daminati, M. G., and Vitale, A., 1985, Glycosylation is not needed for the intracellular transport of phytohemagglutinin in developing *Phaseolus vulgaris* cotyledons and for the maintenance of its biological activities, Physiol. Plant., 65:15.

Bowles, D. J., Marcus, S. E., Pappin, J. C., Findlay, J. B. C., Eliopoulos, E., Maylox, P. R., and Burgess, J., 1986, Posttranslational processing of concanavalin A precursors in jackbean cotyledons, J. Cell Biol., 102:1284.

Brown, W. J., Goodhouse, J., and Farquhar, M. G., 1986, Mannose-6-phosphate receptors for lysosomal enzymes cycle between the Golgi complex and endosomes, J. Cell Biol., 103:1235.

Chrispeels, M. J., 1985, The role of the Golgi apparatus in the transport and posttranslational modification of vacuolar (protein body) proteins, Oxford Surv. Plant Molec. Cell Biol., 2:43.

Craig, S., and Goodchild, D. J., 1984, Golgi-mediated vicilin accumulation in pea cotyledons is redirected by monensin and nigericin, Protoplasma, 122:91.

Dorel, C., Voelker, T. A., Herman, E. M., and Chrispeels, M. J., 1989, Transport of proteins to the plant vacuole is not by bulk flow through the secretory system, and requires positive sorting information, J. Cell Biol., 108:327.

Faye, L., Johnson, K. D., Sturm, A., and Chrispeels, M.J., 1989, Structure, biosynthesis, and function of asparagine-linked glycans on plant glycoproteins, Physiol. Plant., 75:309.

Gonzalez-Noriega, A., Grubb, J. H., Talkad, V., and Sly W. S., 1980, Chloroquine inhibits lysosomal enzyme pinocytosis and enhances lysosomal enzyme secretion by impairing receptor recycling, J. Cell Biol., 85:839.

Higgins, T. J. V., 1984, Synthesis and regulation of major proteins in seeds, Annu. Rev. Plant Physiol., 35:47.

Higgins, T. J. V., Chrispeels, M. J., Chandler, D. M., and Spencer, D., 1983, Intracellular sites of synthesis and processing of lectin in developing pea cotyledons, J. Biol. Chem., 258:9550.

Johnson, L. M., Bankaitis, V. A., and Emr, S. D., 1987, Distinct sequence determinants direct intracellular sorting and modification of a yeast vacuolar protease, Cell, 48:875.

Kelly, R. B., 1985, Pathways of protein secretion in eukaryotes, Science, 230:25.

Khoo, V., and Wolf, J. J., 1970, Origin and development of protein granules in maize endosperm, Am. J. Bot., 57:1042.

Larkins, B. A., and Hurkman, W. J., 1978, Synthesis and deposition of zein in protein bodies of maize endosperm, Plant Physiol., 62:256.

Lauriere, M., Lauriere, C., Johnson, K. D., Chrispeels, M. J., and Sturm, A., 1989, Characterization of a xylose-specific antiserum that reacts with the complex asparagine-linked glycans of extracellular and vacuolar glycoproteins, Plant Physiol., 90:1182.

Matile, P., 1975, "The Lytic Compartment of Plant Cells," Springer Verlag, Heidelberg, FRG, 153 pp.

Matile, P., 1978, Biochemistry and function of vacuoles, Annu. Rev. Plant Physiol., 29:193.

Mauch, F., and Staehelin, L. A., 1989, Functional implications of the subcellular localization of ethylene-induced chitinase and β-1,3 glucanase in bean leaves, Plant Cell, 1:447.

Nelson, N., 1988, Structure, function, and evolution of proton-ATPases, Plant Physiol., 86:1.

Osborn, T. C., Alexander, D. C., Sun, S. S. M., Cardona, C., and Bliss, F. A., 1988, Insecticidal activity and lectin homology of arcelin seed protein, Science, 240:207.

Schekman, R., 1985, Protein localization and membrane traffic in yeast, Ann. Rev. Cell Biol., 1:115.

Sonnewald, U., Studer, D., Rocha-Sosa, M., and Wilmitzer, L., 1989, Immunocytochemical localization of patatin, the major glycoprotein in potato (*Solanum tuberosom* L.) tubers, Planta, 179:176.

Staswick, P., 1988, Soybean vegetative storage protein structure and gene expression, Plant Physiol., 87:250.

Staswick, P., 1989, Developmental regulation and the influence of plant sinks on vegetative storage protein gene expression in soybean leaves, Plant Physiol., 89:309.

Sturm, A., Voelker, T. A., Herman, E. M., and Chrispeels, M. J., 1988, Correct glycosylation, Golgi-processing, and targeting to protein bodies of the vacuolar protein phytohemagglutinin in transgenic tobacco, Planta, 175:170.

Tague, B. W., and Chrispeels, M. J., 1987, The plant vacuolar protein, phytohemagglutinin, is transported to the vacuole of transgenic yeast, J. Cell Biol., 105:1971.

Tartakoff, A. M., 1983, Perturbation of vesicular traffic with the carboxylic ionophore monensin, Cell, 32:1026.

Valls, L. A., Hunter, C. P., Rothman, J. H., and Stevens, T. H., 1987, Protein sorting in yeast: the localization determinant of yeast vacuolar carboxypeptidase Y resides in the propeptide, Cell, 48:887.

Voelker, T. A., Herman E. M., and Chrispeels, M. J., 1989, *In vitro* mutated phytohemagglutinin genes expressed in tobacco seeds: role of glycans in protein targeting and stability, Plant Cell, 1:95.

Walker-Simmons, M., and Ryan, C. A., 1977, Immunological identification of proteinase inhibitors I and II in isolated tomato leaf vacuoles, Plant Physiol., 60:61.

# BIOCHEMICAL ADAPTATIONS TO ANOXIA IN RICE AND ECHINOCHLOA SEEDS

Robert A. Kennedy[1], Theodore C. Fox[1], Leslie D. Dybiec[2] and Mary E. Rumpho[1]

[1]Botany Department
University of Maryland
College Park, MD 20742

[2]Horticulture Department
Ohio State University
Columbus, OH 43210

## INTRODUCTION

The genus Echinochloa contains some of the most well studied flood tolerant plant species. These plants are interesting for their extreme flood tolerance, their economic importance as weeds in numerous crops around the world, particularly in rice (Oryza sativa [L.]), and because they represent a complete spectrum of flood tolerance within one genus (Barrett and Wilson, 1981, Kennedy et al., 1987b). Of the five species studied here, E. phyllopogon (Stev.) Koss and E. oryzoides (Ard.) Fritsch Clayton are flood tolerant and confined to aquatic environments. E. muricata (Beauv.) Fern is semi-tolerant and found along streambanks, whereas E. crus-galli (L.) Beauv. and E. crus-pavonis (H.B.K.) Schult. are intolerant and found only in drier sites (Barrett and Wilson, 1981). In nature, these species can all be found in or around the rice agro-ecosystem. In the laboratory, all of the species except E. crus-pavonis are able to germinate and grow in a strict $N_2$ atmosphere, as does rice.

Our research has concentrated on the tolerant species E. phyllopogon (previously referred to as E. crus-galli var. oryzicola). Contrary to the prevailing view, we demonstrated that glycolysis is not the only metabolic pathway operating in E. phyllopogon under anoxia (Rumpho and Kennedy, 1983a, Kennedy et al., 1987a). In fact, E. phyllopogon exhibits a comprehensive and integrated anaerobic metabolic scheme, which at least qualitatively, appears very similar to that operating in aerobic environments. From these findings we were interested in determining the energy requirements for growth of this plant under both environments, building upon our earlier biochemical and physiological data.

The diversity of the Echinochloa genus provides for a unique system for comparative studies on the mechanisms that impart flood tolerance. Alteration of gene expression in plant tissues subjected to a variety of stresses is common (Kelley and Freeling, 1984, Mocquot et al. 1987, Ramagopal,

*Recent Advances in the Development and Germination of Seeds*
Edited by R.B. Taylorson
Plenum Press, New York

1988). Typically, exposure to low oxygen induces an increase in the activity of several glycolytic enzymes, including alcohol dehydrogenase (ADH), which leads to alcoholic fermentation (Crawford, 1977, Rumpho and Kennedy, 1981). Increased alcohol synthesis is found in both tolerant and intolerant plants (Smith and apRees, 1979), however, including all of the *Echinochloa* species studied here (Rumpho and Kennedy, 1981, Cobb and Kennedy, 1987).

In the present experiments we examined what role, if any, ethanol plays in the germination and growth of tolerant *vs*. intolerant *Echinochloa* species. We also looked at whether anaerobiosis induces differential gene expression, evidenced by *de novo* protein synthesis in tolerant *vs*. intolerant plants. Last, we determined rates of protein synthesis in individual seedling parts of *E. phyllopogon* grown under anaerobic conditions.

## ENERGETICS OF GROWTH IN *E. PHYLLOPOGON*

The energetic requirements for growth of *E. phyllopogon* under aerobic and anaerobic conditions were modeled using growth data and changes in metabolite pools. Since chlorophyll is not synthesized in the absence of oxygen, anaerobically germinated seedlings do not photosynthesize, relying on seed reserves for metabolic activity (Kennedy et al., 1980, Bozarth, 1983). To avoid complicated comparisons between heterotrophic ($N_2$-grown) and autotrophic ($O_2$-grown) grown seedlings, aerobically germinated seeds were maintained in the dark. Consequently, seed reserves provided the only substrate for growth in both air- and $N_2$-grown seedlings, thereby permitting direct comparisons during the first seven days of germination. Here, growth is defined as accumulation of dry weight in shoots and roots. Under anaerobic conditions, roots were not formed; growth occurred only in shoots (Kennedy et al., 1980).

The growth rate of *E. phyllopogon* was 301 and 76 $\mu$g dry weight/seedling/day under aerobic and anaerobic conditions, respectively (Fox and Kennedy, 1988). Thus, anoxia reduced growth to 25% of that in air. Normally, without oxygen, carbohydrate cannot be oxidized via the tricarboxylic acid (TCA) cycle and glycolysis accommodates cellular demand for ATP (Bewley and Black, 1983). However, ATP production from glycolysis is much less efficient than that from the TCA cycle (2 vs. 38 moles ATP per mole glucose, respectively). If glycolysis was the sole energy-transducing pathway in $N_2$-germinated seedlings, then a growth rate of 76 $\mu$g dry weight/seedling/day in shoots would require catabolism of 348 $\mu$g dry weight/seedling/day in the seed. In contrast, we found 268 $\mu$g dry weight/seedling/day was lost from seeds of anaerobically germinated seedlings. Thus, metabolism in anaerobically grown seedlings of *E. phyllopogon* was more efficient than predicted.

We further analyzed the energetic requirements of *E. phyllopogon* by estimating the contribution of metabolic pathways to aerobic and anaerobic metabolism. We estimated the amount of ATP produced by catabolic reactions and consumed by anabolic reactions with a model derived from Penning de Vries and co-workers (Penning de Vries et al., 1974, 1982). To simplify comparisons, NADH was converted to ATP-equivalents assuming three moles ATP per mole NADH. In air-grown seedlings the TCA cycle is the main source of ATP, primarily through oxidative phosphorylation. The difference between the dry weight lost from the seed and the dry weight accumulated in shoots and roots was considered to be oxidized for ATP synthesis. From this value, we calculated that catabolic reactions can provide 120% of the ATP required for anabolic reactions (Fox and Kennedy, 1988).

Without oxygen as the terminal electron acceptor, oxidative phosphorylation cannot occur in plants. As a result, ATP production during anaerobiosis is restricted to the substrate level phosphorylation reactions of glycolysis. Assuming that the difference between the dry weight lost from the seed and that accumulated in the shoot was fermented to ethanol, only 42% of the ATP required for growth could be produced. The total amount of ethanol synthesized during anaerobiosis (i.e., ethanol contained in and excreted from the seedling) was equivalent to a dry weight loss of 24 μg/seedling/day (Rumpho and Kennedy, 1983b). This amount of ethanol would require only 12% of the dry matter metabolized under $N_2$. We looked for accumulation and excretion of other end-products of fermentation, particularly malate and lactate, but found none; ethanol was the sole product of fermentation in *E. phyllopogon* (Rumpho and Kennedy, 1983b). Thus, a major portion of the carbohydrate available for energy production appeared to be metabolized by another route.

Several lines of evidence indicate that *E. phyllopogon* develops functional mitochondria under $N_2$. Whereas mitochondria in other species degenerate or exhibit abnormalities (Ueda and Tsuji, 1971, Oliveira, 1977, Vartapetian et al., 1977), mitochondria of anaerobically germinated seedlings of *E. phyllopogon* were morphologically indistinguishable from those grown in air (VanderZee and Kennedy, 1981). Incorporation of $^{14}C$ from sucrose and acetate demonstrated that TCA cycle intermediates were formed (Rumpho and Kennedy, 1983a) and *in vitro* spectrophotometric assays showed that TCA cycle enzyme activities were present (Fox and Kennedy, 1989). In addition, when seedlings germinated in $N_2$ were briefly exposed to air, isolated mitochondria were capable of reducing oxygen, indicating that the respiratory chain developed during anaerobiosis (Kennedy et al., 1987a).

ATP synthesis derived from TCA cycle activity primarily results from mitochondrial electron transport coupled with oxidative phosphorylation. However, during operation of the TCA cycle a small quantity of ATP is synthesized via substrate level phosphorylation by succinyl-CoA. If 88% (assumes 12% of carbohydrate is metabolized to ethanol, as discussed above) of the carbohydrate available for energy production in $N_2$-germinated seedlings of *E. phyllopogon* was metabolized via the TCA cycle without being coupled to oxidative phosphorylation, sufficient ATP could be synthesized to meet the growth requirements (Fox and Kennedy, 1988). For this pathway to proceed a mechanism to oxidize NADH must also function, otherwise, the TCA cycle would cease for lack of the necessary cofactor, NAD. One such mechanism would be to convert NADH to NADPH via pyridine nucleotide transhydrogenase. In turn, NAD would be produced and NADPH recycled to NADP through lipid synthesis. Under $N_2$, E. phyllopogon accumulates large amounts of lipid in shoots (Knowles and Kennedy, 1984, Everard and Kennedy, 1985). We estimated that lipid synthesis would consume 35% of the NADH generated during TCA cycle operation. In addition, we found that nitrate was depleted from the seeds of E. phyllopogon during germination (Kennedy et al., 1983). Although the pool of endogenous nitrate in E. phyllopogon is much larger than that in rice and wheat, our calculations indicate that nitrate reduction alone could oxidize only 1% of the NADH generated by anaerobic operation of the TCA cycle. If the TCA cycle operates during anaerobic germination as described, at this time, we do not have evidence for the mechanism(s) sufficient to oxidize NADH at rates needed to support the growth rate of E. phyllopogon seedlings under anaerobiosis.

## EFFECT OF ETHANOL ON GERMINATION AND GROWTH OF ECHINOCHLOA

Glycolysis and the formation of ethanol appear to be essential for the anaerobic germination of Echinochloa (Rumpho and Kennedy, 1983b). Despite several studies (Jackson, 1982, Rumpho and Kennedy, 1983b, Kennedy et al. 1987b, Roberts et al., 1989) there is considerable disagreement regarding the relationship of ADH activity and maximal ethanol production to tolerance of anoxia. There is a limit for ethanol accumulation in all tissues, after which membrane damage takes place (Hoek and Taraschi, 1988). Different organisms may possess various means of dealing with ethanol, as we have demonstrated previously for E. phyllopogon (Rumpho and Kennedy, 1983b). ADH activity and ethanol production increased in each of the five Echinochloa species discussed here when exposed to anaerobic conditions. The intolerant species E. crus-pavonis produced the least ethanol (Kennedy et al., 1987b). A low threshold for ethanol tolerance could limit ethanol, and hence energy production. To address this hypothesis, we examined the relationship between ethanol and anoxia tolerance in the five Echinochloa species.

Total germination of any of the Echinochloa species was not affected by the absence of oxygen, except for E. crus-pavonis, which did not germinate under $N_2$ (Figures 1a and c). In all of the other species, except E. oryzoides, the rate of germination was delayed by about one day in $N_2$, compared to air. In E. oryzoides, germination was actually enhanced by anaerobic conditions. The addition of 1% ethanol (concentration approximately equal to the greatest amount produced by any of the species after 7 days anaerobiosis [Kennedy et al, 1987b]) in air or $N_2$ also had no effect on total percent germination, but did delay the rate of germination in both environments (Figures 1b and d). There were no obvious differences between tolerant and intolerant species in their responses to the addition of high ethanol to the medium.

While ethanol had virtually no effect on germination percentage, a differential effect was observed on growth of tolerant vs. intolerant species. Increasing ethanol (from 0 to 3% [v/v]) resulted in decreased shoot and root growth in

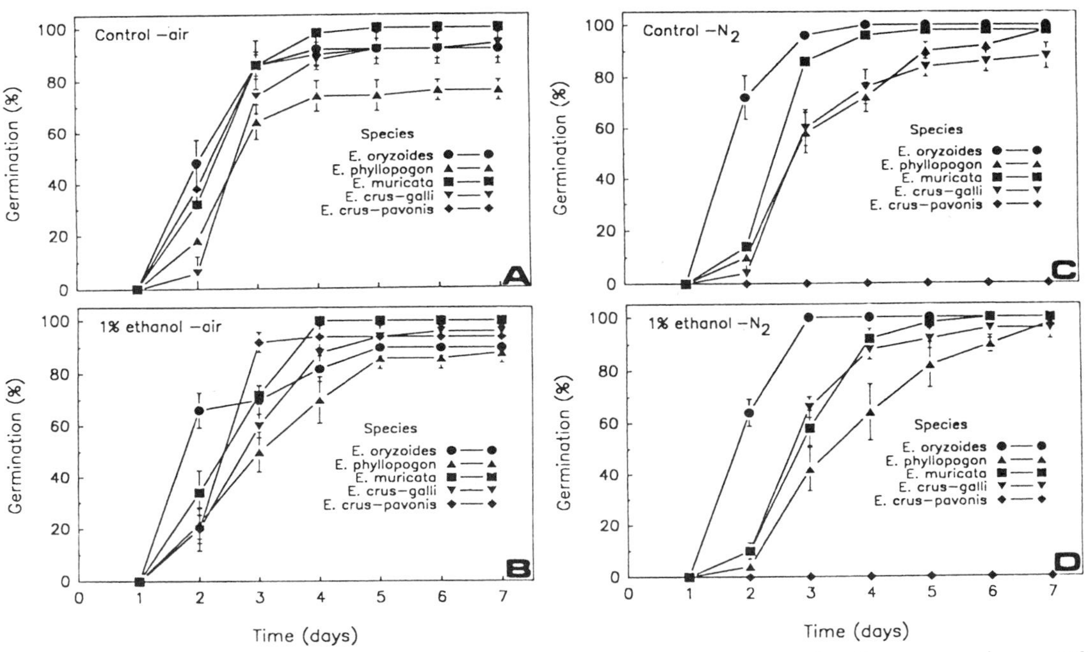


Fig. 1. Germination of <u>Echinochloa</u> species in air <u>vs</u>. $N_2$ and plus or minus 1% ethanol. Ten seeds each of the five <u>Echinochloa</u> species were surface-sterilized and placed on filter paper saturated with sterile distilled water or 1% (v/v) ethanol in plastic petri dishes. Five replicate plates were used for each species. Identical experiments were set up in air and an anaerobic chamber at 25C and in room light. Each day, solutions were removed and replaced with new sterile solutions and germination recorded. Values represent the mean ± standard error.

all species relative to control seedlings (Figures 2a-c). No roots were produced by any species under $N_2$.

Tolerant and intolerant plants did differ in the effect of ethanol on root and shoot growth in air. In tolerant plants (including rice), root growth was inhibited more than shoot growth. The reverse was true in intolerant plants (Figure 2d). With increasing ethanol, the shoot/root ratio increased in the tolerant species E. oryzoides, E. phyllopogon and rice, whereas, it remained the same or decreased in the semi- and non-tolerant species (E. muricata, E. crus-galli and E. crus-pavonis). This observation is significant, particularly for energy conservation in the absence of oxygen. Shoot growth is obviously more important than root elongation in enabling the seedling to reach an aerobic zone. Diversion of limited seed reserves to roots for growth would be detrimental to survival of the seedling.

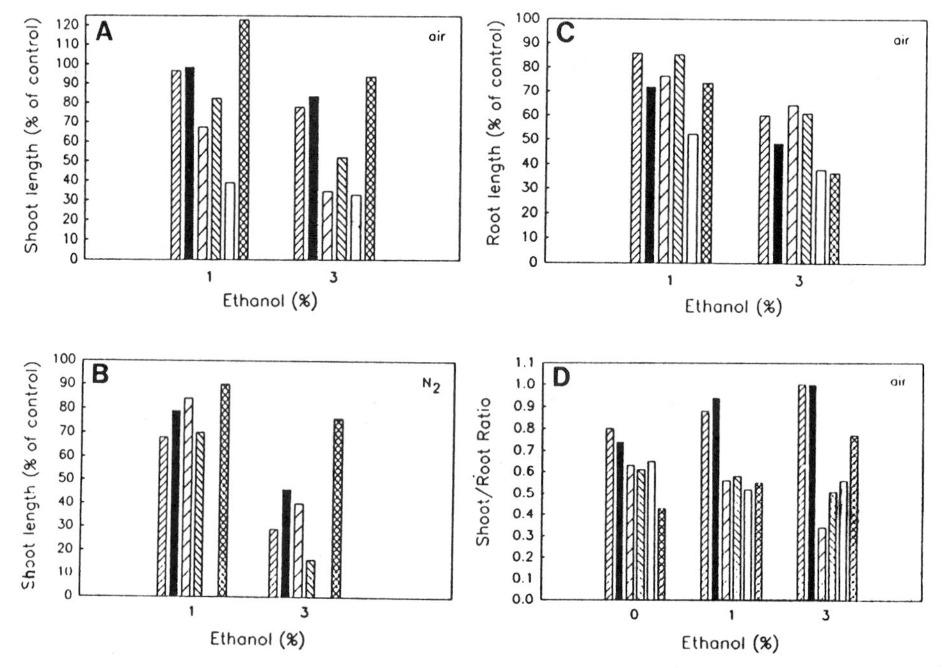


Fig. 2. Growth of Echinochloa species and rice in air vs. $N_2$ and plus or minus added ethanol. Seeds were sterilized and germinated as described for Fig. 1. After 7 days, seedlings were harvested and shoot and root lengths measured. Values represent the means of fifty seedlings and are given as percent of control (no added ethanol) (A-C). (D) Shoot/root ratios in air plus 0, 1 and 3% (v/v) ethanol. Echinochloa species are presented left to right in the order of most tolerant to least tolerant, followed by rice on the far right. The bar representing each species is as follows: E. oryzoides ( ), E. phyllopogon ( ), E. muricata ( ), E. crus-galli ( ), E. crus-pavonis ( ), and rice ( ). Note that E. crus-pavonis does not germinate under $N_2$ and no roots are produced by any species under $N_2$.

## EFFECT OF ETHANOL OR ANAEROBIOSIS ON *DE NOVO* PROTEIN SYNTHESIS

It has been shown previously that short-term exposure of maize seedlings to anaerobic conditions results in cessation of aerobic protein synthesis and selective synthesis of 10 major and 10 minor proteins in the primary root (Kelley and Freeling, 1984). Several of these proteins have been identified and are involved in the glycolytic breakdown of carbohydrates including, sucrose synthase (87 kD), pyruvate decarboxylase (64 kD), phosphoglucoisomerase (55 kD), phosphoglucomutase (42 kD), ADH (40 kD), glyceraldehyde phosphate dehydrogenase (35 kD) and cytosolic fructose-1,6-bisphosphate aldolase (33 kD) (see Bailey-Serres et al., 1988 for review.) In maize, metabolic pathways other than glycolysis are believed to be inoperative without oxygen and the maize primary root tip dies after 70 h anaerobiosis (Sachs et al., 1980).

The only anaerobic tolerant species in which protein synthesis has been studied in the absence of oxygen is rice (Mocquot et al., 1981). Rice responded quite differently to anoxia than did maize; aerobic protein synthesis was not completely halted and aerobic polypeptides were labeled up to nine days under anoxia.

We examined the patterns of *de novo* synthesized polypeptides in *Echinochloa* and rice germinated in air or $N_2$ for three days, and also after germination in air plus 1% ethanol. Several polypeptides ranging in molecular weight from 95 to 12 kD were labeled in air, $N_2$, and air plus ethanol, as demonstrated by the fluorograms (Figs. 3 and 4). The only exception was *E. crus-pavonis* in which no radiolabeled bands were apparent under $N_2$. The approximate molecular weights of the major bands labeled under $N_2$ for the other four *Echinochloa* species included 87, 78, 62, 55, 50, 48, 46, 42, 40, 38 and 31 kD (see arrows in lane 7, Fig. 3). The bands at about 87 and 62 kD are especially enhanced under anaerobic conditions. In air, the major labeled polypeptides were 87, 79, 78, 76, 55, 50, 48, 46, 43, 39 and 22 kD (see arrows in lane 1, Fig. 3). *E. phyllopogon* exhibited the most distinct protein pattern; additional bands were observed at 89, 74, 66, 49, 47 and 29 kD, whereas the bands identified for the other species in air at 55, 48 and 46 kD were absent or greatly reduced in *E. phyllopogon* (lane 2, Fig. 3).

The polypeptide patterns of the fluorograms of all the species labeled in air plus 1% ethanol (Fig. 4) resembled those of the aerobic controls (see Fig. 3, lanes 1-6). However, there was increased labeling of the bands at 87 and 62 kD which was also previously observed for the fluorograms of $N_2$-grown seedlings (Fig. 3, lanes 7-10). Numerous polypeptides were synthesized in the presence of ethanol including major ones at approximately 87, 74, 62, 55, 50, 48, 46, 43, 39, 37, 24 and 22 kD (see arrows in Fig. 4, lane 1). Although there were some quantitative differences, no major qualitative differences in labeling patterns between tolerant and intolerant *Echinochloa* species were observed. However, several different bands were labeled in rice seeds compared to *Echinochloa* as shown by the arrows in lane 6 (Fig. 4).

In $N_2$, all Echinochloa species except E. crus-pavonis synthesized several polypeptides. Both externally added ethanol (in air) and anaerobiosis induced altered de novo polypeptide patterns in all species relative to their respective aerobic control patterns. Thus, in Echinochloa, it does not appear that aerobic polypeptide synthesis is appreciably turned-off under anaerobic conditions, with the exception of E. crus-pavonis. However, upon transfer from $N_2$ to air, E. crus-pavonis is capable of synthesizing several proteins identical to those of the other species (unpublished data).

There are several quantitative differences between air and $N_2$ polypeptide labeling patterns, including an increase in the labeling of the 87 and 62 kD polypeptides under $N_2$ (presumably

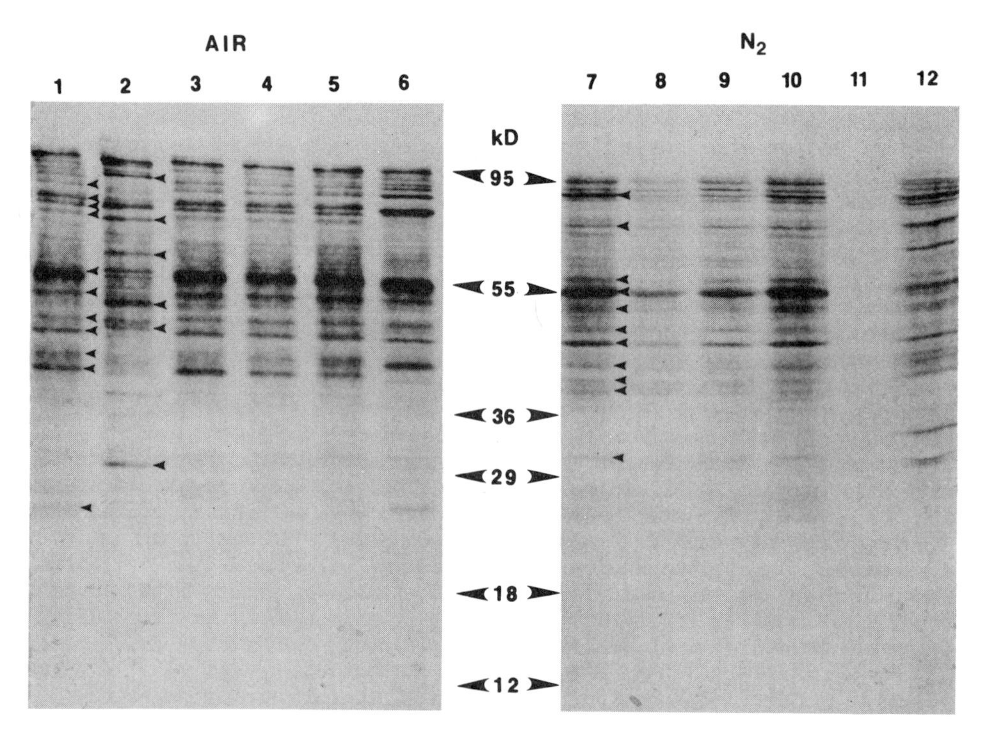


Fig. 3. Fluorogram of Echinochloa species and rice labeled with $^{35}$S-methionine. Seeds were grown in air or $N_2$ for 3 days and incubated with 50 microcuries $^{35}$S-met for 4 hours in the same atmosphere. Sodium dodecyl sulfate-polyacrylamide gel electrophoresis (SDS-PAGE) was carried out on extracts of entire seedlings (Laemmli, 1970). Gels were treated for fluorography and exposed to x-ray film to reveal the labeled polypeptides. Species in lanes 1-6 (air) and repeated in lanes 7-12 ($N_2$) are: E. oryzoides, E. phyllopogon, E. muricata, E. crus-galli, E. crus-pavonis and rice. Relative mobility of molecular weight standards (kilodaltons) are indicated by the center arrows. Arrows in lanes 1, 2 and 7 mark the major labeled polypeptides.

sucrose synthase and pyruvate decarboxylase [Kelley and Freeling, 1984]). E. phyllopogon especially exhibited several distinct differences in its labeled polypeptide profile compared to the other species. Otherwise, within each experimental condition, there were few differences among the different Echinochloa species, despite their widely varying tolerance to anaerobiosis. Several labeled bands of the same molecular weight in both air and $N_2$ possibly corresponded to glycolytic enzymes including glucose-6-P isomerase, phosphoglucomutase and ADH.

Since other metabolic pathways, in addition to glycolysis, have been demonstrated to be active in at least E. phyllopogon under anaerobic conditions (Rumpho and Kennedy, 1983a), it should be expected that some of the polypeptides present under anaerobic conditions represent enzymes of other pathways, e.g., TCA cycle, lipid synthesis and the oxidative pentose phosphate pathway. We have not examined the metabolism of the other species in sufficient detail to generalize this hypothesis to them.

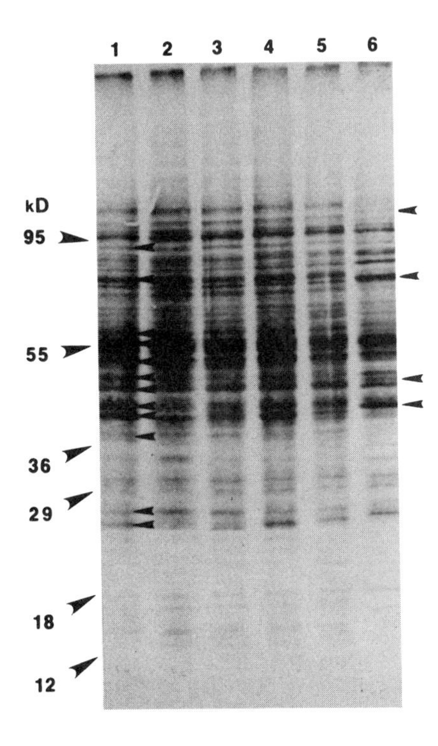


Fig. 4. Fluorogram of Echinochloa and rice grown in air plus 1% ethanol for 3 days and labeled with 50 microcuries $^{35}S$-methionine for 4 hours under the same conditions. SDS-PAGE and fluorography were as described for Fig. 3. Species in lanes 1-6 are: E. oryzoides, E. phyllopogon, E. muricata, E. crus-galli, E. crus-pavonis, and rice. Arrows in lane 1 mark major polypeptides in Echinochloa and in lane 6 polypeptides unique to rice. Relative mobility of molecular weight markers are indicated on the left in kilodaltons (kD).

RATES OF PROTEIN SYNTHESIS IN INDIVIDUAL SEED PARTS OF E. PHYLLOPOGON UNDER ANAEROBIC CONDITIONS

As mentioned above the growth rate of E. phyllopogon under anaerobic conditions is 25% of that in air (Fox and Kennedy, 1988). Total protein synthesis, measured by in vivo incorporation of $^{35}S$-methionine into protein, was 10% of that in air (Table 1). However, protein synthesis was not proportionately reduced in shoot, root and seed tissues. Anaerobiosis reduced protein synthesis to 8% of the aerobic level in the seed. The rate of protein synthesis in shoots of anaerobically grown seedlings was 86% of that in air.

The mass of shoot tissue formed under anoxia is reduced compared to that formed in air. However, the rate of incorporation of $^{35}S$-methionine into protein in the shoots is similar in both air and $N_2$. Hence, the rate of protein synthesis per unit dry matter is actually stimulated by anoxia, i.e. 1.5 vs. 3.7 fmol/mg dry weight/h in air- and $N_2$-grown seedlings, respectively. It is apparent that E. phyllopogon shoots produced under anaerobic conditions have the ability to synthesize proteins required for structural and catalytic processes.

Table 1. Incorporation of $^{35}S$-methionine into proteins of 7 day old seedlings of E. phyllopogon. Seedlings were exposed to 20 mCi $^{35}S$-methionine in the imbibition solution for 1 hour.

| | Incorporation of $^{35}S$-Methionine | | |
|---|---|---|---|
| Tissue | Air | $N_2$ | $N_2$/Air |
| | (fmol/seedling/h) | | (%) |
| Seedling | 22.5 | 2.3 | 10 |
| Shoot | 1.4 | 1.2 | 86 |
| Seed | 13.4 | 1.1 | 8 |
| Root[a] | 7.7 | - | - |

[a]No roots are formed under $N_2$.

CONCLUSIONS

1. Glycolysis alone cannot satisfy the energetic needs of E. phyllopogon under anaerobic conditions. However, if fermentation is combined with an active TCA cycle, substrate level phosphorylation (uncoupled from electron transport) can provide sufficient ATP to maintain the growth rate observed under anoxia in this species.

2. Exogenous ethanol had virtually no effect on germination of tolerant vs. intolerant Echinochloa species.

However, ethanol did result in an increase in the shoot/root ratios of tolerant species and a decrease in intolerant species. Preferential growth of the shoot is important for the plant to reach an aerobic zone before other factors become limiting for its survival.

3. All of the Echinochloa species examined here except E. crus-pavonis synthesized numerous polypeptides under anaerobiosis. Within each environmental treatment there were very few differences in labeled polypeptides between tolerant and intolerant plants. Differential expression was seen when comparing between environments. Both anaerobiosis and exogenously applied ethanol enhanced labeling of the 87 and 62 kD polypeptides.

4. Although protein synthesis in whole seedlings of E. phyllopogon is reduced after seven days anoxia, shoot tissue retained the ability to synthesize proteins at rates approaching those in air.

REFERENCES

Bailey-Serres, J., Kloeckener-Gruissem, B., and Freeling, M., 1988, Genetic and molecular approaches to the study of the anaerobic response and tissue specific gene expression in maize, Plant Cell Environ., 11:351.

Barrett, S. C. H., and Wilson, B. F., 1981, Colonizing ability in the Echinochloa crus-galli complex (barnyardgrass). I. Variation in life history, Can. J. Bot., 59:1844.

Bewley, J. D., and Black, M., 1983, "Physiology and Biochemistry of Seeds in Relation to Germination," Springer-Verlag, Berlin.

Bozarth, C. S., 1983, Greening and photosynthesis in anaerobically grown Echinochloa crus-galli and Oryza sativa after exposure to air. Ph.D. Dissertation, Washington State University, Pullman, WA.

Cobb, B. G., and Kennedy, R. A., 1987, Distribution of alcohol dehydrogenase in roots and shoots of rice (Oryza sativa) and Echinochloa seedlings, Plant Cell Environ., 10:633.

Crawford, R. M. M., 1977, Tolerance of anoxia and ethanol metabolism in germinating seeds, New Phytol., 79:511.

Everard, J. D., and Kennedy, R. A., 1985, Physiology of lipid metabolism during anaerobic germination of Echinochloa crus-galli var. oryzicola, Plant Physiol., 77:S-98.

Fox, T. C., and Kennedy, R. A., 1988, Modeling of energy requirements for growth of barnyard grass seedlings under aerobic and anaerobic conditions, Plant Physiol., 86:S-54.

Fox, T. C., and Kennedy, R. A., 1989, Mitochondrial enzymes in aerobically and anaerobically germinated seedlings of Echinochloa and rice, Planta (in press).

Hoek, J. B., and Taraschi, T. F., 1988, Cellular adaptation to ethanol, TIBS, 13:269.

Jackson, M. B., 1982, An examination of the importance of ethanol in causing injury to flooded plants, Plant Cell Environ., 8:163.

Kelley, P. M., and Freeling, M., 1984, Anaerobic expression of maize fructose-1,6-diphosphate aldolase, J. Biol. Chem., 259:14180.

Kennedy, R. A., Barrett, S. C. H., VanderZee, D., and Rumpho, M.E., 1980, Germination and seedling growth under anaerobic conditions in Echinochloa crus-galli (barnyard grass), Plant Cell Environ., 3:243.
Kennedy, R. A., Fox, T. C., Siedow, J. N., 1987a, Activities of isolated mitochondria and mitochondrial enzymes from aerobically and anaerobically germinated barnyard grass (Echinochloa) seedlings, Plant Physiol., 85:474.
Kennedy, R. A., Rumpho, M. E., Fox, T. C., 1987b, Germination physiology of rice and rice weeds: metabolic adaptations to anoxia, in: "Plant Life in Aquatic and Amphibious Habitats," R. M. M. Crawford, ed., Blackwell Scientific, Oxford.
Kennedy, R. A., Rumpho, M. E., VanderZee, D., 1983, Germination of Echinochloa crus-galli (barnyard grass) seeds under anaerobic conditions. Respiration and response to metabolic inhibitors, Plant Physiol., 72:787.
Knowles, L. O., and Kennedy, R. A., 1984, Lipid biochemistry of Echinochloa crus-galli during anaerobic germination, Phytochem., 23:529.
Laemmli, U. K., 1970, Cleavage of structural proteins during the assembly of the head of bacteriophage T4, Nature, 227:680.
Mocquot, B., Ricard, B., and Pradet, A., 1987, Rice embryos can express heat-shock genes under anoxia, Biochimie, 69:677.
Mocquot, B., Prat, C., Mouches, C., and Pradet, A., 1981, Effect of anoxia on energy charge and protein synthesis in rice embryo, Plant Physiol., 68:636.
Oliveira, L., 1977, Changes in ultrastructure of mitochondria of roots of Triticale subjected to anaerobiosis, Protoplasma, 91:267.
Penning de Vries, F. W. T., Brunsting, A. H. M., van Laar, H. H., 1974, Products, requirements and efficiency of biosynthesis. A quantitative approach, J. Theor. Biol., 45:339.
Penning de Vries, F. W. T., van Laar, H. H., 1982, Simulation of Plant Growth and Crop Production. Centre for Agricultural Publications and Documentation, Wageningen.
Pradet, A., Mocquot, B., Raymond, P., Morisset, C., Aspart, L., and Delsem, M., 1985, Energy metabolism and synthesis of nucleic acids and proteins under anoxic stress, in: "Cellular and Molecular Biology of Plant Stress," J. L. Key and T. Kusuge, eds., Alan R. Liss, Inc., New York.
Ramagopal, S., 1988, Regulation of protein synthesis in root, shoot and embryonic tissues of germinating barley during salinity stress, Plant Cell Environ., 11:501.
Roberts, J. K. M., Chang, K., Webster, C., Callis, J., and Walbot, V., 1989, Dependence of ethanolic fermentation, cytoplasmic pH regulation, and viability on the activity of alcohol dehydrogenase in hypoxic maize root tips, Plant Physiol., 89:1275.
Rumpho, M. E., and Kennedy, R. A., 1981, Anaerobic metabolism in germinating seeds of Echinochloa crus-galli (barnyard grass). Metabolite and enzyme studies, Plant Physiol., 68:165.
Rumpho, M. E., and Kennedy, R. A., 1983a, Activity of the pentose phosphate and glycolytic pathways during anaerobic germination of Echinochloa crus-galli (barnyard grass) seeds, J. Exp. Bot., 155:1.

Rumpho, M. E., and Kennedy, R. A., 1983b, Anaerobiosis in *Echinochloa crus-galli* (barnyard grass) seedlings. Intermediary metabolism and ethanol tolerance, *Plant Physiol.*, 72:44.

Sachs, M. M., Freeling, M., and Okimoto, R., 1980, The anaerobic proteins of maize, *Cell*, 20:761.

Smith, A. M., and apRees, T., 1979, Pathways of carbohydrate fermentation in the roots of marsh plants, *Planta*, 146:327.

Ueda, K., and Tsuji, H., 1971, Ultrastructural changes of organelles in coleoptile cells during anaerobic germination in rice seeds, *Protoplasma*, 73:203.

VanderZee, D., and Kennedy, R. A., 1981, Germination and seedling growth in *Echinochloa crus-galli* var. *oryzicola* under anoxic germination conditions, *Planta*, 155:1.

Vartapetian, B. B., Andreeva, I. N., Kozlova, G. I., and Agapova, L. P., 1977, Mitochondrial ultrastructure in roots of mesophytes and hydrophytes under anoxia after glucose feeding, *Protoplasma*, 91:243.

# SOME ASPECTS OF METABOLIC REGULATION OF SEED GERMINATION AND DORMANCY

Daniel Côme[1] and Françoise Corbineau[2]

[1]Laboratoire de Physiologie Végétale Appliquée
Université Pierre et Marie Curie
Tour 53, 1er étage
4, place Jussieu, 75252 Paris Cedex 05, France
[2]Laboratoire de Physiologie des Organes Végétaux Après Récolte
CNRS
4 ter, route des Gardes, 92190 Meudon, France

## ABBREVIATIONS

ADH, alcohol dehydrogenase (EC 1.1.1.1); ADP, adenosine diphosphate; AMP, adenosine monophosphate; ATP, adenosine triphosphate; FBPase 1, fructose 1,6-bisphosphatase (EC 3.1.3.11); FBPase 2, fructose 2,6-bisphosphatase (EC 3.1.3.46); Fru-6-P, fructose 6-phosphate; Fru-1,6-$P_2$, fructose 1,6-bisphosphate; Fru-2,6-$P_2$, fructose 2,6-bisphosphate; Glu-6-P, glucose 6-phosphate; G6PDH, glucose 6-phosphate dehydrogenase (EC 1.1.1.49); 4-MP, 4-methylpyrazole; NAD, nicotinamide adenine dinucleotide; NADH, nicotinamide adenine dinucleotide reduced; NADP, nicotinamide adenine dinucleotide phosphate; NADPH, nicotinamide adenine dinucleotide phosphate reduced; $P_i$, inorganic phosphate; PFK, 6-phosphofructokinase (EC 2.7.1.11); PFK 1, 6-phosphofructo-1-kinase (EC 2.7.1.11); PFK 2, 6-phosphofructo-2-kinase (EC 2.7.1.105); 6PGDH, 6-phosphogluconate dehydrogenase (EC 1.1.1.44); $PP_i$, pyrophosphate; $PP_i$-PFK, pyrophosphate:fructose 6-phosphate 1-phosphotransferase (EC 2.7.1.90); SHAM, salicylhydroxamic acid.

## INTRODUCTION

Seed germination is subject to a very precise regulation, the complexity of which originates both in the action of various external factors and in characteristics within the seeds themselves. This phenomenon, which leads to the growth of preformed organs of the embryo, can only occur if all environmental components are suitable (water availability, oxygen supply, suitable temperature, and sometimes light). However, many seeds are unable to germinate, or do so with great difficulty, even when placed in the presence of water and oxygen at moderate temperatures. It is precisely in this inability to germinate, commonly known as dormancy, that these seeds display their basic physiological characteristics.

The aim of the present paper is to examine some aspects of the regulation of germination and dormancy from a metabolic point of view including respiration and energetic metabolism, pentose phosphate pathway, activation of glycolysis, and ethanol metabolism.



*Recent Advances in the Development and Germination of Seeds*
Edited by R.B. Taylorson
Plenum Press, New York

## ATTEMPTED DEFINITION OF GERMINATION AND DORMANCY

### The Germination Process

Germination was accurately defined by Evenari (1957) as a process limited by the onset of seed hydration and the start of radicle elongation. This definition has been implicitly adopted by all physiologists, who agree that a seed has germinated when its radicle has pierced the seed coats or, in the case of an isolated embryo, has begun to elongate.

Various authors (Rollin, 1975; Tissaoui and Côme, 1975; Perino, 1987) have shown that the overall germination process takes place in three successive phases : imbibition, germination *sensu stricto* and growth. Imbibition corresponds to a rapid water intake which leads to a regular increase in the respiratory activity. Germination *sensu stricto* is the true germination phase. It is an activation process of the embryo, which is not accompanied by any apparent morphological changes. Growth is marked by the beginning of radicle elongation and by a significant change in the physiological state.

The crucial phase of the germination process is germination *sensu stricto*, since it is on this phase that growth leading to seedling formation depends. It was clearly shown (Côme, 1980/81, 1982; Côme and Thévenot, 1982; Perino, 1987) that the nature of these two phases are quite different phenomena which do not involve the same regulatory mechanisms.

### What is Dormancy ?

Seeds are considered to be dormant when they do not germinate, or germinate poorly, in environmental conditions that are apparently favourable

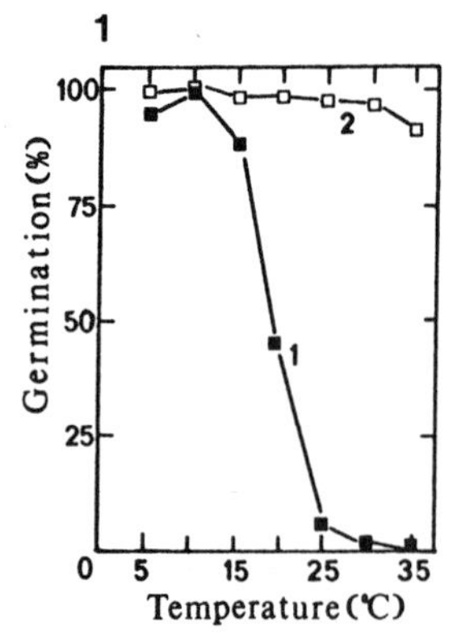


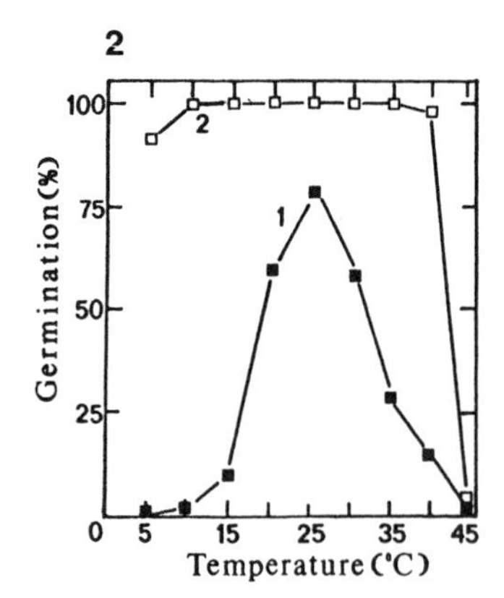


Fig. 1. Influence of temperature on the germination percentage obtained after 7 days with *Avena sativa* seeds (cv. Moyencourt), at harvest time (1) and after 4 months of dry storage at 30°C (2). From Corbineau et al. (1986).

Fig. 2. Influence of temperature on the germination percentage obtained after 7 days with *Helianthus annuus* seeds (cv. Mirasol), at harvest time (1) and after 12 months of dry storage at 5°C (2). From Corbineau (unpublished results).

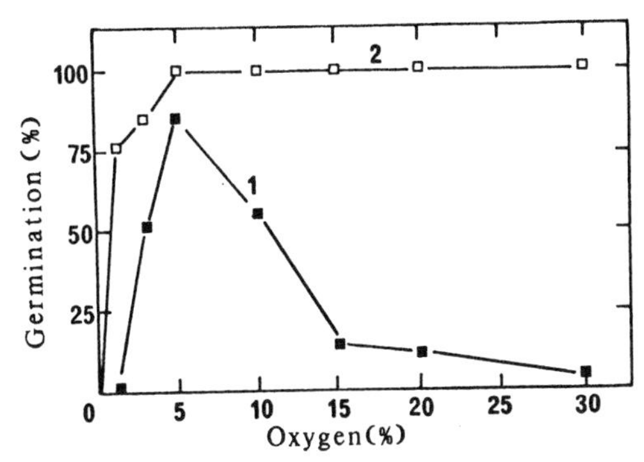


Fig. 3. Influence of oxygen concentration on the germination percentage obtained after 14 days at 15°C with dormant (1) and non-dormant (2) apple (*Pyrus malus*) embryos. From Côme et al.(1985).

to germination (good supply with water, good oxygenation, apparently good temperature). However, a small change in these conditions is enough to obtain good germination of most of the dormant seeds, because dormancies are generally relative phenomena which appear or not depending on the external factors (Côme, 1982). As a matter of fact every dormancy of whatsoever character (embryo dormancy or seed coat imposed dormancy) must be considered as a regulatory mechanism which allows germination only in very strict external conditions. Removal of dormancy is characterized by a widening of the range of conditions in which germination becomes possible. The following examples illustrate these phenomena, but many others could be given.

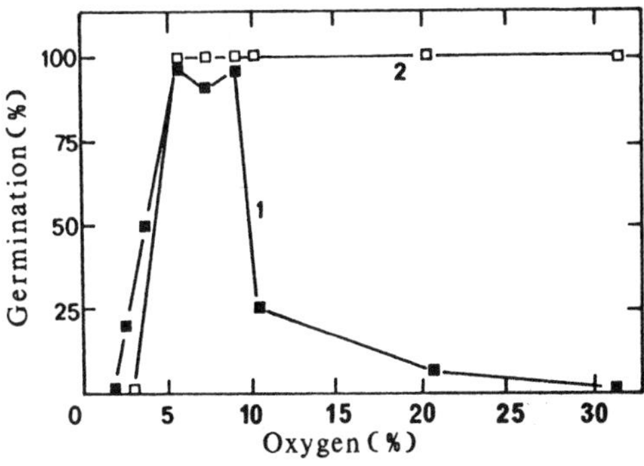


Fig. 4. Influence of oxygen concentration on the germination percentage obtained after 7 days at 40°C in continuous white light, with dormant (1) and non-dormant (2) *Oldenlandia corymbosa* seeds. From Corbineau and Côme (1980/81).

Cereal (barley, oat or wheat) seeds are dormant at harvest time since their germination is very difficult at moderate temperatures, but they germinate easily at 5°C or 10°C (Corbineau and Côme, 1980, 1981; Corbineau et al., 1981, 1986; Côme and Corbineau, 1984; Côme et al., 1984). After some months of dry storage their dormancy is broken, and they become capable of germinating up to 30° or 35°C (Fig.1). On the contrary, dormancy of freshly harvested sunflower seeds corresponds to an inability to germinate at temperatures lower than about 25°C (Corbineau and Côme, 1987), and breaking of dormancy during dry storage renders the seeds able to germinate in a larger range of temperatures (Fig.2). Dormant apple embryos (Côme et al., 1985) or dormant *Oldenlandia corymbosa* seeds (Corbineau and Côme, 1980/81), which cannot germinate in air, are able to do so in atmospheres much poorer in oxygen. When they are no longer dormant they germinate easily in air and even in atmospheres enriched in oxygen (Figs 3 and 4).

## RESPIRATION AND ENERGETIC METABOLISM

In most cases, a seed can only germinate if its respiratory activity, and then if the production of ATP, is sufficient. This is probably the reason why a minimum of oxygen is required. However, these events occur in both dormant and non-dormant seeds. It is not a process which is involved in dormancy. For example, dormant and non-dormant apple embryos exhibit exactly the same respiratory activity as long as the non-dormant embryos have not yet germinated (Perino, 1987). It is radicle elongation of non-dormant embryos which is associated with an increased respiratory activity, but not the absence of dormancy. The same behaviour applies to ATP production and energy charge ((ATP + 0.5 ADP)/(ATP + ADP + AMP)). Dormant oat seeds do not germinate at 30°C, but non-dormant seeds germinate easily. Nevertheless, the ATP/ADP ratio and the energy charge increase in the same way during imbibition at 30°C, until the germination of non-dormant seeds (Fig.5). Similar results were obtained with apple embryos (Tissaoui, 1975) and sunflower seeds (Corbineau, unpublished results). Dormancy is not due, therefore, to an inability to synthesize ATP from ADP and AMP.

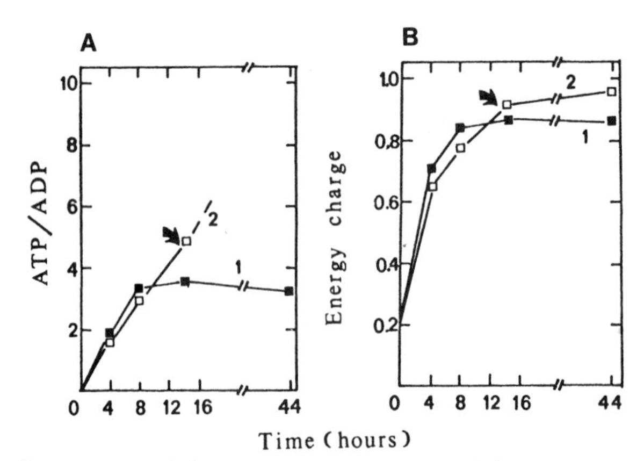


Fig. 5. ATP/ADP ratio (A) and energy charge (B) as a function of time at 30°C, for dormant (1) and non-dormant (2) oat (cv. Moyencourt) seeds. The arrows indicate when the radicle of the non-dormant seeds began to elongate. From Côme et al. (1988).

## INVOLVEMENT OF THE PENTOSE PHOSPHATE PATHWAY

The idea that the pentose phosphate pathway could be involved in seed germination comes from the effects of respiratory inhibitors and other metabolic inhibitors. Respiratory inhibitors (KCN, $NaN_3$, $Na_2S$) greatly stimulate germination of dormant seeds of cereals and many other species (Roberts, 1973; Roberts and Smith, 1977; Bewley and Black, 1982). In the case of oat, these inhibitors are most effective at a concentration of $10^{-3}$ M when they are applied continuously from the start of imbibition (Fig.6), and higher concentrations are soon lethal. However, very high concentrations ($10^{-2}$ M or $5.10^{-2}$ M) strongly stimulate germination if given for a short period at the beginning of germination (Table 1). Similar results were obtained with apple embryos (Perino, 1987).

Different hypotheses have been proposed to explain the stimulatory effect of KCN on germination. According to Taylorson and Hendricks (1973), the cyanhydric gas formed from KCN could react with L-cystein to give $\beta$ L-cyanoalanine necessary for the synthesis of arginine and aspartic acid. These two aminoacids could be limiting factors for germination, and thus their synthesis from KCN could promote it. Apple embryo is indeed able to metabolize cyanhydric acid (Miller and Conn, 1980; Dziewanowska and Lewak, 1982). However, other respiratory inhibitors which are not metabolized, such as $NaN_3$ (Perino, 1987) or $Na_2S$ (Wysinska et al., 1981) have the same effect as KCN. Consequently, the hypothesis of Taylorson and Hendricks is probably not applicable to the apple embryo.

Some studies suggest that the stimulation of germination by respiratory inhibitors may involve the cyanide insensitive pathway (Esashi et al., 1979; Yu et al., 1979; Upadhyaya et al., 1982, 1983). This is not so in barley or oat seeds (Corbineau et al., 1984) nor in apple embryo (Perino, 1987), however, since an inhibitor of this pathway (SHAM) does not alter the stimulatory effect of $NaN_3$ or KCN (Fig.7).

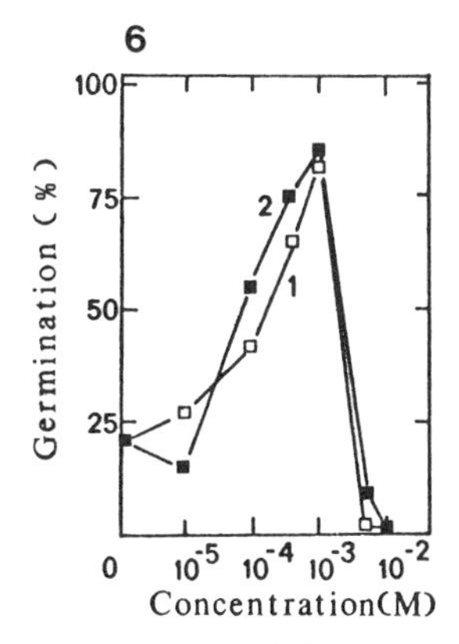


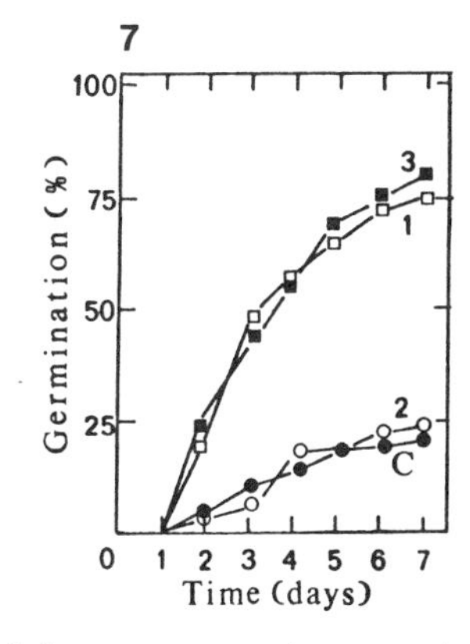


Fig. 6. Effect of KCN (1) and $NaN_3$ (2) concentration on the germination percentage obtained after 7 days at 30°C with dormant oat (cv. Moyencourt) seeds placed in the presence of the respiratory inhibitors from the start of imbibition. From Côme et al. (1988).

Fig. 7. Effect of $10^{-3}$ M $NaN_3$ (1), $5.10^{-3}$ M SHAM (2) and $10^{-3}$ M $NaN_3$ + $5.10^{-3}$ M SHAM (3) on the germination at 25°C of dormant oat (cv. Moyencourt) seeds. Control seeds (C) were placed onto water. From Côme et al. (1988).

Table 1. Influence on dormant oat (cv. Moyencourt) seeds of the duration of treatment with $10^{-2}$ M and $5.10^{-2}$ M $NaN_3$ on the germination percentage obtained at 25°C at the end of the treatment and after transfer onto water for 7 days. Control seeds were placed directly onto water at 25°C for 7 days. From Lecat (1987).

| Germination (%) | $NaN_3$ concentration (M) | Duration of treatment with $NaN_3$ | | | | | | | |
|---|---|---|---|---|---|---|---|---|---|
| | | 0 | 4h | 8h | 16h | 1d | 2d | 3d | 6d |
| Control | 0 | 24 | | | | | | | |
| Before transfer onto water | $10^{-2}$ | - | 0 | 0 | 0 | 0 | 0 | 0 | 0 |
| After transfer onto water | | - | 56 | 58 | 88 | 88 | 90 | 83 | 33 |
| Before transfer onto water | $5.10^{-2}$ | - | 0 | 0 | 0 | 0 | 0 | 0 | 0 |
| After transfer onto water | | - | 28 | 82 | 60 | 30 | 12 | 0 | 0 |

Numerous authors (Roberts, 1969, 1973; Simmonds and Simpson, 1972; Roberts and Smith, 1977; Perino and Côme, 1981; Perino et al., 1984) consider that respiratory inhibitors stimulate germination through an effect on the pentose phosphate pathway. Moreover, inhibitors of Krebs cycle (malonate) or glycolysis (NaF), and stimulators of the pentose phosphate pathway (nitrites, nitrates, methylene blue) improve germination (Roberts, 1973; Roberts and Smith, 1977; Bewley and Black, 1982; Perino, 1987). As a matter of fact, it was shown by Perino (1987) that, in the case of apple embryos, all these substances stimulate germination sensu stricto and inhibit growth (Fig.8).

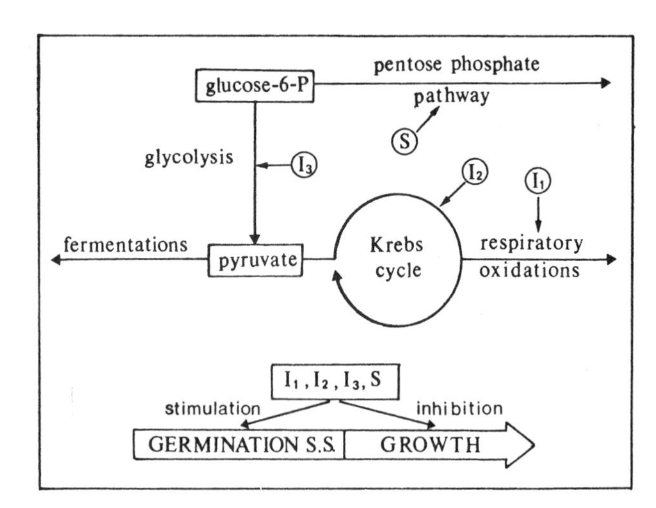


Fig. 8. Effects of various substances on germination sensu stricto and growth, in the case of apple embryos. $I_1$, respiratory inhibitors (KCN, $NaN_3$ or $Na_2S$); $I_2$, inhibitor of Krebs cycle (malonate); $I_3$, inhibitor of glycolysis (NaF); S, stimulators of pentose phosphate pathway (nitrites, nitrates or methylene blue).

Table 2. Activity of glucose-6-phosphate dehydrogenase (G6PDH) and 6-phosphogluconate dehydrogenase (6PGDH) in dormant and non-dormant oat (cv. Prieuré) seeds. The activity was determined *in vitro* at 30°C, by measuring NADPH production spectrophotometrically at 340 nm. From Côme et al. (1988).

| Enzymes | Activity (nM NADPH $min^{-1}$ mg $protein^{-1}$) Dormant seeds | Non-dormant seeds | $\frac{Dormant}{Non\text{-}dormant} \times 100$ |
|---|---|---|---|
| G6PDH | 31.0 | 32.0 | 97 |
| 6PGDH | 37.3 | 33.9 | 110 |

Although the hypothesis implicating the pentose phosphate pathway is the most likely, it depends on indirect arguments and is somewhat contentious (Bewley and Black, 1982). The two main enzymes of the pentose phosphate pathway (G6PDH and 6PGDH) are present in oat seeds, and are located in the embryo and the aleurone layer (Lecat, 1987), but their *in vitro* activities are similar in dormant and non-dormant seeds (Table 2). Both enzymes have also the same *in vitro* activity in dormant and non-dormant apple embryos, and they are much more active in the embryonic axis than in the cotyledons where they are present only in the procambium (Perino, 1987). Nevertheless, such *in vitro* measurements do not give information on *in vivo* enzymatic activities, because the regulation of the pentose phosphate pathway is very complex, as shown by numerous published studies and results obtained with oat seeds (Table 3). Phosphate ions inhibit enzymes in this pathway (Speer, 1974) and also inhibit germination (Xhaufflaire, 1968; Durand, 1974; Corbineau, unpublished results). The activities of G6PDH and 6PGDH are also regulated by ATP (Muto and Uritani, 1972; Grossman and McGowan, 1975; Ashihara and Komamine, 1976), NADH and NADPH (Muto and Uritani, 1972; Ashihara and Komamine, 1974, 1976; Lendzian and Bassham,1975; Pelroy et al., 1976; Apte et al., 1978), and by the NADPH/NADP ratio (Perino, 1987; Côme et al., 1988). Consequently, the activity measured *in vitro* with purified enzymes does not necessarily reflect the *in vivo* activity. It can be the various factors involved in the regulation of the enzymatic activity that are different. Measurements of the $^{14}CO_2$ produced when glucose-6-$^{14}C$ and glucose-1-$^{14}C$ are supplied to the seeds, and calculation of the $C_6/C_1$ ratio did not give also decisive results (Bewley and Black, 1982; Lecat, 1987). The same is true for the oxidised and reduced forms of coenzymes involved in glycolysis (NAD and NADH) and the pentose phosphate pathway (NADP and NADPH) (Roberts, 1973; Bewley and Black, 1982). Thus, the participation of the pentose phosphate pathway in the germination process is not established. However, by measuring the evolution of G6PDH and 6PGDH activities, and NAD, NADH, NADP and NADPH contents in apple embryos during imbibition, Perino (1987) has shown that the transition from germination *sensu stricto* to growth seems to be associated with an increasing participation of this pathway to the glucidic catabolism.

On the basis of the results of Vidal et al. (1980) concerning NADPH-dependent malate dehydrogenase in the chloroplasts, which is activated by NADPH produced in light, a hypothesis has been put forward to explain the possible part taken by the pentose phosphate pathway in the germination process (Côme, 1987; Côme et al., 1988). It is suggested that this pathway is involved through the production of NADPH, which reduces proteins with disulfide bridges of the thioredoxin type, owing to a thioredoxin reductase (NADPH-dependent enzyme). The reduced proteins may activate enzymes necessary for germination. This hypothesis, however, has not yet been verified.

Table 3. Effects of phosphate ions ($K_2PO_4$), ATP, NADH and NADPH on the activity of glucose-6-phosphate dehydrogenase (G6PDH) and 6-phosphogluconate dehydrogenase (6PGDH) of dormant oat (cv. Prieuré) seeds. Measurements made in vitro at 30°C and pH 7.5. From Côme et al. (1988).

| Enzymes | Activity (percentage control) in presence of | | | | |
|---|---|---|---|---|---|
| | Tris-HCl 0.02 M (control) | $K_2PO_4$ 0.4 M | ATP 3 mM | NADH 0.4 mM | NADPH 0.4 mM |
| G6PDH | 100 | 52 | 30 | 47 | 45 |
| 6PGDH | 100 | 42 | 85 | 18 | 17 |

## ACTIVATION OF GLYCOLYSIS THROUGH FRUCTOSE 2,6-BISPHOSPHATE

Phosphofructokinase (PFK1), which is involved in the phosphorylation of Fru-6-P to Fru-1,6-$P_2$ in the presence of ATP, is well established as a control step in glycolysis (Everson and Rowan, 1965; Turner and Turner, 1980). More recently, a $PP_i$-dependent phosphofructokinase ($PP_i$-PFK) was shown to play an important part in the glycolytic pathway in higher plants (Carnal and Black, 1979, 1983; Sabularse and Anderson, 1981; Kruger et al., 1983; Hers, 1984; Smyth and Black, 1984; Cséke et al., 1985).

Fructose 2,6-bisphosphate (Fru-2,6-$P_2$) is a regulator of PFK activity discovered in liver cells (Van Schaftingen et al., 1981), but also present in most eukaryotic cells (Van Schaftingen, 1986). In higher plants, it is a very potent stimulator of $PP_i$-PFK (Sabularse and Anderson, 1981; Van Schaftingen et al., 1982; Hers, 1984) and an inhibitor of fructose 1,6-bisphosphatase (FBPase 1). It is synthesized by 6-phosphofructo-2-kinase (PFK2) from Fru-6-P and ATP and is hydrolysed to Fru-6-P and $P_i$ by FBPase 2. PFK 2 is inhibited by phosphoenolpyruvate and glycerate-3-phosphate. Fig. 9 summarizes the involvement of Fru-2,6-$P_2$ in the control of glycolysis.

The suggestion that Fru-2,6-$P_2$ could play a part in the germination process comes from the fact that its concentration increases manyfold in seeds during imbibition (Hers, 1984) and in fungal spores after induction of their germination by various means (Van Laere et al., 1983). In *Avena sativa* embryos the ratio of potential $PP_i$-PFK activity to ATP-PFK (PFK 1) was shown to be about 1.62 (Corbineau et al., 1989), indicating the probable engagement of tissues in biosynthetic activity (Dennis and Greyson, 1987). In the same species, large differences were observed in changes in the concentrations of Fru-2,6-$P_2$ and other phosphate esters upon imbibition at 30°C of dormant and non-dormant seeds (Larondelle et al., 1987). In non-dormant oat seeds, which began to germinate after about 20 hours, the concentration of Fru-2,6-$P_2$ sharply increased until 16 hours, and then decreased during a period that roughly corresponded to radicle protrusion (Fig.10). In dormant seeds, Fru-2,6-$P_2$ concentration increased during the first few hours, and then remained constant (Fig.10). Phosphoenolpyruvate and glycerate 3-phosphate started to increase markedly only when the radicle of non-dormant seeds elongated, whereas they increased steadily in the dormant ones from the beginning of imbibition, always remaining much higher than in non-dormant seeds. By transferring dormant and non-dormant oat seeds from 10°C to 30°C, or inversely, it was confirmed that germination was associated with a rise in Fru-2,6-$P_2$ content and a simultaneous drop in the concentration of phosphoenolpyruvate (Larondelle et al., 1987). Moreover, when germination of dormant oat seeds was promoted at 30°C by ethanol, Fru-2,6-$P_2$ increased, and phosphoenolpyruvate and glycerate 3-phosphate remained low until radicle elongation, as in non-dormant seeds imbibed with water (Larondelle et al., 1987).

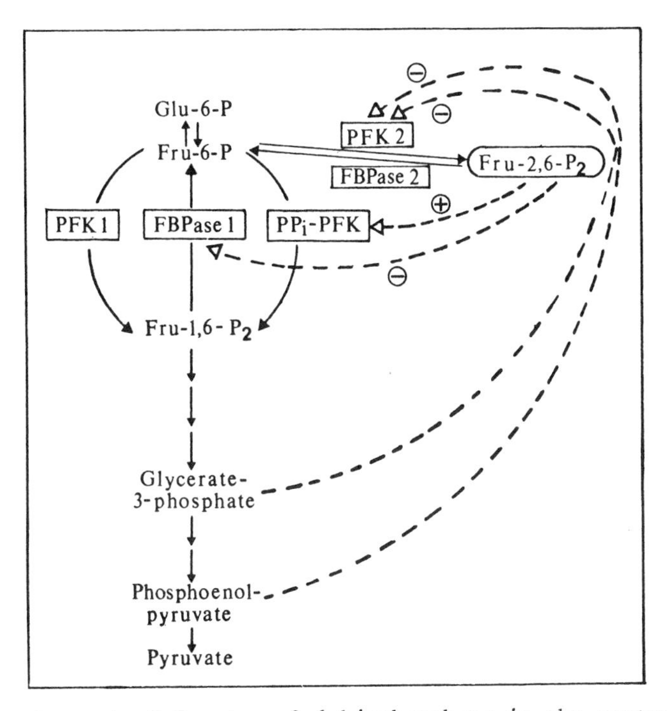


Fig. 9. Involvement of fructose 2,6-bisphosphate in the control of glycolysis.

In oat seeds thus Fru-2,6-$P_2$ always reaches transiently high values in conditions that allow germination to occur. This increase in Fru-2,6-$P_2$ content probably induces a stimulation of glycolysis, but it may also play another, still unknown, function in metabolism. Whatever that may be, it is the earliest metabolic event described in the literature that permits us to distinguish a seed which will or will not germinate.

## ETHANOL METABOLISM

Many organic compounds stimulate germination of numerous seeds (Taylorson and Hendricks, 1979, 1980/81; Bewley and Black, 1982). In particular, ethanol has such an effect with seeds of lettuce (Pecket and Al-Charchafchi, 1978), barley (Le Deunff, 1983), Avena fatua (Adkins et al., 1984 a), Avena sativa (Larondelle et al., 1987; Lecat, 1987) and other species. Taylorson and Hendricks (1979, 1980/81) assume that ethanol may act at the membrane

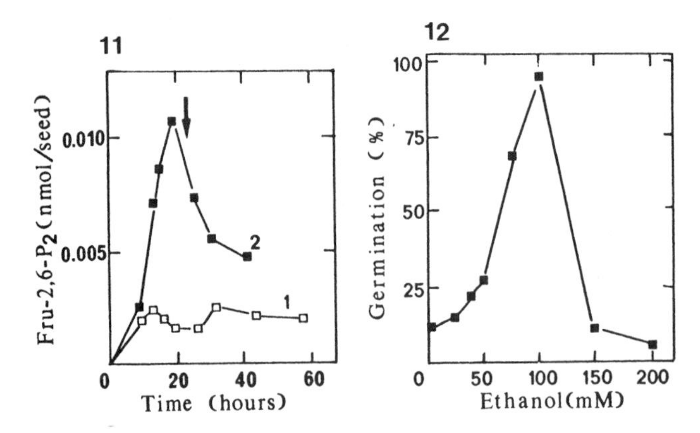


Fig. 10. Changes in Fru-2,6-$P_2$ content of dormant (1) and non-dormant (2) oat (cv. Moyencourt) seeds during their imbibition at 30°C. The arrow indicates when the radicle of the non-dormant seeds began to elongate. From Larondelle et al. (1987), slightly modified.

Fig. 11. Effect of ethanol concentration on the germination percentage obtained after 7 days at 30°C with naked dormant Avena sativa seeds. The seeds were placed in the presence of ethanol from the start of imbibition. From Corbineau and Gouble (unpublished results).

level through its properties as an anesthetic. However, it may also be involved metabolically in the stimulation of germination as respiratory substrate (Fidler, 1968), since it increases oxygen uptake by wild oat seeds (Adkins et al., 1984 b) and Fru-2,6-$P_2$ content in cultivated oat seeds (Larondelle et al., 1987).

Very recent results obtained by Corbineau and Gouble (unpublished data) with dormant Avena sativa seeds have demonstrated that the stimulatory effect of ethanol requires its metabolism through alcohol dehydrogenase (ADH). These results can be summarized as follows. At relatively low concentrations (50-100 mM), ethanol strongly stimulates germination when given continuously (Fig.11), and short treatments with higher concentrations are also very effective (Fig.12). The effect of other alcohols depends on $K_m$ values for their oxidation by ADH (Table 4). Good ADH substrates (ethanol,

Table 4. $K_m$ values for different alcohols as substrates of alcohol dehydrogenase. From Corbineau and Gouble (unpublished results).

| Alcohols | $K_m$ (mM) |
|---|---|
| Ethanol | 9.6 |
| Butanol-1 | 17.2 |
| 2-propen-1-ol | 3.6 |
| Propanol-1 | 30.3 |
| Propanol-2 | non oxidizable |
| Methanol | non oxidizable |

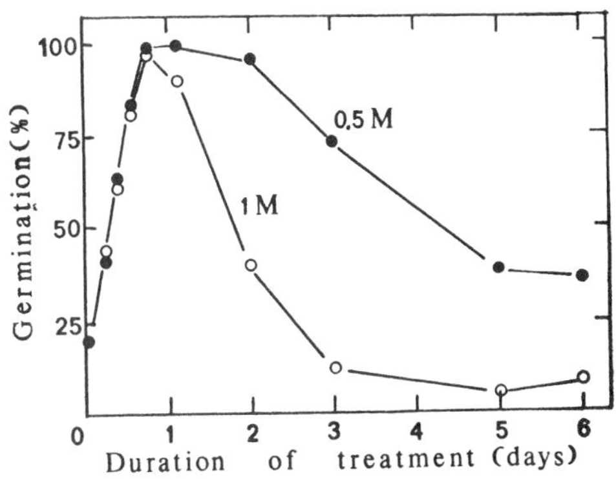


Fig. 12. Effect of duration of treatment at 30°C with 0.5 M and 1 M ethanol on the germination percentage obtained 7 days after transfer onto water at the same temperature, with naked dormant *Avena sativa* seeds. From Corbineau and Gouble (unpublished results).

Table 5. Germination, oxygen uptake and $CO_2/O_2$ ratio in naked dormant *Avena sativa* seeds pretreated with various alcohols or with ethanol + 4-methylpyrazole. All treatments were at 30°C. Germination was determined after a subsequent 7 days on water . Oxygen uptake and $CO_2/O_2$ ratio were measured after 24 hours. Pretreatment with alcohols was for 16 hours except in the case of 2-propen-1-ol (0.5 hour). From Corbineau and Gouble (unpublished results).

| Preincubation medium | Germination (%) | $O_2$ uptake ($\mu$l $h^{-1}$) | $CO_2/O_2$ |
|---|---|---|---|
| Water (control) | 12 | 55.5 ± 4.9 | 0.90 |
| Ethanol (200 mM) | 84 | 80.6 ± 5.1 | 0.76 |
| Butanol-1 (100 mM) | 92 | 78.0 ± 7.1 | - |
| 2-propen-1-ol (50 mM) | 73 | - | - |
| Propanol-1 (50 mM) | 99 | 79.0 ± 8.5 | - |
| Propanol-2 (100 mM) | 32 | 56.3 ± 2.3 | - |
| Methanol (100 mM) | 30 | 53.5 ± 0.7 | - |
| Ethanol (200 mM) + 4-MP (50 mM) | 24 | 50.3 ± 2.2 | - |

butanol-1, 2-propen-1-ol, propanol-1) improve germination, whereas alcohols which cannot be oxidized by ADH (propanol-2, methanol) have practically no effect on germination (Table 5). Moreover, the action of ethanol is abolished in the presence of 4-methylpyrazol (4-MP) which is an inhibitor of ADH (Table 5). The stimulation of germination by ethanol and other alcohols is associated with an increase in oxygen uptake and a decrease in $CO_2/O_2$ ratio (Table 5). These results, and those of Larondelle et al. (1987) concerning the increase in Fru-2,6-$P_2$ induced by ethanol, suggest that alcohols which are substrates of ADH stimulate germination through an activation of glycolysis and Krebs cycle.

## CONCLUSION

The search for metabolic events that could characterize seed germination or dormancy is a very difficult problem. In order that such events may be of real biological meaning they must be detected before radicle protrusion; otherwise they concern the growth of the young seedling but not the true germination process (germination *sensu stricto*). In fact, many studies have shown that noticeable metabolic differences between a seed which germinates and a seed which does not germinate can only be observed after the start of radicle elongation.

The most recent investigations in this field attempt to explain why some substances (such as respiratory inhibitors or ethanol) or some external factors (for example temperature) likely to modify seed metabolism are capable of stimulating germination or of breaking dormancy. These investigations are still fragmentary and the results obtained are difficult to interpret and to relate to each other. Nevertheless, they suggest that some processes involved in the regulation of essential metabolic pathways, such as the pentose phosphate pathway, glycolysis or Krebs cycle, could provoke germination or break dormancy. At the present time, the most remarkable event that has been shown to be distinctive of a seed which will germinate is a large increase in Fru-2,6-$P_2$ during imbibition. This finding is without any doubt worth further attention.

## REFERENCES


Adkins, S.W., Naylor, J.M., and Simpson, G.M., 1984a, The physiological basis of seed dormancy in *Avena fatua* . V. Action of ethanol and other organic compounds, *Physiol. Plant*., 62:18.

Adkins, S.W., Simpson, G.M., and Naylor, J.M., 1984b, The physiological basis of seed dormancy in *Avena fatua*. VI. Respiration and the stimulation of germination by ethanol, *Physiol. Plant*., 62:148.

Apte, S.K., Rowell, P., and Stewart, W.D.P., 1978, Electron donation to ferredoxin in heterocysts of the $N_2$-fixing alga *Anabaena cylindrica*, *Proc. Roy. Soc. London*, B200:1.

Ashihara, H., and Komamine, A., 1974, Regulation of the activities of some enzymes of the pentose phosphate pathway in *Phaseolus mungo*, *Z. Pflanzenphysiol*., 74:130.

Ashihara, H. and Komamine, A., 1976, Characterization of regulatory properties of glucose-6-phosphate dehydrogenase from black gram (*Phaseolus mungo*), *Physiol. Plant*., 36:52.

Bewley, J.D., and Black, M., 1982, "Physiology and biochemistry of seeds in relation to germination. Tome 2, Viability, dormancy and environmental control", Springer-Verlag, Berlin, Heidelberg, New York.

Carnal, N.W., and Black, C.C., 1979, Pyrophosphate-dependent 6-phosphofructokinase, a new glycolytic enzyme in pineapple leaves, *Biochem. Biophys. Res. Commun*., 86:20.

Carnal, N.W., and Black, C.C., 1983, Phosphofructokinase activities in

photosynthetic organisms. The occurence of pyrophosphate-dependent 6-phosphofructokinase in plants and algae, Plant Physiol., 71:150.
Côme, D., 1980/81, Problems of embryonal dormancy as exemplified by apple embryo, Isr. J. Bot., 29:145.
Côme, D., 1982, Germination, in:"Croissance et développement. Physiologie végétale II", P. Mazliak, ed., Hermann, Paris.
Côme, D., 1987, Germination et dormances des semences, in:"Le développement des végétaux. Aspects théoriques et synthétiques", H. Le Guyader, ed., Masson, Paris.
Côme, D., and Corbineau, F., 1984, La dormance des semences des céréales et son élimination. I. Principales caractéristiques, C. R. Acad. Agric. France, 70:141.
Côme, D., Corbineau, F., and Lecat, S., 1988, Some aspects of metabolic regulation of cereal seed germination and dormancy, Seed Sci. & Technol., 16:175.
Côme, D., Lenoir, C., and Corbineau, F., 1984, La dormance des céréales et son élimination, Seed Sci. & Technol., 12:629.
Côme, D., Perino, C., and Ralambosoa, J., 1985, Oxygen sensitivity of apple (Pyrus malus L.) embryos in relation to dormancy, Isr. J. Bot., 34:17.
Côme, D., and Thévenot, C., 1982, Environmental control of embryo dormancy and germination, in:"The physiology and biochemistry of seed development, dormancy and germination", A.A. Khan, ed., Elsevier Biomedical Press, Amsterdam, New York, Oxford.
Corbineau, F., Carmignac, D.F., Gahan, P., and Maple, A.J., 1989, Glycolytic activity in embryos of Pisum sativum and of non-dormant or dormant seeds of Avena sativa L. expressed through activities of PFK and $PP_i$-PFK, Histochemistry, 90:359.
Corbineau, F., and Côme, D., 1980, Quelques caractéristiques de la dormance du caryopse d'orge (Hordeum vulgare L., variété Sonja). C. R. Acad. Sc. Paris, D280:547.
Corbineau, F., and Côme, D., 1980/81, Some particularities of the germination of Oldenlandia corymbosa L. seeds (tropical Rubiaceae), Isr. J. Bot., 29:157.
Corbineau, F., and Côme, D., 1981, La dormance des céréales, qu'est-ce que c'est ? Cultivar, 142:15.
Corbineau, F., and Côme, D., 1987, Régulation de la germination des semences de tournesol par l'éthylène, in:"2 ème colloque sur les substances de croissance et leurs utilisations en agriculture", ANPP, Paris.
Corbineau, F., Lecat, S., and Côme, D., 1986, Dormancy of three cultivars of oat seeds (Avena sativa L.), Seed Sci. & Technol., 14:725.
Corbineau, F., Lenoir, C., Lecat, S., and Côme, D., 1984, La dormance des semences des céréales et son élimination. II. Mécanismes mis en jeu, C. R. Acad. Agric. France, 70:148.
Corbineau, F., Sanchez, A., Chaussat, R., and Côme, D., 1981, La dormance du caryopse de blé (Triticum aestivum L., var. Champlein) en relation avec la température et l'oxygène, C. R. Acad. Agric. France, 9:826.
Cséke, C., Balogh, A., Wong, J.H., and Buchanan, B.B., 1985, Pyrophosphate fructose-6-phosphate phosphotransferase (PFP) : an enzyme relating fructose-2,6-bisphosphate to the control of glycolysis and gluconeogenesis in plants, Physiol. Vég., 23:247.
Dennis, D.T., and Greyson, M.F., 1987, Fructose 6-phosphate metabolism in plants, Physiol. Plant., 69:395.
Durand, M., 1974, Influence de quelques régulateurs de croissance sur la germination et la dormance de l'embryon de pommier (Pirus malus L.), Thesis 3 ème cycle, University Pierre et Marie Curie, Paris.
Dziewanowska, K., and Lewak, S., 1982, Hydrogen cyanide and cyanogenic compounds in seeds. IV. Metabolism of hydrogen cyanide in apple seeds under conditions of stratification, Physiol. Vég., 20:165.
Esashi, Y., Ohhara, Y., Okazaki, M., and Hishinuma, K., 1979, Control of cocklebur seed germination by nitrogenous compounds : nitrite,

nitrate, hydroxylamine, thiourea, azide and cyanide, Plant Cell Physiol., 20:344.
Evenari, M., 1957, Les problèmes physiologiques de la germination, Bull. Soc. Franç. Physiol. Vég., 3:105.
Everson, R.G., and Rowan, K., 1965, Phosphate metabolism and induced respiration in washed carrot slices, Plant Physiol., 40:1247.
Fidler, J.C., 1968, The metabolism of acetaldehyde in plant tissues, J. Exp. Bot., 58:41.
Grossman, A., and McGowan, R.E., 1975, Regulation of glucose-6-phosphate dehydrogenase in blue-green algae, Plant Physiol., 55:658.
Hers, H.-G., 1984, The discovery and the biological role of fructose 2,6-bisphosphate, Biochem. Soc. Trans., 12:729.
Kruger, N.J., Kombrink, E., and Beevers, H., 1983, Pyrophosphate fructose 6-phosphate phosphotransferase in germinating castor bean seedlings, FEBS Lett., 153:409.
Larondelle, Y., Corbineau, F., Dethier, M., Côme, D., and Hers, H.-G., 1987, Fructose 2,6-bisphosphate in germinating oat seeds. A biochemical study of seed dormancy, Eur. J. Biochem., 166:605.
Lecat, S., 1987, Quelques aspects métaboliques de la dormance des semences d'avoine (Avena sativa L.). Etude plus particulière de l'action des glumelles, Thesis, University Pierre et Marie Curie, Paris.
Le Deunff, Y., 1983, Mise en évidence de l'influence bénéfique de l'alcool éthylique en solution aqueuse sur la levée de dormance des orges, C. R. Acad. Sc. Paris, III 296:433.
Lendzian, K., and Bassham, J.A., 1975, Regulation of glucose-6-phosphate dehydrogenase in spinach chloroplasts by ribulose 1,5-diphosphate and $NADPH/NADP^+$ ratios, Biochim. Biophys. Acta, 396:260.
Miller, J.M., and Conn, E.E., 1980, Metabolism of hydrogen cyanide by higher plants, Plant Physiol., 65:1199.
Muto, S., and Uritani, I., 1972, Inhibition of sweet potato glucose-6-phosphate dehydrogenase by $NADPH_2$ and ATP, Plant Cell Physiol., 13:377.
Pecket, R.C., and Al-Charchafchi, F., 1978, Dormancy in light-sensitive lettuce seeds, J. Exp. Bot., 29:167.
Pelroy, R.A., Kirk, M.R., and Bassham, J.A., 1976, Photosystem II regulation of macromolecule synthesis in the blue-green alga Aphanocapsa 6714, J. Bacteriol., 128:623.
Perino, C., 1987, Etude physiologique et métabolique des phases de la germination de l'embryon de pommier, Thesis Doct. Sci. Nat., University Pierre et Marie Curie, Paris.
Perino, C., and Côme, D., 1981, Influence du cyanure de potassium sur la germination de l'embryon de pommier (Pirus malus L.) non dormant, Physiol. Vég., 19:219.
Perino, C., Simond-Côte, E., and Côme, D., 1984, Effets du cyanure de potassium et de l'acide salicylhydroxamique sur la levée de dormance et l'activité respiratoire des embryons de pommier, C. R. Acad. Sc. Paris, III 299:249.
Roberts, E.H., 1969, Seed dormancy and oxidation processes, Symp. Soc. Exp. Biol., 23:161.
Roberts, E.H., 1973, Oxidative processes and the control of seed germination, in:"Seed Ecology", W. Heydecker, ed., Butterworths, London.
Roberts, E.H., and Smith, R.D., 1977, Dormancy and the pentose phosphate pathway, in: "The physiology and biochemistry of seed dormancy and germination", A.A. Khan, ed., Elsevier North-Holland Biomedical Press, Amsterdam.
Rollin, P., 1975, Influence de quelques inhibiteurs sur la respiration et la germination des akènes de Bidens radiata, Physiol. Vég., 13:369.
Sabularse, D.C., and Anderson, R.L., 1981, D-fructose-2,6-bisphosphate : a naturally occurring activator for inorganic pyrophosphate:D-fructose-6-phosphate 1-phosphotransferase, Biochem. Biophys. Res. Commun., 103:848.

Simmonds, J.A., and Simpson, G.M., 1972, Regulation of the Krebs cycle and pentose phosphate activities in the control of dormancy of Avena fatua, Can. J. Bot., 50:1041.

Smyth, D.A., and Black, C.C., 1984, The discovery of a new pathway of glycolysis in plants, What's new Plant Physiol., 15:13.

Speer, H.L., 1974, Activity of glucose-6-phosphate dehydrogenase from lettuce seeds (Lactuca sativa), Can. J. Bot., 52:2225.

Taylorson, R.B., and Hendrick, S.B., 1973, Promotion of seed germination by cyanide, Plant Physiol., 52:23.

Taylorson, R.B., and Hendricks, S.B., 1979, Overcoming dormancy in seeds with ethanol and other anesthetics, Planta, 145:507.

Taylorson, R.B., and Hendricks, S.B., 1980/81, Anesthetic release of seed dormancy. An overview, Isr. J. Bot., 29:273.

Tissaoui, T., 1975, La dormance et la germination de l'embryon de pommier (Pirus Malus L.) en relation avec l'oxygène, la température et le métabolisme des nucléotides adényliques, Thesis Doct. Sci. Nat., University Pierre et Marie Curie, Paris.

Tissaoui, T., and Côme, D., 1975, Mise en évidence de trois phases physiologiques différentes au cours de la "germination" de l'embryon de pommier non dormant, grâce à la mesure de l'activité respiratoire, Physiol. Vég., 13:95.

Turner, J.P., and Turner, D.H., 1980, The regulation of glycolysis and the pentose phosphate pathway, in: " The biochemistry of plants, vol. 2", P.K. Stumpf and E.E. Conn, eds, Academic Press, New York.

Upadhyaya, M.K., Naylor, J.M., and Simpson G.M. , 1982, The physiological basis of seed dormancy in Avena fatua. I. Action of the respiratory inhibitors, sodium azide and salicylhydroxamic acid, Physiol. Plant., 54:419.

Upadhyaya, M.K., Naylor, J.M., and Simpson, G.M., 1983, The physiological basis of seed dormancy in Avena fatua. II. On the involvement of alternative respiration in the stimulation of germination by sodium azide, Physiol. Plant., 58:119.

Van Laere, A., Van Schaftingen, E., and Hers, H.-G., 1983, Fructose 2,6-bisphosphate and germination of fungal spores, Proc. Natl. Acad. Sci. USA, 80:6601.

Van Schaftingen, E., 1986, Fructose 2,6- bisphophate, Adv. Enzymol. Relat. Areas Mol. Biol., 59:315.

Van Schaftingen, E., Jett, M.-F., Hue, L., and Hers, H.-G., 1981, Control of liver 6-phosphofructokinase by fructose 2,6-bisphosphate and other effectors, Proc. Natl. Acad. Sci. USA, 78: 3483.

Van Schaftingen, E., Lederer, B., Bartrons, R., and Hers, H.-G., 1982, A kinetic study of pyrophosphate:fructose-6-phosphate phosphotransferase from potato tuber. Application to a microassay of fructose 2,6-bisphosphate, Eur. J. Biochem., 129:191.

Vidal, J., Jacquot, J.P. and Gadal, P., 1980, Light activation of NADP malate dehydrogenase in a reconstituted chloroplastic system, Phytochemistry, 19:1919.

Wyzinska, D., Perino, C., and Côme, D., 1981, Influence du sulfure de sodium sur la germination de l'embryon de pommier non dormant, C. R. Acad. Sc. Paris, III 292:1029.

Xhaufflaire, A., 1968, Effet du tampon phosphate sur la germination et la croissance radiculaire des plantules de Lens culinaris traitées ou non par différentes phytohormones, Mededelingen Rijksfakulteit Landbouwwetenschappen Gent, 33:1347.

Yu, K.S., Michell, C.A., Yentur, S., and Robitaille, H.A., 1979, Cyanide insensitive, salicylhydroxamic acid-sensitive processes in potentiation of light-requiring lettuce seeds, Plant Physiol., 63:121.

# PHYTOCHROME AND SENSITIZATION IN GERMINATION CONTROL

William J. VanDerWoude

Plant Photobiology Laboratory
Agricultural Research Service, USDA
Beltsville, Maryland 20705, U.S.A.

## INTRODUCTION

The germination behavior of light-dependent seeds in response to temperature change, growth hormones, and chemical regulators displays broad and complex light requirements. Such treatments, as well as the varying soil temperature of the natural environment, induce high levels of germination in darkness, or very high light sensitivity for promotion of germination. One aspect of such behavior has been the absence of clear correlations between germination response and phototransformation of the photoreceptor, phytochrome. Interpretation of behavior in terms of underlying mechanisms has therefore been difficult. Research progress during the past decade in several laboratories now indicates the involvement of sensitization of the phototransduction system to very low levels of the active, far-red absorbing form of phytochrome, $P_{fr}$. Several lines of evidence now demonstrate phytochrome to be a molecular dimer. Studies of sensitization have led to the development and support of a dimeric model for the molecular action of phytochrome. Although additional direct evidence is yet required, these and related studies indicate a role of increased membrane fluidity in sensitization. This report summarizes progress and discusses some of the current questions in this research area.

## SENSITIZATION INDUCED BY LOW TEMPERATURE

This report will focus primarily on our studies of sensitization in positively photoblastic seeds (achenes) of Grand Rapids lettuce (*Lactuca sativa* L.). Treatment with alternating temperatures or low temperature preincubation has long been known to increase the germination in darkness of light-requiring seeds (Stokes, 1965). Our initial studies (VanDerWoude and Toole, 1980) showed that several hours of low temperature (4-16°C), dark incubation of imbibed lettuce seeds increased germination at 20°C. To more clearly define the role of phytochrome in this response all subsequent studies were conducted with seeds depleted of $P_{fr}$. Such depletion may be accomplished by giving imbibed seeds a short far-red (FR) irradiation and a 24 h dark incubation at 20°C. This removes $P_{fr}$ by processes of dark destruction and/or reversion. No "safe" lights may be used after this first FR exposure since they may establish $P_{fr}$ levels sufficient to promote germination in sensitized seeds. After 24 h of subsequent "prechilling" at 4°C, dark germination remains low but brief FR irradiation fully promotes

*Recent Advances in the Development and Germination of Seeds*
Edited by R.B. Taylorson
Plenum Press, New York

the germination of such seeds. Prechilling therefore increases germination sensitivity to the low, usually inhibitory or nonpromotive, $P_{fr}$ levels established by FR.

Sensitization increases with lowered prechilling temperature to a maximum response near 4°C. It is linearly dependent upon the duration of prechilling at 4°C up to a near maximal response at 10 h. Sensitization decays in a converse manner when seeds are returned to 20°C after 10 h at 4°C. The kinetics for the low temperature promotion and subsequent decay of sensitivity to FR are similar to those found for thermally induced changes in membrane lipid composition in plants (Quinn, 1988). The kinetics therefore suggest the induction of increased membrane fluidity by prechilling as follows. The immediate effect of reduced temperature is a decrease in membrane fluidity. A return to initial fluidity levels occurs during low temperature incubation. Return to 20°C then immediately results in high membrane fluidity that subsequently declines. Such "homeoviscous adaptation" of membranes to temperature change has been observed and studied in many organisms (Thompson, 1983). Although direct information on membrane behavior in seeds is yet required, the observed kinetics for prechilling-induced sensitization suggest a direct relationship between increased sensitivity and increased membrane fluidity.

Studies of the influences of alcohols and other anesthetics (Taylorson and Hendricks, 1979; Chadoeuf-Hannel, 1985) on the promotion of phytochrome -dependent seed germination support the importance of membrane function in this process. The importance to biological function of the lateral motion of many membrane proteins is now recognized (Axelrod, 1983).

Prechilling-induction of sensitization does not involve increases in the rate of potentiation of germination by $P_{fr}$. When the germination of sensitized and control seeds is induced by brief red (R) light, followed several hours later by FR, the observed rates of " $P_{fr}$ action" are very similar. This behavior suggests that prechilling modifies the control of potentiation by phytochrome rather than metabolic, biosynthetic, or other processes of potentiation. This is in contrast to the apparent modification of the processes of potentiation by prolonged dark incubation (Duke et al., 1977).

## FLUENCE-RESPONSE STUDIES

Detailed examination of the fluence-response behavior of sensitized seeds has contributed greatly to an understanding of the sensitization mechanism. Such behavior of prechilled lettuce seeds (VanDerWoude, 1985) is distinctly biphasic. Germination is promoted by a range of Very Low Fluences (VLF), that for 660 nm light are in the range of $2 \times 10^{-9}$ to $5 \times 10^{-7}$ mol $m^{-2}$. The magnitude of VLF Responses (VLFR) increases with sensitization by prechilling at 4°C, up to a maximum of 95% germination after 24 h. Unchilled controls display almost no VLFR. Low Fluences (LF) in the range of $5 \times 10^{-5}$ to $3 \times 10^{-4}$ mol $m^{-2}$ promote LF Responses (LFR) above the maximum VLFR. The LFR is responsible for nearly all germination in unchilled controls.

The VLFR and LFR do not represent separate seed populations. The seed population is considered as having an approximately normal distribution of threshold requirements for potentiation of germination. To the degree that a given treatment induces sensitization, all seeds in the population are sensitized. The population of seeds that display the VLFR are those in which the potentiation requirement is satisfied by the VLFR. Mechanisms underlying both the VLFR and the LFR appear to be operating in all seeds in the population. Analysis of the LFR behavior of sensitized seeds indicates responses to be the additive result of activities underlying both the VLFR and the LFR.

Between the VLF and LF ranges exists a response plateau equal to the maximum VLFR. For half-maximal responses, a 24 h, 4°C incubation increases sensitivity to 660 nm light about 10,000-fold relative to unchilled seeds. Irradiation with 730 nm light, given immediately after a saturating 660 nm irradiation, reduces responses in sensitized seeds to levels near that of the maximum VLFR, but not below. Such photoreversal requires the same range of fluences needed in nonsensitized seeds to fully reverse promotion by R. Promotion of the VLFR in sensitized seeds by 730 nm irradiation requires fluences about 50-fold higher than for 660 nm, in good accordance with the difference in extinction coefficients of red-absorbing form of phytochrome, $P_r$, at these two wavelengths.

Other pretreatments also sensitize lettuce seed germination and induce the VLFR. Brief incubation at high temperature (1 h at 28°C) or for a few hours on 1% ethanol results in biphasic fluence-response behavior very similar to that of prechilled seeds. The kinetics for the promotion of sensitivity by 1% ethanol and the subsequent decay of sensitivity are very similar to those for prechilling of seeds (VanDerWoude, unpublished). Ethanol-induced sensitization increases germination only if it is used as a pretreatment. If ethanol is allowed to remain present it inhibits the late, growth phases of the germination process. For all sensitizing treatments the VLFR is promoted over the same general range of fluences.

Several additional sensitizing factors are now known to promote the VLFR and biphasic fluence-response behavior in seed germination. Included are: prolonged high temperature (Small et al., 1979; Cone et al., 1985), gibberellic acid (Rethy et al., 1987; DeGreef et al., this volume), and the pyridazinone-type herbicide, SAN 9789 (Widell et al., 1981). The similar behaviors resulting from these varied treatments suggest the operation of a common underlying mechanism of sensitization.

Seedling growth displays similar biphasic fluence-responses, even in the absence of a sensitizing treatment. Examples include: coleoptile growth in etiolated barley seedlings (Uematsu et al., 1981), and etiolated oat seedlings, where coleoptile growth is promoted while that of the mesocotyl is inhibited (Mandoli and Briggs, 1981). Shinkle and Briggs (1984; 1985) very thoroughly characterized an auxin-induced increase in light sensitivity of phytochrome-mediated growth responses in oat coleoptile sections. Similar to the influence of gibberellic acid on seeds, indole-3-acetic acid establishes a growth response to very low fluences of red light. Establishment of a pH gradient across the plasma membrane by incubation in low pH buffer, or by use of the drug, fusicoccin (which stimulates proton efflux from cells), can substitute for auxin in producing the enhanced sensitivity to very low $P_{fr}$ levels.

## PHYTOCHROME IS A DIMERIC MOLECULE

The VLF range of fluences establish immeasurably low $P_{fr}$ levels, below 0.1% of total phytochrome, as compared to the LF which establishes proportions of $P_{fr}$ greater than 10%. It has long been recognized that many phytochrome-mediated responses do not correspond with photometrically estimated or spectrophotometrically measured levels of $P_{fr}$ (Hillman, 1967; Smith, 1983). One source of difficulty in explaining such response behaviors has been the classical concept that phytochrome exists and acts as a monomer. Biochemical evidence now shows phytochrome to be a dimer and indicates its monomers to be identical, each having one chromophore (Jones and Quail, 1986). Evidence from electron microscopic (Jones and Erickson,

1989) and small-angle X-ray scattering studies (Tokutomi et al., 1989) also demonstrate the dimeric structure of phytochrome. The dimer structure results from association of monomers at a region of the molecule distal to the chromophore.

Studies of light-induced phytochrome pelletability by Pratt and Marmé (1976) provide evidence for the existence dimeric phytochrome *in vivo*. Red fluences that produce low $P_{fr}$ levels induce the subsequent pelletability of nearly equal amounts of $P_r$ and $P_{fr}$, a behavior consistent with the presence of two chromophores per molecule, only one of which need be in the $P_{fr}$ form to elicit pelletability. A similar relationship was demonstrated by Napier and Smith (1987) for the binding of phytochrome dimers to membranes; only one-half of the dimer needs to be in the $P_{fr}$ form for binding to occur. Brockmann et al. (1987) provided additional evidence for the existence *in vivo* of dimeric phytochrome, including measurements indicating that the $P_r$ half of a partially phototransformed dimer is subject to destruction, just as is $P_{fr}$.

## THE DIMERIC MECHANISM OF PHYTOCHROME ACTION

The finding that phytochrome exists as a molecular dimer has provided new opportunities for understanding sensitization and the phytochrome-related mechanisms that underlie biphasic fluence-response behavior. As a dimer, phytochrome can exist in three interphototransformable species; $P_r$:$P_r$, $P_r$:$P_{fr}$, and $P_{fr}$:$P_{fr}$. Based on these species and their proposed specific association with a receptor, X, a dimeric mechanism of phytochrome action (Fig. 1.) was proposed and analyzed by numerical simulation (VanDer-Woude, 1985). Simulations of the mechanism, using established values for the photochemical properties of phytochrome, very closely parallel the observed fluence-response behavior of sensitized seeds, as well as such behavior of other phytochrome-mediated responses. Such simulations have provided physiological evidence for the *in vivo* photomorphogenic action of dimeric phytochrome and support the validity of the dimeric model. Several aspects of the dimeric mechanism and its relationship to the VLFR and LFR are are discussed below:

1. Phytochrome exists as a molecular dimer, having one chromophore on each of two identical monomers. Pending additional information, existing simulations of the dimeric mechanism assume that the reversible phototransformations of each monomer occur independently and have little influence on extinction coefficients and quantum efficiencies for the photoconversion of its partner.

2. The three interphototransformable species of dimeric phytochrome have the following activity relationships:
   $P_r$:$P_r$, photomorphogenically inactive
   $P_r$:$P_{fr}$, variably active, dependent on degree of sensitization
   $P_{fr}$:$P_{fr}$, photomorphogenically active, independent of sensitization.

3. The dimer species $P_r$:$P_{fr}$ and $P_{fr}$:$P_{fr}$ act in association with a specific, photomorphogenically active receptor, X. The simulations indicate X to be present in very low abundance (about 1/1000 the cellular concentration of phytochrome in dark-grown tissues, on the order of 1,000 receptors per cell).

4. In dark-incubated tissues the abundance of phytochrome (on the order of 1,000,000 dimers per cell) serves as a large "antenna" for the very sensitive detection of even only a few photons per cell. Very low photon fluences (VLF) establish very low levels of $P_r$:$P_{fr}$ (less than 0.1% of total dimer), and very little or no $P_{fr}$:$P_{fr}$ because of a very

low statistical probability for the phototransformation of both monomers. Fluences for the threshold promotion of the VLFR are calculated to establish as little as 0.003% of phytochrome as $P_{fr}$ (De Petter et al., 1988). However, such very low levels of $P_r$:$P_{fr}$ are adequate to occupy and activate a portion of the very low population of receptors. The maximum of the VLF portion of a biphasic fluence-response curve represents the saturation of X with $P_r$:$P_{fr}$. Sensitization does not appear to change the fluence requirement for maximum $P_r$:$P_{fr}$-X formation, indicating that the abundance of X and dimer are not changed. There are VLFR behaviors, such as in the germination of thermodormant lettuce seeds (Small et al., 1979), in which FR at VLF levels promotes germination but increased FR inhibits and R at LF is subsequently required to promote. This may be explained by phototransformations that form inactive but stable $P_{fr}$:$P_r$-X and $P_r$:$P_r$-X complexes and reduce levels of $P_r$:$P_{fr}$-X. When such VLFR behavior exists, sensitization may be examined by using defined VLF of R or FR that maximize the VLFR, but not FR fluences that approach phytochrome photoequilibria.

5. Sensitizing treatments, by temperature, hormones, anesthetics, and other chemicals, increase the activity of $P_r$:$P_{fr}$-X dimer-receptor complex and increase the magnitude of VLF responses. Using the example of sensitization by gibberellic acid (Rethy et al., 1987), it now appears that this hormone does not substitute for $P_{fr}$ in promoting germination since at least a very low level of $P_{fr}$ is required. Rather, gibberellic acid appears to sensitize the phytochrome phototransduction mechanism and increase the activity of $P_r$:$P_{fr}$-X.

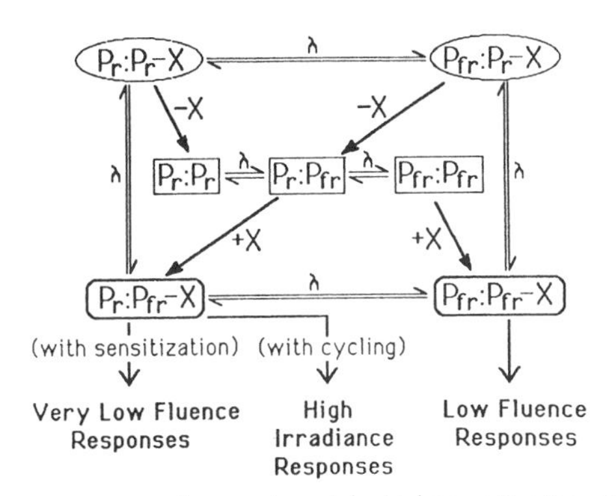


Fig. 1. Primary components and activities of the dimeric mechanism of phytochrome action. The activity of $P_r$:$P_{fr}$-X increases with sensitization or its photochemical cycling.

6. As discussed above, the nature and response kinetics of sensitizing treatments suggest that the mechanism of sensitization involves changes in membrane properties (possibly allowing the increased mobility of membrane components) which alter the behavior and increase the activity of $P_r$:$P_{fr}$-X.

7. The formation of $P_{fr}$:$P_{fr}$ and the establishment of the phytochrome-receptor complex, $P_{fr}$:$P_{fr}$-X, require low fluences (LF). The activity of $P_{fr}$:$P_{fr}$-X is not influenced by sensitizing treatments.

8. Comparisons of observed responses and simulations indicate that the VLFR and the LFR differ greatly with respect to the formation of dimer-X complexes. The VLFR is related logarithmically to the formation of $P_r$:$P_{fr}$-X (increasing levels give reduced increases in response). The VLFR is dependent on the abundance of total phytochrome. In contrast, the LFR is related nearly linearly to the formation of $P_{fr}$:$P_{fr}$-X, most of which occurs by the phototransformation of $P_r$:$P_{fr}$-X. The LFR is therefore relatively independent of the abundance of phytochrome, but is correlated with the proportion of phytochrome in the $P_{fr}$ form. The differences in the apparent behavior of $P_r$:$P_{fr}$-X and $P_{fr}$:$P_{fr}$-X fit a mechanism in which $P_{fr}$:$P_{fr}$-X activates and remains bound to an additional membrane component, Y, while $P_r$:$P_{fr}$-X remains free and may keep many Y active in an appropriate (high fluidity?) membrane environment. Evidence is increasing for the function of a phosphatidylinositol-mediated signal transduction mechanism in plants (Memon et al., 1989) and in the action of phytochrome (Morse et al., 1987). Mobility of the membrane components of such mechanisms in other organisms (Berridge, 1987) is an essential factor in the operation of the signal amplification cascade.

9. At fluences above those necessary to maximize VLF responses ($5 \times 10^{-7}$ mol $m^{-2}$), the availability of $P_r$:$P_{fr}$ is adequate to fully or substantially fill X. Even buried seeds may be exposed to such VLF (Bliss and Smith, 1985). In the natural environment therefore, sensitization and modulation of the activity of $P_r$:$P_{fr}$-X appears to be an important control point in the regulation of germination, as well as in the regulation of plant growth and development, by both endogenous and exogenous factors. Even the few seconds of irradiation received by buried seeds during mechanical disturbance of the soil in normal agricultural practice may promote high germination in weed seed populations. As an example, Hartmann and Nezadal (1989) found that when all soil disturbing operations were conducted at night (between 1 h after sunset and 1 h before sunrise, and without equipment lighting), weed coverage in the absence of herbicide use was reduced from 80% to 2%.

## FUNCTION OF THE DIMERIC MECHANISM IN HIGH IRRADIANCE RESPONSES

Continuous daily irradiation, as in the natural environment, promotes phytochrome-mediated behavior termed the "High Irradiance Response" (HIR). The HIR is exhibited by many phytochrome-mediated responses (Mancinelli and Rabino, 1978) including seed germination, where prolonged irradiation appears to inhibit the late, growth phase of germination, even after seeds are no longer inhibited by brief FR irradiation (Bartley and Frankland, 1982; Frankland, 1986). Simulation of the dimeric mechanism for 12 h of continuous irradiation leads to calculated action spectra very similar to those observed for the HIR (VanDerWoude, 1987). The action maximum in the red region appears due to the action of $P_{fr}$:$P_{fr}$-X. A red action maximum is not generally observed in the HIR inhibition of germination. This may be explained by much greater rates of red light-induced reduction of the levels of "etiolated", Type I phytochrome in seeds, than of the "green", Type II phytochrome that appears on seedling development in the light. Action maxima in the blue and in the far-red near 710 nm appear related to the presence and photochemically-mediated turnover rate, or "cycling", of $P_r$:$P_{fr}$-X. Frankland demonstrated a corresponding empirically-derived relationship showing dependence of the HIR in germination on an optimal balance of $P_{fr}$ abundance and $P_{fr}$ cycling. The dimeric mechanism specifies the major importance of cycling in the HIR to be in the turnover of $P_r$:$P_{fr}$-X.

The dependence of the HIR in the blue and far-red regions of the spectrum on the cycling of $P_r:P_{fr}$-X may provide an insight to the primary mechanism of phytochrome action. Cycling, involving the random dissociation and reformation of $P_r:P_{fr}$-X complexes, may permit $P_r:P_{fr}$-X to repeatedly induce short-lived activity of the next step of the phytochrome signal transduction chain at all membrane domains of X, even though the proportion of X occupied by $P_r:P_{fr}$ at any instant may be low. Consequently, the overall cellular level of such activity may increase with increased levels of $P_r:P_{fr}$ and $P_r:P_{fr}$-X cycling. The dynamic dissociation and reformation of $P_r:P_{fr}$-X that occurs during irradiation in the HIR is the mechanistic equivalent of the mobility of $P_r:P_{fr}$-X that appears to function in the VLFR. A single type of phytochrome receptor and signal transduction mechanism may therefore underlie all VLFR, LFR, and HIR behavior. The demonstrated ability of the dimeric mechanism to account for VLFR, LFR, and HIR behavior indicates that it may provide a unified mechanism for phytochrome-mediated responses.

## SUMMARY

The light-dependent germination behavior of many species is particularly complex after various promotive pretreatments. Such pretreatments include incubation for several hours at reduced temperatures or for a few minutes at elevated temperatures, exposure to ethanol or other anesthetics, and gibberellic acid. Changes in the rates of phytochrome action or in abundance of phytochrome, that would explain such sensitization, are not observed. Sensitized seeds respond to light at very low fluences in addition to the low fluences, on the order of 10,000-fold greater, that are required in the absence of sensitization. Such observed biphasic fluence-response behaviors support a mechanism in which phytochrome exists and acts as a molecular dimer. Several lines of biochemical evidence now indicate the dimeric nature of phytochrome. Simulations of the dimeric mechanism indicate that sensitization leads to VLF promotion of germination by increasing the activity of heterodimers, $P_r:P_{fr}$, bound to receptors, X. Indirect evidence from sensitization studies suggests that the increased activity of $P_r:P_{fr}$-X results from an enhanced mobility in its membrane environment. The dimeric mechanism has been extended to an understanding of responses of growth to very low and low fluences, as well as of the high irradiance responses of both growth and germination. It provides a useful conceptual framework for continued phytochrome investigations including the photoregulation of seed germination.

## REFERENCES

Axelrod, D., 1983, Lateral motion of membrane proteins and biological function, J. Membrane Biol., 75:1.

Bartley, M. R., and Frankland, B., 1982, Analysis of the dual role of phytochrome in the photoinhibition of seed germination, Nature, 300:750.

Berridge, M. J., 1987, Inositol triphosphate and diacylglycerol: Two interacting second messengers, Ann. Rev. Biochem., 56:159.

Brockmann, J., Rieble, S., Kazarinova-Fukshansky, N., Seyfried, M., and Schäfer, E., 1987, Phytochrome behaves as a dimer *in vivo*, Plant Cell Environ., 10:105.

Chadoeuf-Hannel, R., and Taylorson, R. B., 1985, Anesthetic stimulation of *Amaranthus albus* seed germination: Interaction with phytochrome, Physiol. Plant., 65:451.

Cone, J. W., Jaspers, P. A. P. M., and Kendrick, R. E., 1985, Biphasic fluence-response curves for light induced germination of *Arabidopsis thaliana* seeds, Plant Cell Environ., 8:605.

Bliss, D., and Smith, H., 1985, Penetration of light into soil and its role in the control of seed germination, Plant Cell Environ., 8:475.

Duke, S. O., Egley, G. H., and Reger, B. J., 1977, Model for variable light sensitivity in imbibed dark-dormant seeds, Plant Physiol., 59:244.

Frankland, B., 1986, Perception of light quantity, in: "Photomorphogenesis in Plants," R. E. Kendrick and G. H. M. Kronenberg, ed., Martinus Nijhoff Publ., Dordrecht, p. 219.

Hartmann, K. M., and Nezadal, W., 1989, Efficient photocontrol of weed in crop fields, Proc. Eur. Symp. Photomorphogenesis in Plants, Sept. 24-29, Freiburg i. Br., W. Germany, p. 65.

Hillman, W. S., 1967, The physiology of phytochrome, Ann. Rev. Plant Physiol., 18:30.

Jones, A. M., and Erickson, H. P., 1989, Domain structure of phytochrome from *Avena sativa* visualized by electron microscopy, Photochem. Photobiol., 49:479.

Jones, A. M., and Quail, P. H., 1986, Quaternary structure of 124-kilodalton phytochrome from *Avena sativa* L., Biochemistry, 25:2987.

Mancinelli, A. L., and Rabino, I., 1978, The "high irradiance responses" of plant photomorphogenesis, Bot. Rev., 44:129.

Mandoli, D. F,. and Briggs, W. R., 1981, Phytochrome control of two low-irradiance responses in etiolated oat seedlings, Plant Physiol. 67:733.

Memon, A. R., Chen, Q., and Boss, W. F., 1989, Inositol phospholipids activate plasma membrane ATPase in plants, Biochem. Biophys. Res. Commun., 162:1295.

Morse, M. J., Crain, R. C., and Satter, R. L., 1987, Light-stimulated inositolphospholipid turnover in *Samanea saman* leaf pulvini, Proc. Natl. Acad. Sci. U.S.A., 84:7075.

Napier, R. M., and Smith, H., 1987, Photoreversible association of phytochrome with membranes II. Reciprocity tests and a model for the binding reaction, Plant Cell Environ., 10:391.

Pratt, L. H. and Marmé, D., 1976, Red light-enhanced phytochrome pelletability: Re-examination and further characterization, Plant Physiol. 58:686.

Quinn, P. J., 1988, Regulation of membrane fluidity in plants, in: Advances in Membrane Fluidity, Vol. 3: "Physiological Regulation of Membrane Fluidity", R. C. Aloia, C. C. Curtain, and L. M. Gordon, ed., Alan R. Liss, Inc., New York, p. 293.

Rethy, R., Dedonder, A., De Petter, E., Van Wiemeersch, L., Frederricq, H., De Greef, J., Stevaert, H., and Stevens, H., 1987, Biphasic fluence-response curves for phytochrome-mediated *Kalanchoe* seed germination sensitized by gibberellic acid, Plant Physiol., 83:126.

Shinkle, J. R., and Briggs, W. R., 1984, Indole-3-acetic acid sensitization of phytochrome-controlled growth of coleoptile sections, Proc. Nat. Acad. Sci. USA, 81:374.

Shinkle, J. R. and Briggs, W. R., 1985, Physiological mechanism of the auxin-induced increase in light sensitivity of phytochrome-mediated growth responses in *Avena* coleoptile sections, Plant Physiol. 79:349.

Small, J. G. C., Spruit, C. J. P., Blaauw-Jansen, G., and Blaauw, O. H., 1979, Action spectra for light-induced germination in dormant lettuce seeds, Planta, 144:125.

Smith, H., 1983, Is $P_{fr}$ the active form of phytochrome?, Phil. Trans. R. Soc. Lond. B, 303:443.

Stokes, P., 1965, Temperature and seed dormancy, in: "Encyclopedia of Plant Physiology", Vol. XV/2, Springer-Verlag, Berlin, p. 746.

Taylorson, R. B., and Hendricks, S. B., 1979, Overcoming dormancy in seeds with ethanol and other anesthetics, Planta, 145:507.

Thompson, G. A., Jr., 1983, Mechanisms of homeoviscous adaptation in membranes, in: "Cellular Acclimatization to Environmental Change," A. R. Cousins and P. Sheterline, ed., Cambridge University Press, p. 33

Tokutomi, s., Nakasako, M., Sakai, J., Kataoka, M., Yamamoto, K. T., Wada, M., Tokunaga, F., and Furuya,, M., 1989, A model for the dimeric molecular structure of phytochrome based on small-angle X-ray scattering, FEBS Lett., 247:139.

Uematsu, H., Hosoda, H., and Furuya, M., 1981, Biphasic effect of red light on the growth of coleoptiles in etiolated barley seedlings. Bot. Mag. Tokyo, 94:273.

VanDerWoude, W. J., 1985, A dimeric mechanism for the action of phytochrome: Evidence from photothermal interactions in lettuce seed germination, Photochem. Photobiol., 42:655.

VanDerWoude, W. J., 1987, Application of the dimeric model of phytochrome action to high irradiance responses, in: "Phytochrome and Photoregulation in Plants," M. Furuya, ed., Academic Press, New York, p. 249.

VanDerWoude, W. J,. and Toole, V. K., 1980, Studies on the mechanism of enhancement of phytochrome-dependent lettuce seed germination by prechilling. Plant Physiol. 58:686.

Widell, K.-O., Sundquist, C., and Virgin, H. I., 1981, The effects of SAN 9789 and light on phytochrome in the germination of lettuce seeds, Physiol. Plant., 52:325.

# THE ROLE OF LIGHT AND NITRATE IN SEED GERMINATION

Henk W. M. Hilhorst and Cees M. Karssen

Dept Plant Physiology, Agricultural University,
Arboretumlaan 4, NL-6703 BD Wageningen, The Netherlands

## INTRODUCTION

The minimal requirement for seed germination is the imbibed state. If seeds germinate independently of external factors such as light and soil components, and without a need for pretreatments, either dry or imbibed, at certain temperatures, they can be considered non-dormant in an absolute sense. However, most seeds require one or more external stimulants to germinate. This requirement depends on the 'ecological history' of the seed, including the developmental period on the motherplant. By far the most important component of the seed's ecological history is temperature. The 'temperature history' determines the number and magnitude of environmental factors, including temperature, that are required for germination. In other words, it determines the dormancy status of the seeds. Therefore, over the year, dormancy of seeds in the field may alternate, giving rise to dormancy cycles (Bouwmeester and Karssen, 1989). Figure 1 shows that the expression of dormancy not only depends on the temperature history but also on the germination test conditions. Dormancy of seeds of *Sisymbrium officinale* was induced by a prolonged dark incubation in water at 15 °C. At a germination temperature of 24 °C the seeds did not germinate in the dark, irrespective of the pretreatment period. Thus, under these conditions a change in dormancy could not be expressed. However, when the seeds were irradiated with red light (R) and germinated at 24 °C, the influence of the pretreatment became visible by the changing germination level. Addition of nitrate to the germination medium in combination with R irradiation again changed the germination pattern: induction of dormancy appeared to be delayed. This pattern has been shown for a number of species in our laboratory (Bouwmeester and Karssen, 1989) and can be deduced from several germination studies described in the literature.

It may be argued that the effectiveness of germination stimulants is always a reflection of the dormancy status of the seeds. Thus, the activity of a stimulant has to be defined in relation to other test conditions and the 'temperature history' of the seeds.

The present paper mainly deals with two extensively studied germination stimulants: light and nitrate.

*Recent Advances in the Development and Germination of Seeds*
Edited by R.B. Taylorson
Plenum Press, New York

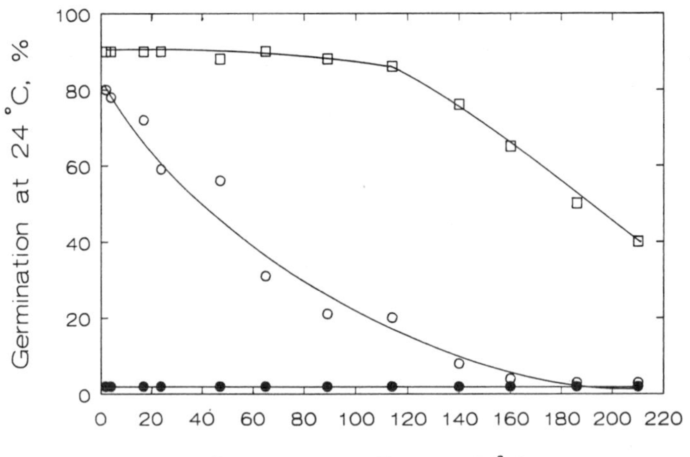


Fig. 1. Germination at 24 $^{o}C$ of seeds of S. officinale, in the dark (●), after R (○), or after R in 25 mM $KNO_3$ (□). Seeds were pre-incubated in water in the dark for the indicated periods at 15 $^{o}C$. Original data.

LIGHT

The stimulating effect of light on seed germination of many wild species has been known for a long time. Since the experiments of Borthwick et al. (1952), it has generally been agreed that red/far-red reversibility of a response is the result of the action of the plant pigment phytochrome. Many factors play a role in phytochrome controlled germination: the amount of pre-existing Pfr, conversions of phytochrome intermediates in dry, partially hydrated and fully imbibed seeds, fluence and wavelength of the applied irradiation, number of irradiations, and interaction with other factors, especially temperature (for extensive review see Bewley and Black, 1982).

Despite the many reports on phytochrome controlled germination few attempts have been made to assess the role of phytochrome in the transduction chain that leads to germination. Evidently, the first step of this chain is the interaction of Pfr with a receptor, resulting in a phytochrome independent intermediate. Evidence for this assumption comes from the phenomenon of escape from the antagonistic effect of far-red irradiation (FR). If imbibed seeds are irradiated with R under conditions that favour germination, and irradiated with FR after variable periods, the seeds escape from FR inhibition. The slope of the escape curve was assumed to be an expression of Pfr activity and high Pfr activities coincided with high final germination levels (Duke et al., 1977). This led to a model in which a Pfr reaction partner (or receptor), X, was introduced. The formation of the PfrX complex was shown to be the rate-limiting step in the induction of germination, and not the total amount of phytochrome or the reversion rate of Pfr to Pr. VanDerWoude (1985) incorporated a dimeric form of phytochrome in the model. This model could explain the biphasic fluence response curves found for some species under specific conditions (Blaauw-Jansen, 1983; Kendrick and Cone, 1985; Mandoli and Briggs, 1981). In this model it is assumed that phytochrome can form two different active complexes with X: Pr:PfrX and Pfr:PfrX. The first form results from irradiation with R of fluences $< 10^{-6}$ mol $m^{-2}$ and the latter from irradiation with R of fluences $>10^{-6}$ mol $m^{-2}$.

Direct examination of the interaction of phytochrome with its receptor X, and the subsequent steps to germination, is still not possible because of extremely low levels and unknown distribution of the active components. Therefore, other methods have to be found to assess the control mechanisms of the germination transduction chain.

One such method is the analysis of dose-response relations whereby the response is the germination of a sub-population of seeds induced by one or more stimuli. For a reliable analysis of the resulting dose-response curves a calculation method is required which takes into account that seed germination is a quantal or 'all-or-none' response. In other words a response at a certain applied dose represents the complete or partial germination of a sub-population of a given seed lot. Whether or not a seed will germinate at a given dose depends on its threshold or tolerance for the applied stimulus. The tolerances of individual seeds are normally distributed over the seed population around a log-dose value for half-maximal germination. Using these assumptions, a calculation method has been proposed to calculate the curve parameters by a linear weighted regression analysis of the curve in a log-dose probit diagram (DePetter et al., 1985). The parameters determining shape and position of the dose-response curve are: m, the log-dose for half-maximal germination; B, the slope of the log-dose probit line; $R^-$, the response at zero dose; $R^+$, maximal response. These parameters are very useful in comparing curves obtained by different treatments but it is also possible to link the parameters with specific biochemical or physical properties of the factors that contribute to the response (Firn, 1986; Weyers, 1987). In Fig.2 some of the possible curve shapes are shown, where
[1] is the standard sigmoidal curve.
[2] is a curve with high response at zero-dose. This may be the result of high endogenous promoter levels. This type of change in fluence-response curves has been described for the light induced germination of *Arabidopsis thaliana* seeds (Cone and Kendrick, 1985). It was indeed shown that a high level of pre-existing phytochrome (Pfr) resulted in this type of response (Cone and Kendrick, 1985)). It is important to note that although the shape of the curve has changed dramatically, the response of the sub-population that was not induced by the pre-existing level of phytochrome is essentially similar to that of curve [1]; neither B nor m have changed. A shift from curve type [1] to type [2] was seen when fluence-response curves of gibberellin (GA)-deficient seeds of *Arabidopsis thaliana* were obtained in a range of $GA_{4+7}$ concentrations (Hilhorst and Karssen, 1988). The curves showed an increased zero response with increasing external GA concentrations. Because these seeds were not able to synthesize GAs the increasing zero response could be ascribed to a response to exogenous GAs that was independent of light. Similar results were obtained with seeds of *Sisymbrium officinale* (Hilhorst and Karssen, 1988) which strengthened the idea that those seeds are not able to synthesize endogenous GAs in the absence of nitrate (see also below). However, in both species the curves also shifted to lower fluence values at increasing GA concentrations. Thus, besides a change of the zero-response $R^-$, m also changed, in a manner similar to the shift of curve [3] to curve [1]. A decrease of the maximal response $R^+$ (curve [4]) may be explained as a decrease in the number or availability of receptors (Firn, 1986). A related change in curve shape not shown here, is one where the response follows the standard curve up to a certain promoter concentration. Above this concentration the curve shows a sharp 'cut-off', indicating that from that level onwards another factor has become limiting for the response. Parameters that change are m, B and $R^+$. We have not been able to find examples in the literature of this type of changing response although it was mentioned by Duke et al. (1978) in an attempt to fit their data to a calculated curve. The last type of response change is expressed by a change in the slope of the curve (curve type [5]). The only example of this type of response change in seed germination was found in the nitrate modulated, light-induced germination of *S. officinale*

and A. thaliana (Hilhorst and Karssen, 1988). Fluence response curves became significantly steeper at increasing nitrate concentrations, even when $R^+$ was already 100 %. It was shown that the germination response was a function of the product of fluence value and nitrate concentration, indicating first order reaction kinetics.

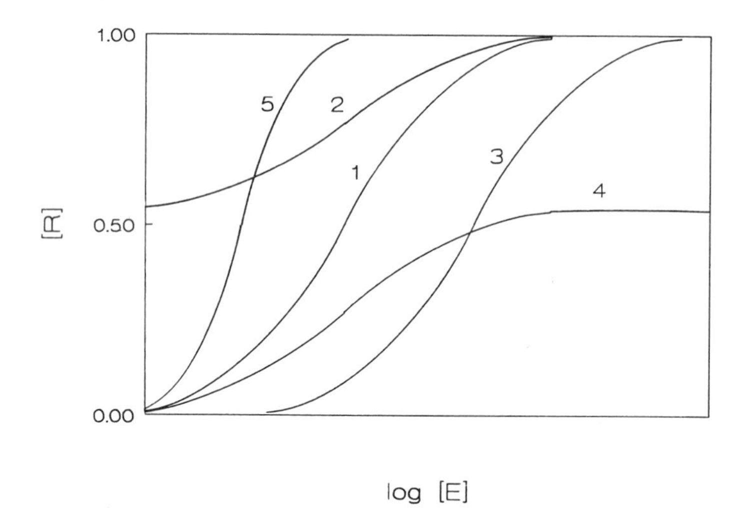


Fig. 2. Set of theoretical sigmoidal curves. [R] = response, [E] = effector concentration. Drawn after Firn (1986).

What do these changes in shape and positions of the curves tell us? By analogy with enzyme kinetics Weyers et al. (1987) derived an equation to describe dose-response relations for plant hormones:

$$R = Rmax[H]^p/([H]^p + K_d) \quad (1)$$

where R = initial rate of response observed, Rmax = the maximal initial rate of response, [H] = hormone concentration, p = interaction coefficient, and $K_d$ = the dissociation constant of the hormone-receptor complex. Equation (1) is analogous to the Michaelis-Menten equation but modified by raising [H] to a power p, to account for the deviations from Michaelis-Menten kinetics which are observed in many concentration-response data. This coefficient is an index of cooperativity between ligands when $p > 1$, and an index of antagonism when $p < 1$. If $p = 1$ we have the Michaelis-Menten type of kinetics.

We have used equation (1) to describe fluence-response curves obtained during induction of dormancy of seeds of Sisymbrium officinale (see Fig.3, original data). The equation we used was of the form:

$$Y = (P)^p.Rmax/[(P)^p + (P_m)^p] \quad (2)$$

with: Y = responding fraction

P = Pfr/Ptot = $[1-e^{(S1 + S2).a.Nt}.\varphi]$

S1 and S2 = photoconversion constants

a = attenuation factor based on seedcoat transmission

$\varphi$ = photo-equilibrium = S1/(S1 + S2)

Nt = fluence (mol $m^{-2}$)

Rmax = fraction of seed population that can be induced by R

Pm = Pfr/Ptot for half-maximal response

p = interaction coefficient

For the observed data points a best fit was calculated by the method described above. The parameters p and (Pm) of equation (2) were substituted by the calculated parameters B (slope) and m (log dose for half maximal germination), respectively. Moreover, a pre-existing level of Pfr of 2% was incorporated to account for the observed low 'dark' germination. In Fig.3 fluence-response curves obtained after 4 different pre-incubation periods at 15 °C are shown. The 24 and 120 h curves nicely fitted equation (2) when Rmax = 1 and a = 1. However, for the 192 and 264 h curves Rmax had to be adjusted to 0.93 and 0.66, respectively, while for the 264 h curve the attenuation factor a had to be lowered to 0.6. Since B was similar for the 24 and 120 h curves, the shift was the result of an increasing m (Tab.1). The 192 and 264 h curves had slightly higher B values but m still increased. Reduction of Rmax values to obtain a good fit must be the result of a growing fraction of seeds whose thresholds became too high for induction of germination. If we assume that the formation of the PfrX complex is the limiting step for germination we may conclude that induction of dormancy in seeds of *Sisymbrium officinale* is characterized by an initial increase of the dissociation constant of the PfrX complex, followed by a decrease of the number of active receptors (X). The remaining seed fraction that germinates after prolonged pre-incubation is characterized by a slightly higher cooperativity (Table 1, B) and, surprisingly, by a lower transmission of the seed coat, which as yet cannot be explained.

This example shows that equation (2) is not only valid for describing general concentration-response curves (Weyers et al., 1987) but also for fluence-responses, a special case of concentration-response. It also shows that changes in these curves may be linked with physico-chemical events

Table 1. Values of parameters from the observed curves of Fig.3. B = slope of log-dose probit line; m = log fluence for half-maximal germination; $R^-$ = germination at zero-dose (dark); $R^+$ = maximal germination.

| Pre-incubation time (h) | B | m | $R^-$ | $R^+$ |
|---|---|---|---|---|
| 24 | 1.88 | -4.48 | 4.96 | 92.80 |
| 120 | 1.84 | -4.16 | 1.59 | 95.06 |
| 192 | 2.50 | -3.97 | 1.04 | 83.11 |
| 264 | 2.20 | -3.38 | 0.20 | 43.71 |

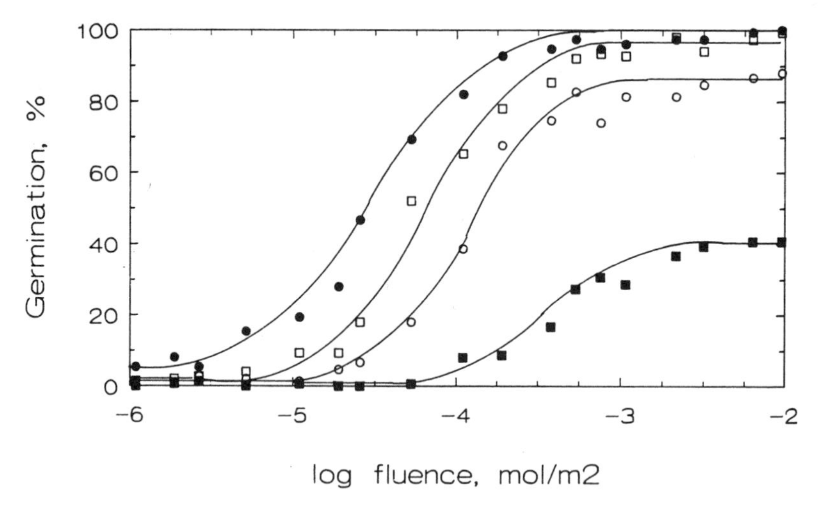


Fig. 3. Fluence-response curves of R-induced germination of seeds of *S. officinale* in 25 mM $KNO_3$ at 24 $^oC$. Seeds were pre-incubated in the dark at 15 $^oC$ for 24 (●), 120 (□), 192 (○) and 264 h (■). Original data.

although it should be kept in mind that certain assumptions had to be made to use the model. The most important ones are that we are dealing with steady state kinetics and that the relation between applied and actual hormone concentrations is known.

As indicated before, the slope of the curve (or promoter range over which the response occurs) is related to the binding characteristics of the receptor (Weyers et al., 1987). The apparent role of nitrate seemed to be the enhancement of the binding characteristics of the Pfr receptor. There is now evidence that the requirement for nitrate is absolute (see below). In this respect nitrate may be regarded as a cooperative co-effector.

NITRATE

Nitrate has long been known to stimulate germination in many species (e.g. Roberts and Smith, 1977; Vincent and Roberts, 1977). In most cases nitrate did not act on its own but in combination with other factors such as temperature and light. Some of these interactions will be discussed later. We will now deal with the mode of action of nitrate on a metabolic and molecular level.

In the recent past two proposals for the mechanism of action of nitrate have been described (Hendricks and Taylorson, 1975; Roberts and Smith, 1977). One of these theories (Roberts and Smith, 1977) assumes that nitrate acts as an alternative hydrogen acceptor that re-oxidizes $NADPH_2$ to NADP, thus stimulating the operation of the pentose phosphate pathway which is believed to be the alternative oxygen requiring process essential for germination. The other hypothesis (Hendricks and Taylorson, 1975) also acknowledges the importance of the pentose phosphate pathway but assumes that reoxidation of $NADPH_2$ proceeds through peroxidase action. The substrate for this enzyme, $H_2O_2$, is made available by the inhibition of catalase activity by nitrite, thiourea or hydroxylamine. These compounds are believed to inhibit enzyme activity by direct binding to the catalase heme protein.

Both theories have been critisized extensively (Bewley and Black, 1982). In both theories a shift in respiratory metabolism is thought to be essential for the transition from the dormant to the non-dormant state. This is supported by several authors who described a stimulation of seed germination by inhibitors of the cytochrome pathway of respiration, such as sodium azide, cyanide and hydroxylamine, thus stimulating an alternative oxygen-requiring process like alternative respiration. The hypothesis of an alternative oxygen requiring process for induction of germination has been tested for *Avena fatua* (Adkins et al., 1984). It was shown that azide, an inhibitor of cytochrome oxidation, could stimulate $O_2$ uptake and also induce germination of dormant seeds. Both azide-induced $O_2$ uptake and germination were inhibited when SHAM was applied. It was concluded that in the absence of the cytochrome pathway of respiration, alternative respiration was required to stimulate germination. If, however, the alternative pathway was blocked by SHAM, while the cytochrome pathway was not inhibited, germination was induced. Thus the authors concluded that either one of both pathways was required for germination. Similarities between the effects of azide and nitrate, especially in their stimulation of $O_2$ uptake, led Adkins et al. (1984) to the conclusion that both components could overcome the same block to germination, in that they were both able to induce reoxidation of NADH. However, as stated by the authors themselves, the results of oxygen uptake and germination were obtained 13 days apart. It is therefore not clear if a causal relationship exists between nitrate action and oxygen uptake. Moreover, though stimulation of oxygen uptake was recorded 4 days before visible germination, it may be questioned whether oxygen uptake triggered germination. Although germination experiments were conducted in the dark, the influence of pre-existing Pfr was not ruled out. Moreover, it was not stated in the paper whether oxygen uptake measurements were also performed in the dark. It may be argued that in the presence of nitrate, germination and $O_2$ uptake were induced by Pfr during the period in which the measurements took place. However, the fact that azide stimulated germination remains intriguing and the conclusion that stimulation of the Krebs cycle via nitrate or inhibition of the cytochrome pathway by azide may point to the requirement of specific intermediates of the Krebs cycle for germination of *A. fatua* (Adkins et al., 1984; Adkins et al., 1988). In *Sisymbrium officinale*, however, very low concentrations of azide inhibited nitrate-stimulated germination (data not shown). In this species it is therefore likely that the cytochrome pathway is essential for germination.

Hilton & Thomas (1986) studied $O_2$ uptake of seeds of 5 species during the period before visible germination, in the presence or absence of exogenous nitrate. $O_2$ uptake patterns differed widely between species. $O_2$ uptake rates increased, decreased or remained at the same level during the period of imbibition while the effect of nitrate on $O_2$ uptake was rather poor compared to the water controls. Also in this study it may be questioned whether a causal relationship exists between action of nitrate and $O_2$ uptake since the moment of induction of germination was not determinded, nor were the involving factors. Even in the water controls both Pfr and nitrate may have played a role. The minor effect of exogenous nitrate on $O_2$ uptake may have been the result of the presence of a certain level of endogenous nitrate which already satisfied the nitrate requirement. A crucial experiment would be to monitor $O_2$ uptake during breakage of dormancy under conditions that seeds will not germinate. Subsequently the effect of germination stimulating factors can be studied, thereby excluding possible masking effects of high initial $O_2$ uptake due to lipoxygenase activity (Parrish and Leopold, 1978). Moreover, caution should be taken when using respiration inhibitors, since it is now known that inhibitors like

SHAM may have considerable side effects at higher concentrations (Moller et al., 1988).

Whether or not a causal relationship exists between the action of nitrate and respiration, all proposals for the stimulating action of nitrate presume that its effectiveness is the result of its reduction, thereby re-oxidizing NAD(P)$H_2$ to NAD(P). It may be questioned whether reduction is required if one thinks of a regulatory role rather than a source of assimilatory nitrogen. Several direct effects of nitrate on cellular processes have been described, including inhibition of tonoplast ATPase and induction of changes in plasmamembrane potential. Moreover, it was suggested that nitrate may play a role in gene activation (Campbell, 1988). In a study of the promotion of seed germination by nitrate and cyanide, Hendricks and Taylorson (1972) already gave indications for a role of nitrate in the unreduced state. They were unable to detect nitrate reductase activity and found that sodium tungstate, an inhibitor of the synthesis of active nitrate reductase, did not influence the nitrate induced germination. They concluded that very low, undetectable levels of nitrate reductase activity were sufficient for the induction of germination. In seeds of _Sisymbrium officinale_ also, no effect of either sodium tungstate or sodium chlorate (competitive inhibitor of nitrate reduction) on the nitrate induced germination was found (Hilhorst and Karssen, 1989). After establishing the period in which nitrate (with Pfr) induced germination, endogenous nitrate levels in the presence or absence of the inhibitors were measured. It appeared that during induction of germination levels of nitrate remained constant. Only after the onset of visible germination, nitrate levels declined, indicating the presence of an active nitrate reducing system. These results, together with results obtained from studies on the light and nitrate induced nitrate reductase activity in seeds of _Sinapis alba_ (Campbell, 1988) suggest a dual role for nitrate in seed germination and early seedling growth. Firstly, nitrate may act as a regulating factor in the induction of seed germination, and secondly, at the onset of true growth, it induces (with Pfr) a nitrate reducing system during seedling growth which is required to start nitrogen assimilation.

However, the actual mechanism of nitrate action in the induction of germination remains obscure. A hypothesis of its role in seeds of _Sisymbrium officinale_ is presented in the following paragraph.

## LIGHT AND NITRATE

For a number of wild species seed germination has been shown to depend on the simultaneous presence of Pfr and nitrate (Vincent and Roberts, 1977; Hilton, 1984; Hilhorst et al., 1986), while for other species nitrate-stimulated dark germination was reported although the possible effects of pre-existing Pfr were not excluded (Roberts and Smith, 1977). In this paragraph we will focus attention on the influence of light and nitrate on germination of _Sisymbrium officinale_.

By now it has been well established that seeds of this species are absolutely dependent on the presence of both factors. Under normal conditions seeds never germinate in darkness, indicating that pre-existing Pfr levels are well below the threshold required for germination. Under non-limiting light conditions we found that germination in water depended on the level of endogenous nitrate. Of 20 seedlots grown in the field in several different years, endogenous nitrate levels were measured and plotted against the light-induced germination in water (Fig. 4). There appeared to be a linear relation between the logarithm of nitrate level and the probit of germination, indicating a normal distribution of mean nitrate levels over different seedlots. Below very low nitrate levels no germination occurred. We were also able to meaure in an indirect way the nitrate levels of individual seeds without destructing the seeds. This could be done because there was a linear relationship between the initial amount of nitrate in the seeds and the amount that leached out during a certain

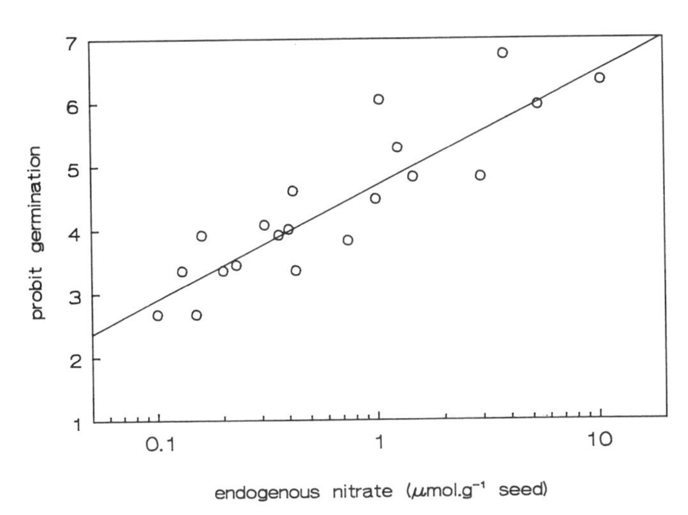


Fig. 4. Plot of nitrate content against probit of germination of seeds from 20 different seed lots of S. officinale. Seeds were pre-incubated for 48 h at 15 °C in the dark and germinated at 24 °C after R irradiation (HJ Bouwmeester and HWM Hilhorst, original data).

incubation period. Thus we obtained from the same seed an indication of its nitrate content and its capacity to germinate in water. We found that within a seed lot nitrate levels of individual seeds were also normally distributed but that germination of these seeds only partially correlated with the nitrate levels. It may be concluded that besides a normal distribution of nitrate levels also the nitrate threshold levels for germination are normally distributed and possibly linked to the distribution of Pfr threshold levels.

As has been shown for other species, the light requirement could be overcome by application of gibberellins. In Sisymbrium the requirement for nitrate also was overcome (Hilhorst et al., 1986). On the other hand, inhibitors of GA-biosynthesis, such as CCC (2-chloro-ethyl trimethylammonium chloride) or tetcyclacis, may inhibit germination (e.g. Adkins et al., 1984; Hilhorst et al., 1986), while GA-deficient mutants of Arabidopsis thaliana failed to germinate in the absence of exogenous GAs (Karssen and Lacka, 1986). This strongly suggests a key role for endogenous GAs in the control of seed germination (Karssen et al., 1989).

Although direct evidence for the stimulation of GA-biosynthesis in mature seeds by external factors is still lacking, there are strong indications that in the absence of germination stimulants there is no biosynthesis of active GAs. Hilhorst et al. (1986) showed that upon induction of secondary dormancy in S. officinale, the stimulative effect of light and nitrate was gradually lost, while the sensitivity to exogenous GAs remained intact. Moreover, escape from the inhibiting effect of tetcyclacis did not start until the stimulants were applied. This indicates that GA-biosynthesis rather than GA-sensitivity is limiting for germination.

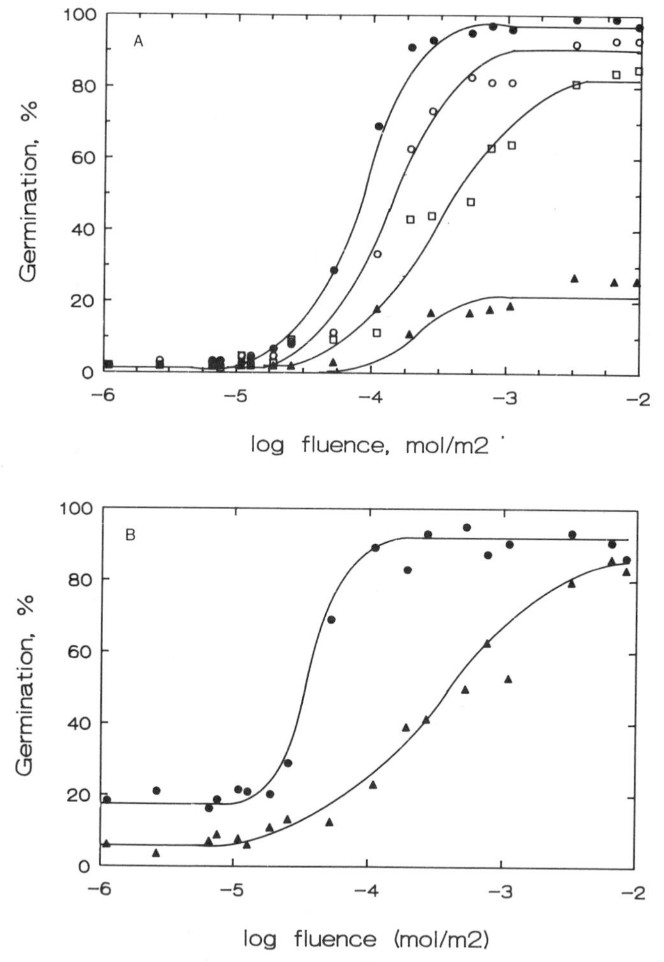


Fig. 5. Fluence-response curves of R-induced germination of seeds of S. officinale (A) and A. thaliana (B). Seeds were pre-incubated in the dark at 15 $^{o}$C for 40 h and germinated at 24 $^{o}$C after R irradiation, in water (▲), 1 (□), 2 (○), and 25 (●) mM $KNO_3$. Redrawn from Hilhorst and Karssen (1988).

In the past, several attempts have been made to clarify the role of Pfr and GAs, and their possible interactions in seed germination (e.g. Bewley et al., 1986; Taylorson and Hendricks, 1976; Carpita and Nabors, 1981). Although the general conclusion was that Pfr and GAs showed some kind of interaction, conclusions about the precise nature of these interactions were equivocal. The main problem in the interpretation of these results is the incompleteness of the dose-response curves. If the minimum and maximum of the curves and their inflection points are not known, or lack sufficient detail, it is difficult to assess the effect of an external factor on the dose-response relation of another factor. In Fig. 2 we showed in what ways a dose-response curve can change its shape and position, already indicating that a simple distinction between 'additive' and 'synergistic' does not always answer our question: is there an interaction between the two factors? Detailed studies of the influence of nitrate on the fluence-response curves for the R-stimulated germination in seeds of Sisymbrium officinale and Arabidopsis thaliana (Hilhorst and Karssen, 1988) revealed that nitrate increased the slopes of the curves for both species (Fig.5). The influence was so strong that the initial gradual response

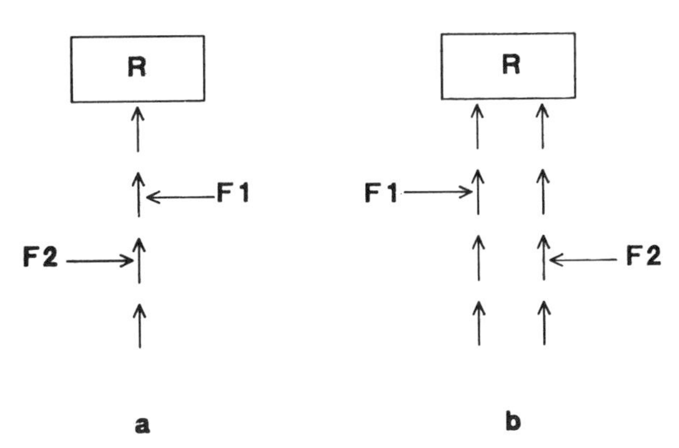


Fig. 6. Descriptive models of co-actions of effectors F1 and F2 on the response [R]. Arrows indicate steps in the pathways leading to the response. Drawn after Schopfer (1986).

changed into a threshold response. It appeared that the response was a function of the product of nitrate concentration and fluence value. Since the slope of the curve is an index for cooperativity (see above) the interaction between Pfr and nitrate may be considered a cooperative coaction. In other words, nitrate enhances the binding of Pfr to its receptor X. This type of interaction can be described by a model shown in Fig. 6a. In the same species we also studied the influence of exogenous GAs on the fluence-response. Again the similarities between both species were striking (Fig.7). Besides a shift upwards the curves were also shifted to lower fluence values at increasing GA concentrations. The enhancement of the dark germination indicates that GAs may act independently of Pfr.

The observed parallel shifts of the fluence-response curves may be interpreted as a change in the dissociation constant of the PfrX complex, as discussed above. This interpretation was given to the observation that exogenous GAs induced a very low fluence response in seeds of Kalanchoe blossfeldiana (DePetter et al., 1985). However, since GA-biosynthesis first has to be induced by R before GAs can act on the photoreceptor it is very unlikely that they do so. The moment sufficient GAs are synthesized, most of the Pfr has been reversed to Pr because the escape time for inhibition by tetcyclacis is 16h, against 8h for FR inhibition. Moreover, GA-response curves can also be shifted to lower concentrations by R irradiation (Hilhorst and Karssen, 1988). It is therefore more likely that Pfr influences the binding characteristics of the GA-receptor in Sisymbrium officinale. Fig.6b gives a possible model for this type of co-action.

Based on detailed dose-response experiments and considerations given in this chapter, a model for the light and nitrate stimulated germination in S. officinale was developed (Fig. 8). The numbers in the model refer to:
1. Cooperative co-action between R and nitrate in which the response is the

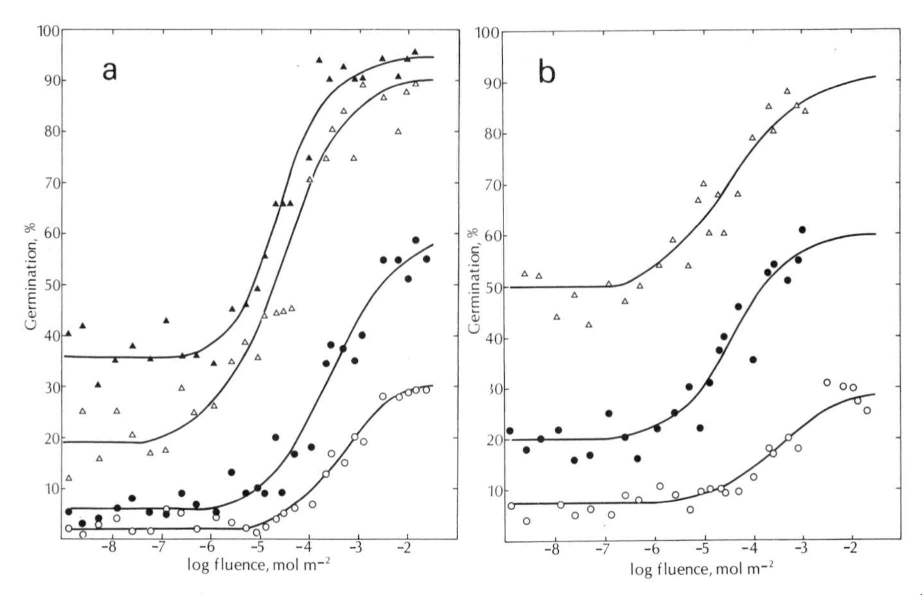


Fig. 7. Fluence-response curves of the R + GA induced germination at 24 °C of seeds of S. officinale (a) and A. thaliana (b). Seeds of S. off. were pre-incubated in water in the dark at 15 °C for 40 h and seeds of A. thal. for 7 days at 10 °C. Seeds were germinated in 2 (○), 5 (●), 10 (△), or 20 (▲) uM $GA_{4+7}$. From Hilhorst and Karssen (1988).

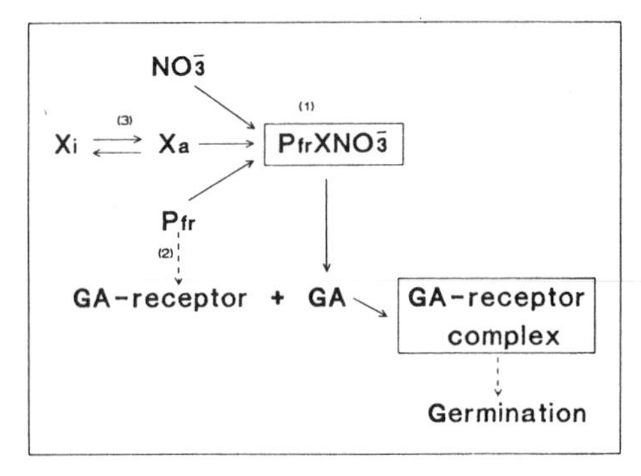


Fig. 8. Hypothetical model for the action of Pfr and nitrate in the induction of seed germination of S. officinale. See text for explanation.

result of the product of red light fluence and nitrate concentration. Both factors are essential.

2. Independent additive co-action of R and GA, with GA as essential factor. Independent, although the effect of R (enhancement of affinity of the GA-receptor) can evidently only be expressed in the presence of GA.

3. Activation of the Pfr receptor X may be an expression of dormancy breaking of light-sensitive seeds (Duke et al., 1977).

The assumption that both Pfr and nitrate bind to the same receptor X is based on experiments in which dose-response relations for R and nitrate were obtained during induction of dormancy. The changes of the m values for both stimulants were highly correlated (Fig.9), suggesting a similar influence of binding characteristics of the receptor on the binding of nitrate and Pfr.

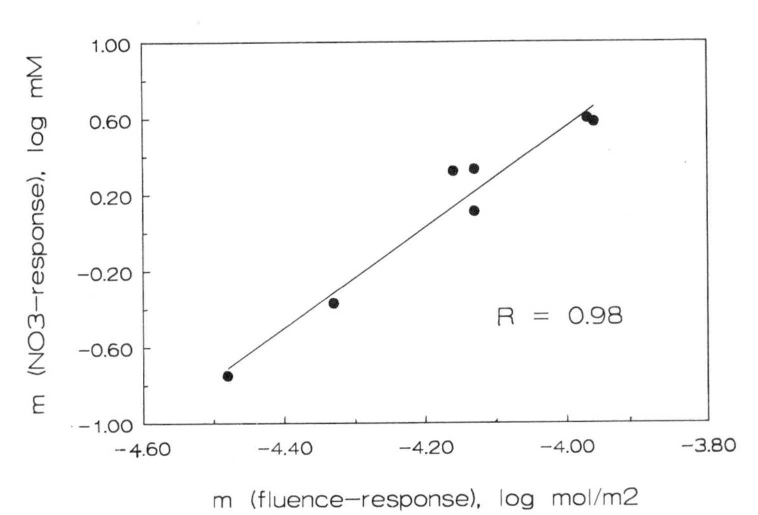


Fig. 9. Correlation diagram of the m values (log-dose for half-maximal germination) obtained from dose-response curves for the nitrate- and R-induced germination of S. officinale after several pre-incubation periods at 15 $^{0}$C in the dark. Original data.

## GENERAL CONCLUSIONS

After this limited survey of the roles of two well known stimulants of seed germination, light and nitrate, the most obvious conclusion that can be drawn is that still very little is known about their mechanisms of action. Since methods to study effector-receptor interactions at a molecular level in plants are only beginning to develop, it will probably take considerable time before these methods can be applied to the systems described in this paper. However, in the meantime, 'classical' seed physiology can play an important role. We believe that a 'dose-response approach' to old problems in seed physiology can give us information on, for example, interactions between environmental factors such as temperature, light and soil components, and their influence on dormancy and germination. It may also provide a sound physiological basis for future research at the molecular level.

REFERENCES


1. Adkins, S. W, Simpson, G. M. and Naylor, J. M., 1984, The physiological basis of seed dormancy in Avena fatua. III. Action of nitrogenous compounds, Physiol. Plant., 60: 227.
2. Adkins, S. W., Simpson, G. M. and Naylor, J. M., 1984, The physiological basis of seed dormancy in Avena fatua. IV. Alternative respiration and nitrogenous compounds, Physiol. Plant., 60: 234.
3. Adkins, S. W., Symons, S. J. and Simpson, G. M., 1988, The physiological basis of seed dormancy in Avena fatua. VIII. Action of malonic acid, Physiol. Plant., 72:477.
4. Bewley, J. D. and Black, M., 1982, "Physiology and Biochemistry of Seeds (Vol.2). Viability, dormancy and environmental control," Springer-Verlag, Berlin.
5. Bewley J. D., Negbi, M. and Black, M., 1968, Immediate phytochrome action in lettuce seeds and its interaction with gibberellins and other germination promoters, Planta, 78: 351.
6. Blaauw-Jansen, G., 1983, Thoughts on the possible role of phytochrome destruction in phytochrome controlled responses, Plant Cell Environ., 6: 173.
7. Borthwick, H. A., Hendricks, S. B., Parker, M. W., Toole, E. H. and Toole, V. K., 1952, A reversible photoreaction controlling seed germination, Proc. Natl. Acad. Sci., (USA), 38:662.
8. Bouwmeester, H. J. and Karssen, C. M., 1989, Environmental factors influencing the expression of dormancy patterns in weed seeds, Ann. Bot, 63:113.
9. Campbell W. H., 1988, Nitrate reductase and its role in nitrate assimilation in plants, Physiol. Plant., 74:214.
10. Carpita, N. C. and Nabors, M. W., 1981, Growth physics and water relations of red-light-induced germination in lettuce seeds, Planta, 152: 131.
11. Cone, J. W., and Kendrick, R. E., 1985, Fluence response curves and action spectra for promotion and inhibition of seed germination in wildtype and long-hypocotyl mutants of Arabidopsis thaliana L, Planta, 163:43.
12. DePetter, E., VanWiemeersch, L., Rethy, R., Dedonder, A., Fredericq, H., DeGreef, J., Steyaert, H. and Stevens, H., 1985, Probit analysis of low and very-low fluence responses of phytochrome-controlled Kalanchoe blossfeldiana seed germination, Photochem. Photobiol., 42: 697.
13. Duke, S. O., 1978, Significance of fluence-response data in phytochrome-initiated seed germination, Photochem. Photobiol., 28: 383.
14. Duke, S. O., Egley, G. H. and Reger, B. J., 1977, Model for variable light sensitivity in imbibed dark-dormant seeds, Plant. Physiol., 59: 244.
15. Firn, R. D., 1986, Growth substance sensitivity: The need for clearer ideas, precise terms and purposeful experiments, Physiol. Plant., 67: 267.
16. Hendricks, S. B. and Taylorson, R. B., 1972, Promotion of seed germination by nitrates and cyanides, Nature, 237: 169.
17. Hendricks, S. B. and Taylorson, R. B., 1975, Breaking of seed dormancy by catalase inhibition, Proc. Natl. Acad. Sci. (USA), 72:306.
18. Hilhorst, H. W. M. and Karssen, C. M., 1988, Dual effect of light on the gibberellin- and nitrate stimulated seed germination of Sisymbrium officinale and Arabidopsis thaliana, Plant. Physiol., 86:591.
19. Hilhorst, H. W. M. and Karssen, C. M., 1989, Nitrate reductase independent stimulation of seed germination in Sisymbrium officinale L. (hedge mustard) by light and nitrate, Ann. Bot., 63:131.
20. Hilhorst, H. W. M., Smitt, A. I. and Karssen, C. M., 1986, Gibberellin-biosynthesis and -sensitivity mediated stimulation of seed germination of Sisymbrium officinale by red light and nitrate, Physiol. Plant., 67:285.

21. Hilton, J. R., 1984, The influence of light and potassium nitrate on the dormancy and germination of Avena fatua L. (wild oat) seed and its ecological significance, New Phytol., 96:31.
22. Hilton, J. R. Thomas, J. A., 1986, Regulation of pregerminative rates of respiration in seeds of various weed species by potassium nitrate, J. Exp. Bot., 37:1516.
23. Karssen, C. M. and Lacka, E., 1986, A revision of the hormone balance theory of seed dormancy: Studies on gibberellin and/or abscisic acid deficient mutants of Arabidopsis thaliana, in: "Plant Growth Substances 1985", M. Bopp, ed., Springer-Verlag, Berlin.
24. Karssen, C. M., Zagorski, S., Kepczynski, J. and Groot, S. P. C., 1989, Key role for endogenous gibberellins in the control of seed germination, Ann. Bot., 63:71.
25. Kendrick, R. E. and Cone, J. W., 1985, Biphasic fluence response curves for induction of seed germination, Plant. Physiol., 79:299.
26. Mandoli, D. F., and Briggs, W. R., 1981, Phytochrome control of two low-irradiance responses in etiolated oat seedlings, Plant. Physiol., 67:733.
27. Moller, I. M., Berczi, A, van den Plas, L. H. W. and Lambers, H., 1988, Measurement of the activity and capacity of the alternative pathway in intact plant tissues: Identification of problems and possible solutions, Physiol. Plant., 72:642.
28. Parrish, D. J. and Leopold, A. C., 1978, Confounding of alternative respiration by lipoxygenase activity, Plant. Physiol., 62:470.
29. Roberts, E. H. and Smith, R. D., 1977, Dormancy and the pentose phosphate pathway, in, "The Physiology and Biochemistry of Seed Dormancy and Germination", A.A. Khan, ed., Elsevier Biomedical Press, Amsterdam.
30. Schopfer, P., 1986, "Experimentelle Pflanzenphysiologie", Springer-Verlag, Berlin.
31. Taylorson, R. B. and Hendricks, S. B., 1976, Interactions of phytochrome and exogenous gibberellic acid in germination of Lamium amplexicaule L. seeds, Planta, 132:65.
32. VanDerWoude, W. J., 1985, A dimeric mechanism for the action of phytochrome: evidence from photothermal interactions in lettuce seed germination, Photochem. Photobiol., 42:655.
33. Vincent, E. M. and Roberts, E. H., 1977, The interaction of light, nitrate and alternating temperature in promoting the germination of dormant seeds of common weed species, Seed Sci. Technol., 5:659.
34. Weyers, J. D. B., Paterson, N. W. A'Brook, R., 1987, Towards a quantitative definition of plant hormone sensitivity, Plant Cell Environ., 10:1.

# WATER-IMPERMEABLE SEED COVERINGS AS BARRIERS TO GERMINATION

G.H. Egley

USDA-ARS, Southern Weed Science Laboratory
P.O. Box 350
Stoneville, MS. 38776

## INTRODUCTION

Some seeds have water-impermeable coverings that prevent water entry into seeds and thereby prevent germination until the impermeability breaks down. Historically, such seeds have been termed "hard" because they do not imbibe water after a day or two and remain hard to the touch whereas nonhard seeds rapidly imbibe and become soft (Assoc. Off. Seed Anal., 1978). This type of physical "coat-imposed dormancy" prevents germination even though water is externally available.

Impermeability breakdown may function as a timing mechanism in that it influences when seed senses that water is available to the embryo (Mayer, 1986). The process distributes germination of seeds of wild plants over a period of time and increases the chances that some seeds will successfully germinate to complete the life cycle. This survival mechanism perpetuates continual weed problems in agricultural situations (Egley and Chandler, 1983; Egley, 1986). Nonsynchronous germination of crop seeds due to water impermeability can create problems when uniform crop emergence and growth are desired. Conversely, some water-impermeability may be beneficial for maintaining vigor of crop seeds during delayed harvest or during storage, particularly under conditions of high humidity (Christiansen and Justus, 1963; Patil and Andrews, 1985; Potts et. al., 1978). Also, some resistance to water diffusion may slow the rate of water entry and protect the embryo against imbibitional injury (Woodstock, 1988). The plant families Leguminosae and Malvaceae contain many species with water-impermeable seeds and have received much attention due to their economic importance (Christiansen and Justus, 1963; Barton, 1965; Quinlivan, 1971; Rolston, 1976). However, many other plant families also have species with water-impermeable seeds (Barton, 1965, Ballard, 1973; Rolston, 1978).

There are several reviews or discussions on the subject of water-impermeability in seeds (Ballard, 1973; Barton, 1965; Quinlivan, 1971; Tran and Cavanagh, 1984; Rolston, 1978; Werker, 1980/81). The present discussion is an attempt to assimilate results of significant investigations on the subject and present a current view on the biology of the development and breakdown of water-impermeability in seeds.

## LOCATION OF WATER-IMPERMEABILITY IN SEEDS

All sites on the seed (e.g., hilum, micropyle, chalazal pore) that were open early during seed development, must seal and must develop water-impervious barriers to effectively insulate the mature embryo from external water (Werker, 1980/81; Tran and Cavanagh, 1984). There are many research reports concerning the location and nature of barriers to water entry into seeds. Most evidence points to the nonliving seed

*Recent Advances in the Development and Germination of Seeds*
Edited by R.B. Taylorson
Plenum Press, New York

coat as the single most significant barrier (Ballard, 1973; Quinlivan, 1971; Tran and Cavanagh, 1984; Rolston, 1978; Werker, 1980/81). The bulk of the evidence has come from studies on hard seeds of the Leguminosae and Malvaceae families. Coats of seeds of these families contain a layer of tightly packed, thick-walled columnar cells that except for a few locations (hilum, micropyle, chalazal pore) completely enclose the embryo (Figures 1,2). The palisade cell layer (Malpighian, macrosclereids) of the coat varies in length, wall thickness and size of cell lumina, depending upon the specific location on the seed. Thus the palisade cell layer of the seed coat appears to fulfill some of the requirements of a major barrier to water diffusion. Werker (1980/81) emphasized that a layer of tightly packed cells is not by itself a sufficient water-impervious barrier because water could permeate the cellulosic fibrils of the cell walls. A water-resistant substance must be on and/or impregnated into the cell walls. However, the tightly packed arrangement would help to prevent water passing between cells that have water-impervious walls.

Figure 1. Longitudinal section through a prickly sida seed (Malvaceae). The palisade (P) cell layer of the seed coat surrounds the cotyledons (CO) and radicle (R) of the embryo. Chalazal cap (CC) cells are beneath the chalazal area. Bar = 1$\mu$m.

Several locations in the seed coat have been proposed as sites of water impermeability. The light line which extends across the upper portion of the seed coat of many seeds, has been historically considered as a barrier (Ballard, 1973; Barton, 1965). The nature of the line is not well understood, but it may be due to an unique arrangement of materials in radial walls of the Malpighian cells that impart different light refractive properties (Barton, 1965; Rolston, 1978). However, impermeability in some seeds exists well beneath the line (Ballard, 1973; Bhalla and Slattery, 1984; Tran and Cavanagh, 1984). Even so, the arrangement of materials in the radial walls that form the line may contribute some resistance to water diffusion through the seed coat, particularly if the palisade cell end walls and cell interior also impede diffusion. The latter is quite likely, because Werker et al. (1979) concluded that impermeability of wild pea (Pisum sativum) seeds was due to a combination of a continuous layer of hard, pectinaceous caps on the palisade cells of the seed coat, quinones in the cells and the tightly-packed arrangement of the cells. The caps of the Malpighian cells were also concluded to impart impermeability to crownvetch (Coronilla varia) seeds (Brant et al., 1971). In hard soybean (Glycine max) seeds, Harris (1987) reported that a prominent light line, abundant tannins in the Malpighian cells, a lack of pores in the coat and cutin in the hilum, all contributed to impermeability. The light line in the hard-seeded malvaceous species, prickly sida (Sida spinosa) contained dense material that was not digested with cellulase and that restricted the diffusion of some water-soluble salts and dyes (Egley and Paul, 1986; Egley et al., 1986). Thus the light line, at least in some hard seeds, may contribute to impermeability but it is not solely responsible. Barriers in hard seeds of velvetleaf (Abutilon theophrasti), another Malvaceae, must have extended into the testa because abrasions and various organic solvents did not overcome impermeability (Horowitz and Taylorson, 1985). Graff and Van Staden (1983) concluded that barriers in the seed coats of two Sesbania species (Leguminosae) extended deeply into the Malpighian layer. Tran and Cavanagh (1984) listed several references that reported impermeability in several species to exist well below the light line, at least to the base of the Malpighian cells.

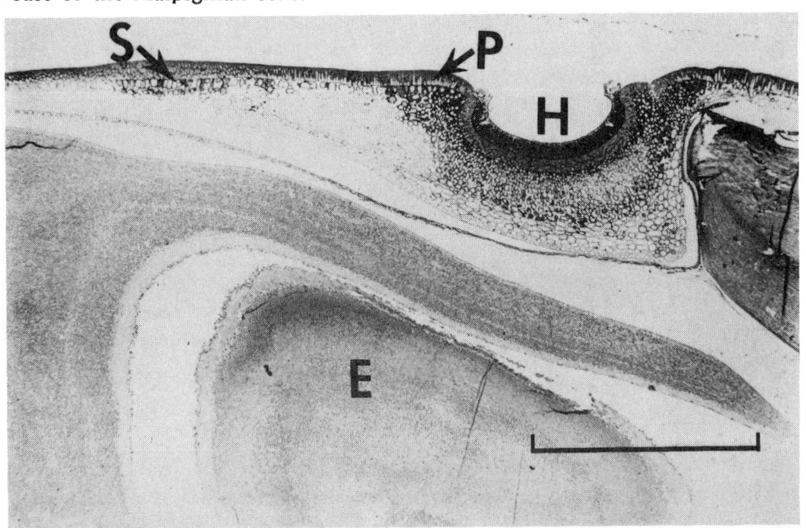


Fig. 2. Longitudinal section through a hemp sesbania seed (Sesbania exaltata). The palisade cell layer surrounds the embryo (E). The strophiolar cells are visible just below the palisade cells of the strophiole (S). The hilar region is also noted (H). Bar = 1$\mu$m.

The cuticle on the seed surface is no longer considered to be the sole or major barrier to water entry into the seed, because removal or disruption of the cuticle of many hard seeds does not always result in imbibition of water (Ballard, 1973; Rolston, 1978; Tran and Cavanagh, 1984). However, the surface cuticle may function as a first-line barrier to water diffusion and may act in combination with other barriers deeper within the seeds to impede movement of water to the embryo. Both the surface cuticle and the palisade cell layer in the seed coat blocked water uptake by Albizia lophantha seeds (Dell, 1980). An impervious seed coat and a noncellular lipid layer at the edge of the hypodermis apparently were jointly responsible for impermeability of Cercis siliquastrum seeds (Riggio-Bevilacqua et al., 1985). Palisade cell walls that contained callose deposits were proposed as causing impermeability in rattlebox (Sesbania punicea) seeds (Riggio-Bevilacqua et al., 1987).

The tight adherance of the seed coat to underlying tissues may prevent lateral movement of water from sites of entry to the embryo (Powell, 1989). Similarly, the characteristics of tissues beneath the coat may be a factor in soybean impermeability because the coat has been observed to "wrinkle" early during imbibition of water (Pereira and Andrews, 1985; McDonald et al., 1988). The loss of impermeability may be due to increased access of water to areas between the coat and cotyledons. In hard cotton (Gossypium hirsutum) seeds, a seal between the chalazal cap located near the base of the chalazal pore, contributed to impermeability by preventing water penetration into the seed's interior (Christiansen and Moore, 1959). Similarly, impermeability beneath the hilar area also restricted lateral movement of water to areas of entry to the embryo of showy crotalaria (Crotalaria spectabilis) seeds (Egley, 1979). Apparently in some seeds, adherence of the seed coat to underlying cells prevents movement of water from areas of penetration into the coat laterally to entry sites into the embryo. This supports the idea that barriers exist below the seed coat and restrict the path of water to the embryo.

Barriers to water diffusion in grapefruit (Citrus paradisi) seeds were located at the inner seed coat where an amorphous cuticular layer encircled the seed at all areas except for the chalazal region (Espelie et al., 1980). Several layers of cells with suberized cell walls filled the gaps in the cuticular layer at the chalazal region (Espelie et al., 1980). Bhalla and Slattery (1984) proposed that parenchyma cell layers just below the seed coat were responsible for impermeability in hard clover seeds. Also, collapsed parenchyma cells below the seed coat of Acacia auriculiformis seed were proposed to form a dense barrier to water diffusion (Pukittayacamee and Hellum, 1988). A hole drilled into hard seeds of huisache (Acacia farnesiana) nearly to the end of the Malpighian cell layer, did not induce imbibition of water (Tran and Cavanagh, 1980). The hole was 155 $\mu$m deep whereas the Malpighian cell depth was 168 $\mu$m and the light line was 88 $\mu$m deep.

Therefore, barriers to water diffusion were detected at various seed coat locations in several different species. The locations ranged from the caps to the bases of the palisade cells. In some instances, the cuticle on the seed's surface and the parenchyma cell layers beneath the seed coat were implicated as barriers. More research to trace the path of water entry into various seeds as impermeability breaks down will help to locate the barriers.

## CHEMICAL NATURE OF WATER-IMPERMEABLE BARRIERS

Phytochemical substances suggested as responsible for restricting diffusion of water into seeds include wax, cutin, suberin, lignin, callose, quinones, and tannins (Rolston, 1978; Werker, 1980/81). Most of the evidence on the chemistry of these substances have come from studies with other plant parts. Wax consists of fatty acids esterified to monohydric and some dihydric alcohols of over 20 carbons in length. It may also contain free fatty acids and high molecular weight ketones and hydrocarbons (Kolattukudy, 1980 b). Wax is soluble in non-polar solvents and is usually extracted from plants with chloroform. Wax frequently occurs with a matrix of the phenolic polymers, cutin or suberin (Kolattukudy, 1980 a). Cutin is a biopolyester consisting mainly of hydroxy and epoxy fatty acids and normally occurs on aerial plant surfaces. However, it has been detected in some internal parts. Cutin often is imbedded in waxes on surfaces of land plants. Cutin synthesis is a special function of plant epidermal cells (Kolattukudy, 1980 b; 1980 a). Suberin is another type of polymer with polyester similarities to that of cutin and with a phenolic matrix somewhat similar to that of lignin. Suberin almost exclusively occurs in internal plant parts and is deposited at extra cellular locations on the plasma membrane side of cell walls. In electron micrographs, suberized regions show a lamellar structure of light and dark bands due to alternate layers of wax and suberin (Kolattukudy, 1980 a, 1984). The composition of suberin is not as well known as that of cutin or wax, because suberin occurs in internal structures and is difficult to extract without contamination with lignin or similar materials. Waxes and cutin on plant surfaces are easier to obtain in pure form. However, the chemistries of all three substances are quite complex (Kolattukudy, 1980 a).

Soliday et al. (1979) reported that the wax associated with suberin was responsible for the resistance to water diffusion in potato tubers. If this result also applies to seed coats, it is probable that wax is the material actually responsible for water-proofing plant tissues and that suberin, and perhaps cutin, only form the matrix for the wax. Due to their hydrophobic properties, suberin and cutin associated with cell walls may slow diffusion of water through tissues but may not completely block water movement.

Phenolics, such as quinones, may be oxidized to form insoluble products that plug cells in the seed coat (Werker et al., 1979). Heavy layers of pectinaceous materials in the seed coat of some seeds have also been implicated in impermeability (Werker et al., 1979). It is doubtful that lignin or callose are primarily responsible for water-proofing the permeability barriers, unless the barriers are heavily impregnated with the materials. It is most likely that wax in association with cutin or suberin, are the substances most responsible for water impermeability in seeds.

More research of the type conducted by Espelie et al., (1980) with grapefruit seeds, in which the chemistry and structure of barriers and the water uptake status of the seeds were studied, are needed to determine the significance of specific tissues and substances in water impermeability of seeds. It is important to correlate the proposed barriers with observed blockage of water entry into the seeds.

## DEVELOPMENT OF WATER-IMPERMEABILITY

### Genetic

The hard seed condition has a genetic basis (Quinlivan, 1971; Tran and Cavanagh, 1984; Rolston, 1978). Because seed impermeability is a varietal characteristic, it is possible to breed or select for a level of impermeability in crop seeds that is appropriate for a particular environment or ecosystem. For example, a slight degree of impermeability may help to maintain seed vigor during seed harvest or storage under conditions that may be more humid than desired. However, the impermeability should not persist and interfere with uniform germination after the seeds are planted.

Lee (1975) determined that two genes were involved in the development of hard cotton seeds but dry conditions during seed maturation enhanced expression of the genes. A single recessive gene was involved with the development of hard seeds of the cultivated lentil (Lens culinaris) (Ladizinsky, 1985). Two or more genes controlled expression of the semi-hard condition in seeds of garden bean (Phaseolus vulgaris) (Dickson and Boettger, 1982). Seeds of garden bean and soybean, each differing in only one gene for seed coat color, had different water imbibition rates. Other white-coated seeds imbibed water more rapidly than did dark-colored seeds (Wyatt, 1977; Tulley et al., 1981). Thus, water-impermeability of many seeds can be altered by manipulating only a few genes.

### Environmental Conditions

Environmental conditions during seed development can modify the expression of water impermeability in seeds (Bewley and Black, 1982; Rolston, 1978). Evanari et al. (1966) found that exposure of the mother plant to long days during the last eight days of ripening of Ononis sicula seeds resulted in seeds with thick coats, a well-developed cuticle and a low level of permeability. Environmental factors can alter cutin composition (Kolattukudy, 1980 a) and thereby may affect water impermeability of some seeds. Impermeability in papilionoid seeds of Legumes is normally associated with seed dehydration during the final maturation stages. For example, as garden bean seeds were reduced to moisture contents below 15%, the number of hard seeds increased (Quinlivan, 1971). Hyde (1954) described the valve-like action of the palisade-counter palisade cell layers in the hilum that permitted papilionoid legume seeds to dehydrate while the seed coat remained impermeable to water loss or uptake (Figure 3). The differential responses of the two cell layers to external moisture level caused the hilum to open when atmospheric moisture was less than that within the seed and caused it to close when external moisture level was greater than the internal moisture level. This valvular action is a physical process that prevents water entry into the seed when the external moisture level is greater than that within the seed. It is not known if a similar phenomenon occurs in seeds that do not have the palisade-counterpalisade cell arrangement. However, it is known that development of impermeability is influenced by moisture levels both internal and external to the seed.

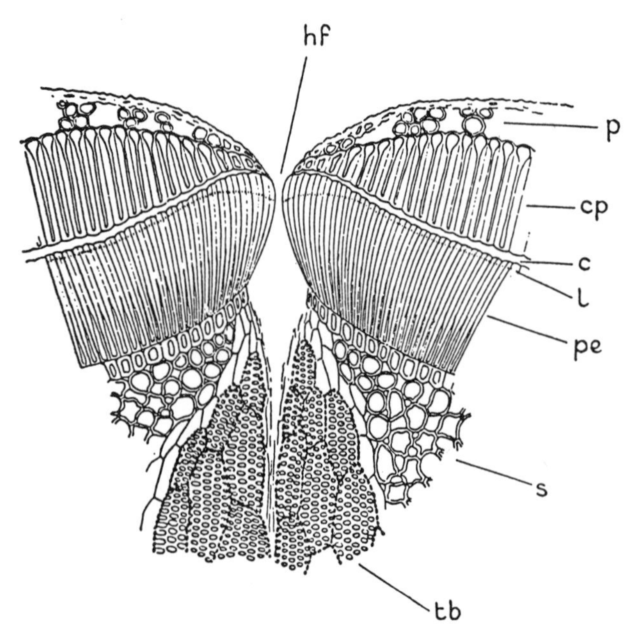


Fig. 3. Transmedian section through the hilum of Lupinus arboreus. hf = hilar fissure, cp = counter-palisade cell layer, l = light line, p = palisade cell layer, Tb = Tracheid bar. From Hyde (1954).

Oxidation of phenolics during seed maturation enhanced the development of water impermeability in coats of wild pea seeds (Marbach and Mayer, 1974). The impermeability did not develop in the absence of oxygen. This tanning reaction darkened the seed coat color and formed quinones that reacted with cellular proteins in the seed coat and contributed to impermeability. However, in subterranean clover (Trifolium subterraneum), Slattery et al. (1982) concluded that darkening of the coat was associated with oxidation of phenolics by catechol oxidase, but development of impermeability was an independent process.

The water and nutrient status of the mother plant may also affect water impermeability of the progeny. Subterranean clover seeds produced by plants growing under soil moisture stress were less impermeable to water than were seeds from plants growing in soil of adequate soil moisture (Aitken, 1939). On the other hand, high soil moisture during soybean seedfill reduced the expression of water impermeability in the seeds. In the latter instance, the enlarged cotyledons disrupted the seed coat integrity of mature seeds and water entered through the stressed coats when the seeds were planted (Hill et al., 1986). In other instances, larger soybean seeds were also less water-impermeable than were smaller, lighter seeds (Smith and Nash, 1961).

Thicker and more water-impermeable seed coats resulted when soybean seeds developed during a deficiency of mineral nutrients and cytokinin (Noodèn et al., 1985). The impermeability which resulted in a germination delay was suggested as a response to stresses such as drought. Also the flowering sequence on the mother plant can influence the order of seed impermeability breakdown. Salisbury and Halloran (1983) found that subterranean clover seeds that developed in fruits from the last flower to be produced, were the first to become permeable. The basis for the sequential effect on seed impermeability was not determined.

There is a need for more information on the modification of the development of seed impermeability by the environment. Effects of light, daylength, temperature, and humidity during different stages of seed development need further study. The moisture and nutritional status of the mother plant also seems important to development of impermeability and needs further evaluation. Such information would also help to explain environmental influences upon survival of seeds in nature.

### Physiology

Many seeds can germinate if removed from the fruit a few days before they dehydrate to less than 20% water dry weight (Table 1; Egley et al., 1983; 1985). Thus the seeds became impermeable to water during late stages of dehydration. Seed coat impermeability of prickly sida and showy crotalaria correlated with seed coat lignification and darkening but not with seed coat dehydration (Egley, 1979; Egley, et al., 1983). The coats became impermeable before they dried to less than 15% water. After the coats became impermeable, the chalazal slit of prickly sida and the hilum of showy crotalaria remained open to water passage until the seed moisture level decreased to below 15%. Imbibition of water was then totally blocked. Apparently, the chalazal slit functions in prickly sida to allow water passage after the coat becomes impermeable as does the hilum in some legume seeds. The chalazal slit appears to become sealed with dried cellular secretions rather than closing due to valvular action (Christiansen and Moore, 1959).

Egley et al. (1985) determined that peroxidase and not polyphenol oxidase was involved in lignification of the seed coat during development of impermeability in some malvaceous and leguminous weed seeds. The time of intense lignification correlated with the onset of impermeability and darkening of the seed coats. The precise role of lignin in impermeability was not determined, however, the least lignified seed coat region of prickly sida was the site where impermeability broke down first (Egley et al., 1983; 1986). Kolattukudy (1980 a; 1981; 1984) pointed out that suberin synthesis is a general response to wounding stresses and that abscisic acid can trigger events leading to suberization. Thus, it is tempting to postulate that the stresses imposed on cells in the seed coat during severe dehydration induce suberin synthesis and that abscisic acid enhances the process. Peroxidase occurs in the seed coat and can calalyze synthesis of suberin as well as lignin (Kolattukudy, 1981).

Additional research is needed to determine if inhibitors of synthesis of wax, cutin, suberin, lignin etc. during seed development will result in inhibition of water impermeability in mature seeds. Such information will help to determine roles of specific substances in impermeability.

## BREAKDOWN OF SEED IMPERMEABILITY

### Artificial methods

Water impermeability in seeds can be broken down artifically by methods that disrupt barriers on or near the seed surface, cause fissures through barriers deeper within the seed or induce loosening of seed coats at predetermined weak points. Common methods of inducing imbibition in hard seeds are soaking of seeds in concentrated sulfuric acid or abrasion of seed surfaces by tumbling seeds in containers with rough surfaces (Rolston, 1978). Other treatments having varied degrees of success in overcoming impermeability include soaking in organic solvents, impaction against hard surfaces, soaking in hydrolytic enzymes, heating under wet conditions and freezing at ultra low temperatures (Rolston, 1978).

Soaking in concentrated sulfuric acid reduces or overcomes impermeability by damaging the cuticles, osteosclereids, hilum and strophiole of *Aspalathus linearis* (Kelly and Van Staden, 1985), disrupting the strophile and testa of rattlebush (Manning and Van Staden, 1987 b) and eroding caps of macrosclereids of crownvetch (Brant et al., 1971). Concentrated sulfuric acid may also rapidly dessicate tissues and cause stress resulting in cell separations and fissures in the seed coat (Duran et al., 1985).

Table 1. Weed seed properties at five stages of development

| Seed Property | Seed developmental stage | | | | |
|---|---|---|---|---|---|
| | 1 | 2 | 3 | 4 | 5 |
| Days after anthesis | | | | | |
| velvetleaf | 12-13 | 14-15 | 16-17 | 18-19 | 20-25 |
| spurred anoda | 13-15 | 16 | 17-18 | 19-21 | 22-25 |
| showy crotalaria | 15-21 | 22-26 | 27-28 | 29-32 | 35-40 |
| hemp sesbania | 15-21 | 22-27 | 28-29 | 30-32 | 35-40 |
| Coat colour | | | | | |
| velvetleaf | white | white | tan | brown | drk-br. |
| spurred anoda | white | white | lt-br. | brown | drk-br. |
| showy crotalaria | white | yel-wt. | lt-gr. | green | drk-gr. |
| hemp sesbania | green | green | lt-gr. | gr-yel. | gr-br. |
| Coat water (% dry wt.) | | | | | |
| velvetleaf | - | - | - | - | 16* |
| spurred anoda | - | 254 | 202 | 155 | 18* |
| showy crotalaria | - | 245 | 200 | 180 | 12* |
| hemp sesbania | - | 185 | 175 | 153 | 17* |
| Germination (%) | | | | | |
| velvetleaf | - | 53 | 47 | 23 | 0 |
| spurred anoda | - | 0 | 9 | 45 | 8 |
| showy crotalaria | - | 0 | 85 | 100 | 28 |
| hemp sesbania | - | 25 | 90 | 93 | 100 |
| hemp sesbania† | - | 67 | 25 | 37 | 53 |
| Germination after coat puncture (%) | | | | | |
| velvetleaf | - | - | - | 67 | 85 |
| spurred anoda | - | - | - | 70 | 88 |
| showy crotalaria | - | - | - | - | 85 |
| hemp sesbania† | - | - | - | 57 | 100 |

* Whole seeds.

†After drying for 2 d to 10% water. From Egley et al. (1985).

Abrasion of seeds on rough surfaces can also erode or damage impermeable areas on or near the seed surface (Rolston, 1978). Mechanical scarification may also impact upon weak areas on the seed coat and cause cell separation of and breaks in impermeable barriers (Ballard, 1973; Kelly, and Van Staden, 1987; Quinlivan, 1971). These weak areas occur at the strophiole (also termed lens by some investigators) of many leguminous seeds and near the chalaza of some malvaceous seeds (Figure 4,5). Impermeability in prickly sida seeds was lost when a heart-shaped portion of the seed coat near the chalazal slit separated from the underlying layer of columnar subpalisade cells (Figure 1). The subpalisade cells occurred only beneath the separation area and had radial cell walls with thin portions without secondary wall thickening near the tops of the cells (Egley and Paul, 1981). When stressed, the radial walls of the subpalisades broke at the thin portions resulting in the separation of the seed coat segment. The contents of the ruptured subpalisades appeared hygroscopic and attracted water to the newly exposed area. Seed coat regions elsewhere on the seed remained impermeable. The thin portions of the subpalisade walls were observed in developing prickly sida seeds and were predetermined weak sites that broke in mature seeds under appropriate conditions (Egley and Paul, 1982) (Figure 6). The thin-walled subpalisade cells of both impermeable and permeable mature dry seeds were intact. However, the walls broke more readily in permeable seeds than in impermeable seeds. Impermeable seeds of velvetleaf and spurred anoda (Anoda cristata), two other species of Malvaceae, also lost impermeability in the chalazal region and had subpalisade cells with thin cell wall portions (Egley et al., 1985).

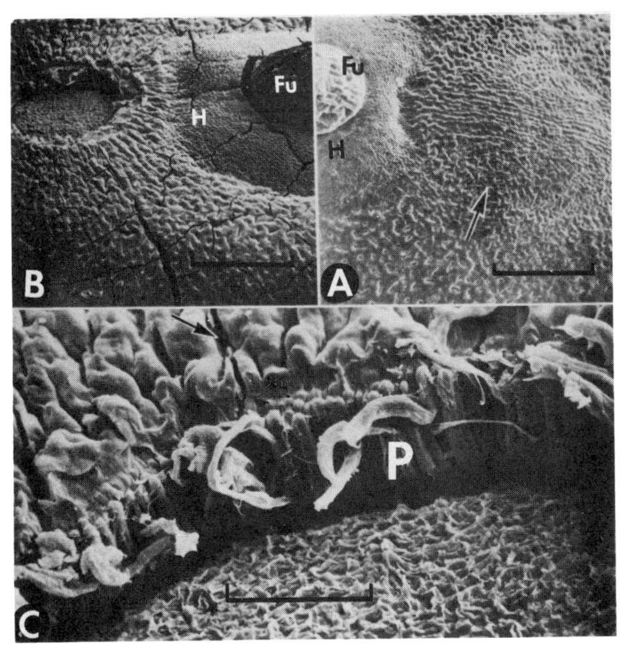


Fig. 4. Strophiolar region of Albizia lophantha seed. Heat-treated seed showing cavity where the strophiolar plug has erupted. A. Strophiolar region prior to eruption of plug (arrow) adjacent to the finiculus (Fu). Bar = 250$\mu$m. B. The cavity is adjacent to hilar tissue (H) funiculus (Fu). Bar = 250 $\mu$m. C. Detail of cavity. Palisade cells (p) of seed coat are visible. Broken, cell walls are evident on cavity floor. Bar = 50 $\mu$m. From Dell (1980).

The strophiole is a weak site in the coat of papilionoid legume seeds where impermeability also breaks down (Hagon and Ballard, 1970). The long, narrow Malpighian cells in the strophiole have a greater tendancy to split when stressed by heat, percussion or scarification, than do the shorter, thicker-walled cells at other seed coat areas (Figure 2). The thin-walled parenchyma cells beneath the Malpighian cells in the strophiole of Albizia lophantha and Acacia kempeana may function as do the subpalisades in prickly sida to facilitate loss of impermeability (Dell, 1980; Hanna, 1984). The strophiole region differentiates early during seed development and is near the hilum. It appears that the strophiole, like the chalazal area, also performs a special function in

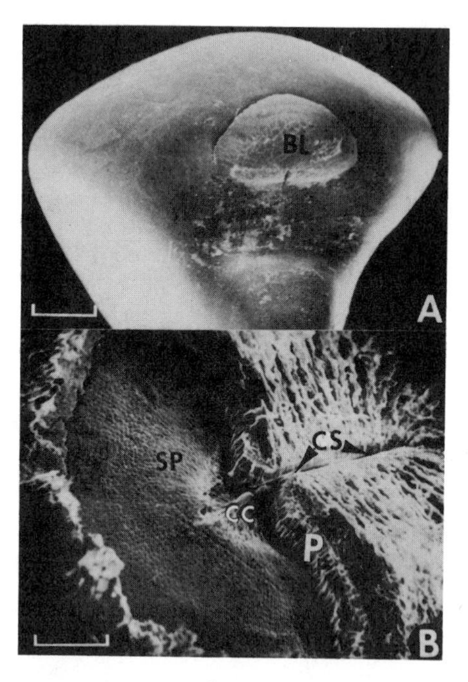


Fig. 5. A. Scanning electron micrograph (SEM) of mature Sida spinosa seed showing "blister" (BL) that formed near chalazal slit. Bar = 250 μm; B. SEM of S. spinosa seed with "blister" removed, exposing subpalisade (SP) layer with chalazal cap (CC) cells in chalazal slit (CS), and palisade layer (P) outside area defined by the blister. Bar = 100 μm. From Egley and Paul, (1981).

the loss of seed impermeability to water (Kelly and Van Staden, 1987). The strophiolar hypodermis may also function by regulating water movement further into the seed (Manning and Van Staden, 1987 a). Methods to artificially break down impermeability were most effective when directed upon these weak areas on the seeds. Applications of hot water, low temperature, pressure and mechanical or chemical scarification induced imbibition of water at the strophiole in legumes (Egley, 1979; Dell, 1980; Hanna, 1984; Manning and Van Staden, 1987; Miklas et al., 1987; Pritchard et al., 1988), and at the chalaza in Malvaceae (Christiansen and Moore, 1959; Manning and Van Staden, 1987 b; Egley and Paul, 1981). Pressure changes caused by plunging seeds in hot water (Brant et al., 1971; Dell, 1980; Hanna, 1984) or in liquid nitrogen (Brant et al., 1971; Pritchard et al., 1988) may break cell walls at weak sites and disrupt seed coat integrity.

Hot water or some organic solvents dissapated a disk-like deposit within the chalazal slit and broke down impermeability in cotton seeds. Water entered the seed through the unblocked slit and moved laterally between the chalazal cap and the inner surface of the palisade layer (Christiansen and Moore, 1959). The palisade cells absorbed water, enlarged and separated from the underlying chalazal cap cells. This separation caused buckling of the coat in the chalazal area, resulting in splits of the seed coat and rapid entry of water through the openings. In permeable cotton seeds, this process occurred in water at normal germination temperatures (Christiansen and Moore, 1959).

Graff and Van Staden (1983) concluded that impermeability is a physical process that is overcome by methods that disrupt the tightly-packed arrangement of the macrosclereids in seeds of legumes. It appears that a similar disruption of the tightly-packed cells in the seed coats or in sealants between cell layers, also overcomes impermeability in many other seeds.

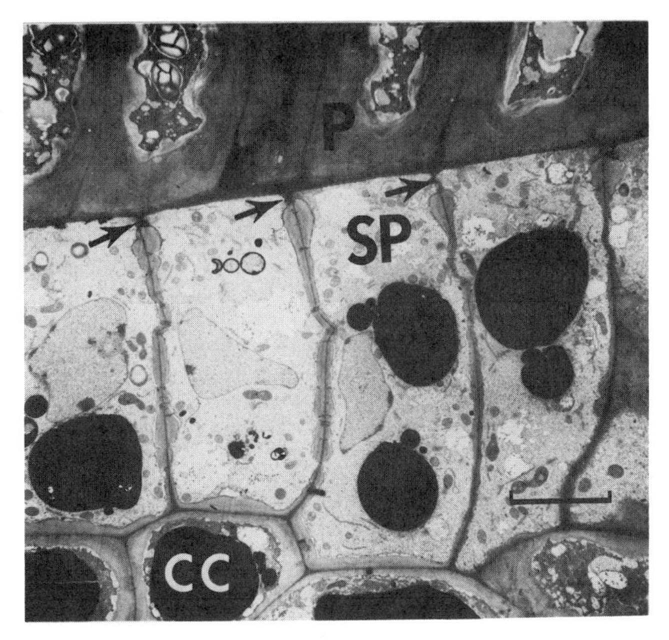


Fig. 6. Transmission electron micrograph of subpalisade region of Sida spinosa seed at 6 days post anthesis. Arrows point to thin cell wall portions where future breakage occurs prior to imbibition of water. The subpalisade layer (SP) is between palisades (P) and chalazal cap (CC) cells. Bar = 5 $\mu$m, From Egley and Paul, (1982).

## Natural methods

There is little confirmed information as to how seeds naturally lose impermeability. Microbial action upon barriers to water uptake has been suggested. Lester (1985) promoted germination of hard seeds of wild tomato (Lycopersicon chessmanii) and two species of Solanum by etching the spermoderm with cellulytic enzymes. He also discussed other investigations in which soil extracts decreased the impermeability of hard seeds. Brant (1971) slightly reduced the impermeability of crownvetch seeds by incubating the seeds in hemicellulase or pectinase. Loss of impermeability through enzymatic activity upon hard seeds by soil organisms is an attractive possibility, but proof is lacking for significant effects upon seed impermeability in the field.

High temperatures due to fire at or just above the soil surface, could produce stresses that break down impermeability (Dell, 1980). Also in certain instances, some hard seeds may be scarified and their impermeability decreased by passage through the digestive system of animals, particularly birds (Krefting, 1949; Rolston, 1978; Potter et al., 1984). Soil tillage may provide sufficient impaction or abrasion to disrupt barriers or break weak sites in coats of some hard seeds.

Exposure of hard seeds to several weeks of diurnal temperatures in the soil has been proposed as a factor in the breakdown of impermeability (Quinlivan, 1971). In Australia, several weeks of alternating daily temperatures (about 15°/60°C) reduced impermeability of hard seeds of subterrenean clover (Quinlivan, 1971). However, a low seed moisture content was necessary. Seeds on or near the soil surface would most likely be affected by hot and dry conditions. Taylor (1981) suggested that softening of subterranean clover was influenced by two different temperature dependant processes.

High temperatures weakened the cells in the strophiole and then temperature fluctuations caused gradual expansion and contraction of cells that eventually disrupted the strophiole and broke down impermeability. Aging of dry seeds at room temperature for several months was sufficient to break down impermeability of hard prickly sida seeds, whereas impermeability was maintained in dry seeds held at -20°C (Egley, 1976). It appears that temperature fluctuations, the seed moisture levels and length of time that seeds are in the soil will influence loss of impermeability in nature.

Based on current evidence, I propose the following model for natural breakdown of impermeability in prickly sida seeds. Over a period of time, stresses such as temperature- induced contractions and expansions of the lightly-lignified, hemicellulose-abundant, tall palisade cells in the chalazal region (Figure 7) may cause some of the underlying subpalisade cells (Figure 6) to break at the thin-walled portions. The fragility of the thin sites on the walls may be enhanced by low seed moisture (<15%) as the seeds age in the soil. The broken cell walls result in separation of part of the coat from the underlying subpalisade cells, and allow some moisture to contact the hygroscopic hemicellulose in the palisades. The hydration of the materials results in rapid cellular expansion, which breaks more subpalisade cell walls and causes extensive buckling of the seed coat where the breakage occurred. Subsequently, the seed coat segment is completely disrupted, the chalazal area is exposed and water enters unimpeded. The broken subpalisades and the hygroscopic material that is extruded from the cells provide an area favorable for absorption of water. Hydration of the embryo and germination soon follow completion of the events that cause the breakdown of impermeability. Even though water impermeability in seeds has received much attention the chemical and structural factors responsible for the condition are not fully understood. There is a need for comprehensive studies that relate the cytochemistry and morphology of proposed barriers with observed prevention of water entry into seeds and movement to the embryo of selected species. Studies comparing these factors in seeds of different species are also needed.

## CONCLUSIONS

Genetic control over the development of seed impermeability to water is influenced by some environmental conditions, particularly those occuring during the last stages of seed maturation. Most evidence implicates the continuous layer of tightly packed palisade cells in the seed coat as containing the major barrier to water entry into seeds. However, a combination of sites, including cuticles, the seed coat and thick-walled cells beneath the seed coat, may all contribute to impermeability. No one barrier is universally responsible for impermeability in seeds. The nature of impermeability apparently varies among species although some similarities exist among closely related species. Waxes, in association with the matrix of suberin or cutin, are the chemical substances most likely as responsible for water-proofing of structures in seeds. However, other substances such as lignin, may also impart resistance to water diffusion, particulary where cell walls are heavily impregnated with the material.

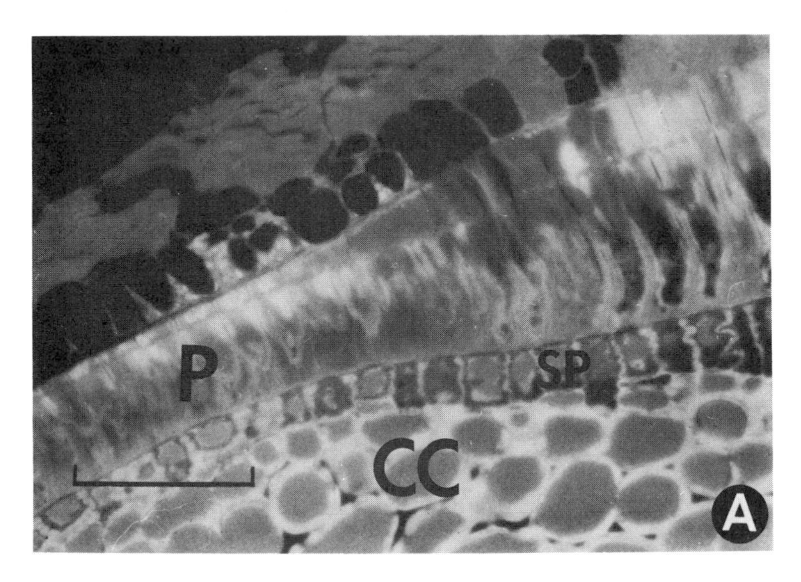


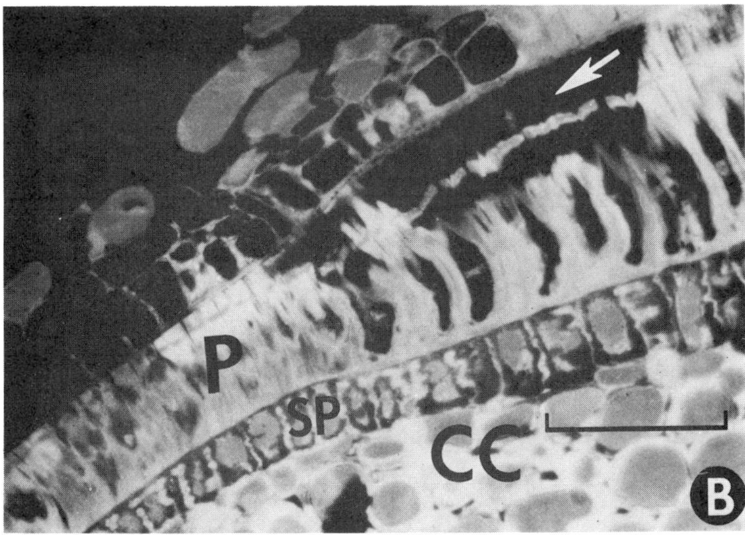


Fig. 7. Section through seed coat of the chalazal region of immature prickly sida seed subsequent to incubation in denatured cellulase control (A) and active cellulase (B). Sections were stained with acridine orange. Sections were made prior to separation of palisade (P) from the subpalisade cell (SP). Note that cellulase digested large amounts of material (arrow, dark spaces) from the cell lumena and distal portions of the taller palisade cells but did not digest large amounts of material (dark spaces) from the shorter cells . The light line (below arrow) remained after cellulase digestion of the surrounding material from the taller cells. Taller palisades are in the region of palisade separation from subpalisade cells. Bar = 50 $\mu$m. From Egley et al. (1986).

The breakdown of water-impermeability is a physical process that is influenced by temperature, moisture and time and culminates in the disruption of barriers to water diffusion to the embryo. In some seeds the disruption occurs at weak sites in or below specific areas (strophiole, chalaza) of the seed coat, when stresses imposed on the weak sites cause breakage of thin-walled cells. It is proposed that once breakage of the cell walls begin, the integrity of the barrier to water diffusion is lost, water enters the seed and contacts cells that subsequently hydrate, expand and cause massive breakage of the thin-walled cells. Buckling of the seed coat then occurs and further separation of the coat from underlying cells results in exposure of the seed's interior to external water and the hydration of the embryo soon follows.

Water-impermeability in seeds has received much attention but the chemical and structural factors responsible for the condition are still not fully understood. Further research is necessary to determine factors that influence the development and breakdown of water-impermeability in seeds. Further information is important to improve our understanding of this type of structure - imposed seed dormancy.

ACKNOWLEGEMENT

The assistance of R.N. Paul, Jr. in the preparation of the figures is gratefully appreciated. His contributions as a co-worker in the author's previous publications on water-impermeable weed seeds are gratefully acknowledged.

LITERATURE CITED

Aitken, Y., 1939, The problem of hard seeds in subterranean clover, Proc. Roy. Soc. Vict., 51:187.

Assoc. of official seed analysts, 1978, Rules for seed testing, J. Seed Technol., 3:29.

Ballard, L. A. T., 1973, Physical barriers to germination, Seed Sci. and Technol., 1:285.

Barton, L. V., 1965, Dormancy in seeds imposed by the seed coat, in: "Encyclopedia of Plant Physiol.," XV; Part 2, W. Ruhland, ed., Springer-Verlag, New York.

Bevilacqua, L. R., Fossati, F., and Dondero, G., 1987, Callose in the impermeable seed coat of Sesbania punicea, Ann. Bot., 59:335.

Bewley, J. D., and Black, M., 1982, "Physiology and Biochemistry of seeds in relation to germination Vol. 2: Viability, Dormancy, and Environmental Control," Springer-Verlag, New York.

Bhalla, P. L., and Slattery, H. D., 1984, Callose deposits make clover seeds impermeable to water, Ann. Bot., 53:125.

Brant, R. E., McKee, G. W., and Cleveland, R. W., 1971, Effect of chemical and physical treatment on hard seed of Penngift crownvetch, Crop Sci., 11:1.

Christiansen, M. N., and Justus, N., 1963, Prevention of field deterioration of cottonseed by an impermeable seedcoat, Crop Sci., 3:439.

Christiansen, M. N., and Moore, R. P., 1959, Seed coat structural differences that influence water uptake and seed quality in hard seed cotton, Agron. Jour., 51:582.

Dell, B., 1980, Structure and function of the strophiolar plug in seeds of Albizia lophantha, Amer. J. Bot., 67:556.

Dickson, M. H., and Boettger, M. A., 1982, Heritability of semi- hard seed induced by low seed moisture in beans (Phaseolus vulgaris L.), J. Amer. Soc. Hort. Sci., 107:69.

Duran, J. M., Estrella, M., and Tortosa, M., 1985, The effect of mechanical and chemical scarification on germination of charlock (Sinapis arvensis L.) seeds,

Egley, G. H., 1986, Stimulation of weed seed germination in soil, Rev. of Weed Sci., 2:67.
Egley, G. H., and Chandler, J. M., 1983, Longevity of weed seeds after 5.5 years in the Stoneville 50-year buried seed study, Weed Sci., 31:264.
Egley, G. H., and Paul, Jr., R. N., 1981, Morphological observations on the early imbibition of water by Sida spinosa (Malcaceae) seed, Amer. J. Bot., 68:1056.
Egley, G. H., and Paul, Jr., R. N., 1982, Development, structure, and function of subpalisade cells in water impermeable Sida spinosa seeds, Amer. J. Bot., 69:1402.
Egley, G. H., and Paul, Jr., R. N., 1986, Detection of barriers to penetration of soluble salts into weed seeds, Plant Physiol. Suppl., 80:21.
Egley, G. H., and Paul, Jr., R. N., Duke, S. O., and Vaughn, K. C., 1985, Peroxidase involvement in lignification in water-impermeable seed coats of weedy leguminous and malvaceous species, Plant. Cell. and Environ., 8:253.
Egley, G. H. and Paul, Jr., R. N. and Lax, A. R., 1986, Seed coat imposed dormancy: Histochemistry of the region controlling onset of water entry into Sida spinosa seeds, Physiol. Plant., 67:320.
Egley, G. H., Paul, Jr., R. N., Vaughn, K. C., and Duke, S. O., 1983, Role of peroxidase in the development of water-impermeable seed coats in Sida spinosa L., Planta., 157:224.
Espelie, K. E., Davis, R. W., and Kolattukudy, P. E., 1980, Composition, ultrastructure, and function of the cutin- and suberin- containing layers in the leaf, fruit peel, juice-sac and inner seed coat of grapefruit (Citrus paradisi Macfed.), Planta., 149:498.
Evenari, M., Koller, D., and Gutterman, Y., 1966, Effects of the enviroment of the mother plant on germination by control of seed- coat permeability to water in Ononis sicula Guss.), Aust. J. Bot. Sci., 19:1007.
Graff, J. L., and Van Staden, J., 1983, The effect of different chemical and physical treatments on seed coat structure and seed germination of Sesbania species, Z. Pflanzenphysiol., 112:221.
Hagon, M. W., and Ballard, L. A. T., 1970, Reversibility of strophiolar permeability to water in seeds of subterranean clover (Trifolium subterranean L.), Aust. J. Bot. Sci., 23:519.
Hanna, P. J., 1984, Anatomical features of the seed coat of Acacia kempeana (Mueller) which relate to increased germination rate induced by heat treatment, New Phytol., 96:23.
Harris, W. M., 1984, Comparative ultrastructure of developing seed coats of "hard-seeded" and soft-seeded" varieties of soybean, Glycine max (L.), Merr. Bot. Gaz., 148:324.
Hill, H. J., West, S. H., and Hinson,, K., 1986, Effect of water stress during seedfill on permeable seed expression in soybean, Crop Sci., 26:807.
Horowitz, M., and Taylorson, R. B., 1985, Behavior of hard and permeable seeds of Abutilon theophrasti Medic. (velvetleaf), Weed Res., 25:363.
Hyde, E. O., 1954, The function of the hilum in some Papilionaceae in relation to the ripening of the seed and the permeability of the testa, Ann. Bot., 18:241.
Kelly, K. M., and Van Staden, J., 1985, Effect of acid scarification on seed coat structure, germination and seedling vigour of Aspalathus linearis, J. Plant. Physiol., 121:37.
Kelly, K. M., and Van Staden, J., 1987, The lens as the site of permeability in the Papilionoid seed Aspalathus linearis, J. Plant. Physiol., 128:395.
Kolattukudy, P. E., 1980, Cutin, suberin, and waxes, in: "The Biochemistry of Plants, Vol. 4, Lipids: Structure and function," P.K. Stumpf, ed., Academic Press, New York.
Kolattukudy, P. E., 1980, Biopolyster membranes of plants: Cutin and suberin, Sci., 208:990.
Kolattukudy, P. E., 1981, Structure, biosynthesis, and biodegradation of cutin and suberin, in: "Ann. Rev. Plant Physiol.," W.R. Briggs, P.B. Green, and R.L. Jones, eds., Ann. Reviews Inc., Palo Alto.
Kolattukudy, P. E., 1984, Biochemistry and function of cutin and suberin, Can. J. Bot., 62:2918.

Krefting, L. W., and Roe, E. I., 1949, The role of some birds and mammals in seed germination, Ecol. Monographs., 19:269.
Ladizinsky, G., 1985, The genetic of hard seed coat in the genus Lens, Euphytica., 34:539.
Lee, J. A., 1975, Inheritance of hard seed in cotton, Crop Sci., 15:149.
Lester, R. N., 1985, Seed germination stimulated by enzyme etching, Biochem. Physiol. Pflanzen., 180:709.
Manning, J. C., and Van Staden, J., 1987, The functional differentiation of the testa in seed of Indigofera paraviflora (Leguminosae: Papilionoideae), Bot. Gaz., 148:23.
Manning, J. C., and Van Staden, J., 1987, The role of the lens in seed imbibition and seedling vigour of Sesbania punicea (Cav.) Benth. (Leguminosae: Papilionoideae), Ann. Bot., 59:705.
Marbach, I., and Mayer, A. M., 1974, Permeability of seed coats to water as related to drying conditions and metabolism of phenolics, Plant. Physiol., 54:817.
Mayer, A. M., 1986, How do seeds sense their environment? Some biochemical aspects of the sensing of water potential, light, and temperature, Isr. J. Bot., 35:3.
McDonald, M. B., Vertucci, C. W., and Roos, E. W., 1988, Seed coat regulation of soybean seed imbibition, Crop Sci., 28:987.
Miklas, P. N., Townsend, C. E., and Ladd, S. L., 1987, Seed coat anatomy and the scarification of Cicer Milkvetch seed, Crop Sci., 27:766.
Noodèn, L. D., Blakley, K. A., and Grzybowski, J. M., 1985, Control of seed coat thickness and permeability in soybean - a possible adaption to stress, Plant Physiol., 79:543.
Patil, V. N., and Andrews, C. H., 1985, Development and release of hardseeded dormancy in cotton (Gossypium hirsutum), Seed Sci. and Technol., 13:691.
Pereira, L. A. G., and Andrews, C. H., 1985, Comparison of non-wrinkled and wrinkled soybean seedcoats by scanning electron microscopy, Seed Sci. and Technol., 13:853.
Potter, R. L., Petersen, J. L., and Ueckert, D. N., 1984, Germination responses of Opuntia spp. to temperature, scarification, and other seed treatments, Weed Sci., 32:106.
Potts, H. C., Duangpatra, J., Hairston, W. G., and Delouche, J. C., 1978, Some influences of hardseededness on soybean quality, Crop Sci., 18:221.
Powell, A. A., 1989, The importance of genetically determined seed coat characteristics to seed quality in grain legumes, Ann. Bot., 63:169.
Pritchard, H. W., Manger, K. R., and Prendergast, F. G., 1988, Changes in Trifolium arevense seed quality following alternating temperature treatment using liquid nitrogen, Ann. Bot., 62:1.
Pukittayacamee, P., and Hellum, A. K., 1988, Seed germination in Acacia auriculiformis: development aspects, Can. J. Bot., 66:388.
Quinlivan, B. J., 1971, Seed coat impermeability in legumes, Jour. Austral. Inst. Agric. Sci., 37:283.
Riggio-Bevilacqua, L., Roti-Michelozzi, G., and Serrato-Valenti, G., 1985, Barriers to water penetration in Cercis silquastrum seeds, Seed Sci. and Technol., 13:175.
Rolston, M. P., 1978, Water impermeable seed dormancy, Bot. Rev., 44:365.
Salisbury, P. A., and Halloran, G. M., 1983, Flowering sequence and the ordering of impermeability breakdown in seed of subterranean clover (Trifolium subterraneum L.), Ann. Bot., 52:679.
Slattery, H. D., Atwell, B. J., and Kuo, J., 1982, Relationship between colour, phenolic content and impermeability in the seed coat of various Trifolium subterraneam L. genotypes, Ann. Bot., 50:373.
Smith, A. K., and Nash, A. M., 1961, Water absorption of soybeans, J. Amer. Oil Chem. Soc., 38:120.

Soliday, C. L., Kolattukudy, P. E., and Davis, R. W., 1979, Chemical and ultrastructural evidence that waxes associated with the suberin polymer constitute the major diffusion barrier to water vapor in potato tuber (Solanum tuberosum L.), Planta., 146:607.

Taylor, G. B., 1981, Effect of constant temperature treatments followed by fluctuating temperatures on the softening of hard seeds of Trifolium subterraneum L., Aust. J. Plant Physiol., 8:547.

Tran, V. N., and Cavanagh, A. K., 1980, Taxonomic implications of fracture, load and deformation histograms and the effects of treatments on the impermeable seed coat of Acacia species, Aust. J. Bot., 28:39.

Tran, V. N., and Cavanagh, A. K., 1984, Structural aspects of dormancy, in: "Seed Physiology Vol. 2: Germination and Reserve Mobilization," D.R. Murray, ed., Academic Press, Orlando.

Tulley, R. E., Musgrave, M. E., and Leopold, A. C., 1981, The seed coat as a control of imbibitional chilling injury, Crop Sci., 21:312.

Werker, E., 1980/81, Seed dormancy as explained by the anatomy of embryo envelopes, Isr. J. Bot., 29:22.

Werker, E., Marbach, I., and Mayer, A. M., 1979, Relation between the anatomy of the testa, water permeability and the presence of phenolics in the genus Pisum, Ann. Bot., 43:765.

Woodstock, L., 1988, Seed imbibition: A critical period for successful germination, J. Seed Technol., 12:1.

Wyatt, J. E., 1977, Seed coat and water absorption properties of seed of near-isogenic snapbean lines differing in seed coat colour, J. Amer. Soc. Hort. Sci., 102:478.

# CONTROL OF GERMINATION AND EARLY DEVELOPMENT IN PARASITIC ANGIOSPERMS

Michael P. Timko, Christa S. Florea, and James L. Riopel

Department of Biology
University of Virginia
Charlottesville, Virginia 22901 USA

## INTRODUCTION

By definition, parasitic plants are a highly specialized group of organisms that derive all, or part, of their nutrition from other plants. There are both foliar and root parasite species. Combined, these plants represent at least eight different families and include mistletoe (Arceuthobium and related genera), dodder (Cuscuta), gerardia (Agalinis), sandalwoods (Buckleya and Thesium), broomrape (Orobanche) and witchweed (Striga) species. Most parasitic plants are hemiparasites that like Agalinis purpurea, photosynthesize, flower and mature to seed-set without a host. In contrast, species such as beechdrops (Epifagus virginiana) and squawroot (Conopholis americana) are holoparasites. These plants lack chlorophyll and are host-dependent for completion of seed-set. Yet other holoparasites, such as Striga asiatica, are photosynthetic, but remain dependent upon their host for other reasons.

Hemiparasites often have a very broad host range, whereas the holoparasites usually are more restricted in their range. The holoparasites Epiphagus and Conopholis parasitize only beech and red oak trees, respectively. Orobanche, a serious problem in a number of Mediterranean and European countries (Parker, 1986), attacks broadleaf plants such as faba bean, sunflower, tomato, squash, and carrot. Striga asiatica, and its relatives S. hermonthica, S. gesnerioides, and S. euphrasioides, are generally restricted to members of the Gramineae (grass) family. Combined, these parasites are responsible for tremendous losses of yield from agronomically important crops.

All of these parasitic plants are characterized by a common organ, the haustorium. Job Kuijt's description (Kuijt, 1969) of this structure as the physiological bridge providing a specialized nutrient channel from host to parasite is still the most valid. Haustoria exhibit a wide range of morphological types. In some root parasite species the haustoria are quite specialized and conspicuous even before host contact. In others, as in the case of Orobanche or Conopholis, the haustorium is difficult to detect prior to intrusive growth (Fig. 1).

In this paper we review the most recent advances in our understanding of the control of germination and early development in parasitic angiosperms. We focus primarily on work involving Striga asiatica, a species for which we know there exists highly specialized requirements for germination and early development. Over the past several years we have gained new insight into how host chemistry influences development in these plants. We have also begun to document the

*Recent Advances in the Development and Germination of Seeds*
Edited by R.B. Taylorson
Plenum Press, New York

biochemical and molecular genetic changes associated with these developmental processes. Studies on Striga serve as a model for further examination of the chemistry of host/parasite relationships. This information is important in providing possible new directions for chemical control of troublesome parasite species.

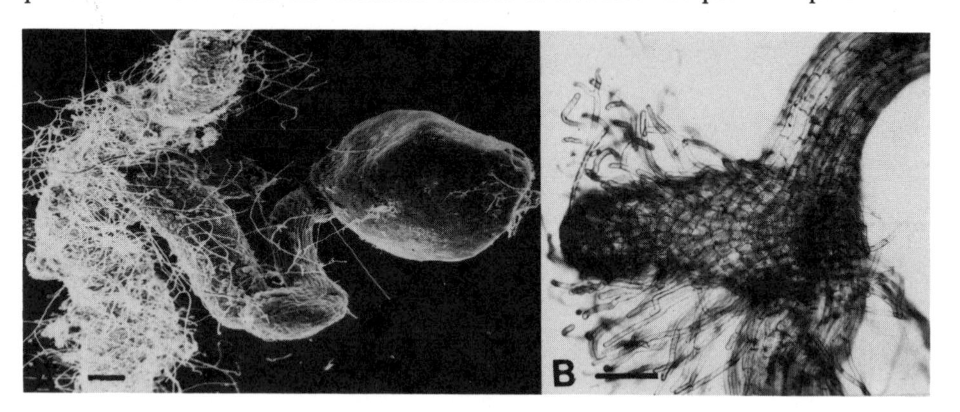


Figure 1. Haustorium of Agalinis purpurea and Conapholis americana.

Scanning electron micrograph of (panel A) seedling of Conopholis at an early invasion stage on oak root. Note the lack of a prominent haustorium. Bar = 2 mm; Panel B the haustorium of Agalinis with abundant haustorial hairs. Bar = 100$\mu$m.

## SEED DEVELOPMENT, WATER CONDITIONING, AND GERMINATION

It is important to recognize that the seeds of most hemiparasites have no special requirements for germination, or if they are temperate species, require only vernalization. For some species germination does not require the presence of a host plant; for others, the requirements for germination are more complex. The requirement for a host derived factor or factors for germination promotion has been reported for Orobanche (Vaucher, 1823; Koch, 1887; Chabrolin, 1938; and others), Tozzia (Heinricher, 1901), Aegenetia (Kusano, 1908), Striga (Saunders, 1933), and Alecta (Botha, 1948; 1950). We presently know the most about the signals used by Striga and Orobanche because of the world-wide impact of these plants on agriculture (Sherif, et al., 1987; Doggett, 1984; Parker, 1986). The detailed aspects of germination induction in Striga and other species will not be covered here since this topic has already been the subject of several other reviews (Kadry and Twefic, 1956; Sunderland, 1960; Brown, 1965; Riopel, 1983; and Worsham, 1987). However, some points warrant our consideration.

The seed of some parasitic plant species are large and contain enough storage metabolites to sustain radicle and shoot growth until a suitable host is contacted. In contrast, species such as Orobanche and Striga produce exceptionally large numbers of small seed. A single Striga plant, for example, produces up to 500,000 seeds, each approximately 0.25 - 0.3 mm long (Pavlista, 1981). As a consequence, these seeds contain extremely limited amounts of storage reserves and can sustain seedling growth for only a short period of time (Worsham, 1987).

In the field, Striga seeds retain their viability for up to 10 years and likely longer (Saunders, 1933). Studies of the age, size and weight of Striga seeds in relation to viability, germination, and subsequent parasitic growth have been conducted by Eplee and his coworkers (Bebawi, et al., 1984). Seed viability and germination have been shown to increase with increases in seed size and weight. Viability was also observed to progressively decline with the age of the seed. Associated with this, germination rates were shown to increase to a maximum at 3 year post-harvest and declined steadily in seed more than 4 years old. Studies with seeds of non-parasitic plant species have failed to demonstrate a consistent correlation between seed size and germination or field emergence. For example, while germination rates of larger seeds of alfalfa, (Medicago sativum) were found superior to that of smaller seeds (Erickson, 1946), there was no correlation in seed size and germination in radish (Brassica rapa) (Lamp, 1962). Doggett (1970) and later Bebawi, et al. (1984) have shown that Striga emergence was related to total number of plants present and a function of seed numbers. Data on the relationship of seed size and weight to emergence is more difficult to interpret. The relative mass of host root tissues in the soil appears to be of greater importance to emergence than seed size. A positive and high correlation has been shown to exist between the total number of Striga plants (emerged and submerged) parasitizing a host root and host root weight. There appears to be a synergistic effect of the parasite upon the host root as manifested by the increased development of the root. This may be part of the survial strategy of the parasite in which greater capacity for host contact results in greater reproductive numbers (see discussion by Bebawi, et al., 1984).

Completion of the entire germination process in seeds of most Striga species requires a period of after-ripening or post-harvest ripening of one to several months, followed by a preconditioning period under moist conditions of about one week. Both processes can be accelerated by elevated temperatures (Hsiao, et al., 1988a). Seeds conditioned in low levels ($10^{-8}$ M) of dl-strigol (a known germination stimulant) instead of water had lower germination rates. In addition, seeds conditioned in strigol required much longer conditioning times and higher concentrations of strigol as a terminal treatment to induce the same percentage of germination as water-conditioned control seeds (Hsiao, et al., 1988b). It is safe to say that the physiological mechanisms involved in the conditioning process of Striga seed are still not well understood. It is increasingly evident that some level of pre-germination conditioning must take place in order for the seed to be responsive to a germination stimulant of either host, or non-host origin. It has been suggested previously that preconditioning may act through one or a combination of mechanisms including the synthesis of an endogenous germination stimulant, leaching of inhibitory compounds from the seed, or increasing the permeability of seed to uptake of exogenous stimuli (Worsham, 1987). Biochemical changes have been noted in Striga seed during pre-germination and germination. Triacylglycerol, an abundant energy-rich storage component used by many oil crop seeds is also abundant in Striga seeds (Menetrez, et al., 1988). Triacylglycerol has been examined for its utility in energy formation during the germination and early development of this parasite. In contrast to those oil crop seeds examined, Striga did not significantly metabolize its triacylglycerol during germination and during the early stages of development post-germination, but appeared to reserve a significant portion of this storage component for the process of haustorial differentiation (Menetrez, et al., 1988). It is interesting to speculate on the basis for this conservation of energy by the parasite. Perhaps since germination is not the only event required to ensure its survival, the holding back of a certain portion of stored metabolic energy might be an evolved mechanism to ensure an energy supply for later critical developmental activities such as haustorial formation. Preconditioning of Striga in the field has been associated with changes in the levels of free amino acids, total proteins, total phenolics, sugars, RNA and DNA (Bkrathalakshmi, 1982). The significance of these changes are as yet unknown.

To date there has been no reported isolation of an endogenous germination stimulant from Striga seeds and only little progress in attempts to generalize on the importance of endogenous inhibitors. It is increasingly clear that the final requirement for germination of Orobanche and Striga seeds is exposure to an exogenous signal. This requirement has been known for for many years (Heinricker, 1898; Pearson, 1912; and Saunders, 1933). A broad range of compounds have been documented over the years as having either stimulatory or inhibitory effects on germination (see Riopel, 1983; Worsham, 1987). Especially noteworthy among the germination stimulants are ethylene, which has been effectively used in this country to promote suicidal germination as a control measure for Striga, and dl-strigol, a stable sesquiterpene isolated from the roots of cotton, a non-host plant species for Striga. Until recently, efforts to identify a germination stimulant from host plant roots have been unsuccessful.

Recently, David Netzley working in the laboratory of Larry Butler at Purdue University observed small colored droplets on the root hairs of Sorghum grown on moistened filter paper in petri plates. Bioassays of the hydrophobic droplet material carried out in our laboratory at the University of Virginia showed the presence of a potent germination stimulant among the constituents of this material. Extracts prepared from these droplets were shown to consist of almost pure concentrations of four para-benzoquinones (collectively termed sorgoleones) (Fig. 2) which in the dihydroquinone configuration are labile, but potent germination stimulants (Chang, et al., 1986; Netzley, et al., 1988).

To demonstrate that the production of the hydrophobic sorgoleone-rich exudate was not a culture artifact but was present on roots under field conditions, Sorghum and corn were grown in potting soil under greenhouse conditions. Soil used to support Sorghum root growth yielded an average of 160 to 230 mg of sorgoleones (equivalent to $10^{-4}$ to $10^{-5}$ mol/l in the soil), whereas soil supporting corn root growth contained no sorgoleones (Netzley, et al., 1988). Since corn is a host for Striga, it is possible that either the germination stimulant from

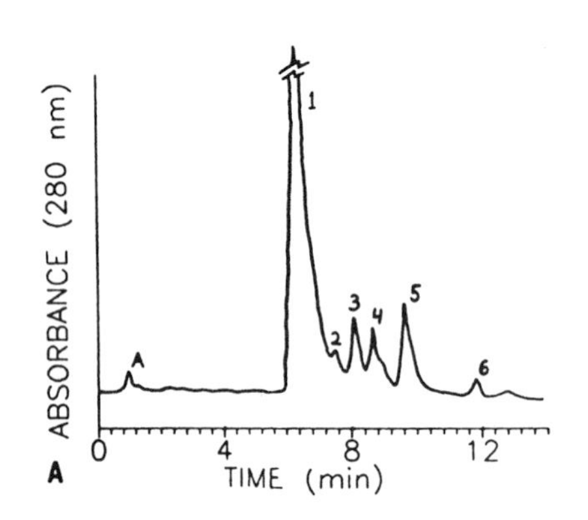


OH
OH
$CH_3O$
OH
B

Figure 2. Isolation and structure of the sorgoleones.

The upper panel of the figure shows an absorbance tracing from an HPLC separation of hydrophobic extract from Sorghum root hairs. The major peaks listed represent (1) sorgoleone-358, (2) dihydroquinone of sorgoleone-358, (3) sorgoleone-360, (4) sorgoleone 386, (5) sorgoleone-362, (6) benzoquinone 6; and A, a labile component. The lower panel show the molecular structure of the dihydroquinone of sorgoleone-358.

this species is more labile than that of its Sorghum counterpart, or is unrelated chemically.

The discovery of the sorgoleones has provided the basis for a theory of distance regulation for germination of Striga. Smith, et al. (1989) suggest that the Striga germination stimulant must convey spacial proximity of the host. They argue that if the hydroquinone is continually exuded from the Sorghum root and is auto-oxidized by dioxygen to the stable quinone, competition between the rates of production by the root and auto-oxidation define a steady state concentration gradient of diminishing signal from the root surface. The precise concentration dependence for germination combined with a concentration gradient of the hydroquinine could provide Striga with a mechanism for very precise distance control. Although intriguing there is currently little data to support this theory. Indeed, two arguments can be made against such a model.

There is evidence for many root components that promote germination. Both Sorghum and corn roots have been shown to exude water soluble germination stimulants (Brown, et al., 1952; Sunderland, 1960; Worsham, et al., 1964). In addition, exudates from roots of a broad range of plants also contain water soluble germination stimulants with high soil mobility. The general effectiveness of trap crops for control methods in Striga infested soils (Yaduraju and Hosmani, 1979) testifies to the existence of exuded chemicals of considerable abundance and soil mobility.

Further, direct measurements of diffusion distances for Sorghum germination promoters both in agar and soil have been made. In agar, Striga seeds germinate at distances up to 2 cm from the host surface (Riopel, 1983) and up to 1 cm from the root surface in soil tests (Linke and Vogt, 1987). These observations suggest that signal chemistry for germination is not especially effective in conserving Striga seeds. The maximum length of radicles growing close to the host root surface where elongation is inhibited is seldom more than 1.5 mm. This length of radicle elongation (1.5 mm) corresponds to a seedling approximately 3-4 days post-germination. Haustorial formation at this age of radicle development is significantly reduced (Riopel and Baird, 1987).

## ISOLATION AND CHARACTERIZATION OF HAUSTORIAL INITIATION FACTORS

The haustorium is a unique multicellular organ unlike any other both in its origin and function. The development of this organ signals the beginning of the parasitic phase of the plant's life-cycle. The association with a suitable host and haustorial development are critical events. It is not surprising that, like germination, a system of chemical signalling has evolved by which the processes of haustorial initiation and development and host penetration are accomplished.

In response to the perception of an exogenous signal, radicle elongation stops abruptly and rapid cellular differentiation within the root tip is initiated. Haustoria arise at endogenous positions of the parent root much like lateral roots. However, they differ from lateral roots in that they arise initially from cells of the inner cortex and in the majority of cases, the stimulus for differentiation is exogenous rather than endogenous.

Haustoria are always present on field collected specimens, but are frequently absent on both hemi- and holoparasites grown in sterile culture without a host (Riopel, 1979; Riopel and Musselman, 1979). In studies of Agalinis purpurea, a typical hemiparasite, we have documented the requirement for an exogenous host signal for initiation of haustorial development. The finding that gum tragacanth, a commercially available foliar exudate of Astragalus, contained an especially potent haustorial inducer was particularly helpful for this work. Two related molecules, Xenognosin A and B (Fig. 3), were isolated from this exudate and structurally characterized (Lynn, et al., 1981).

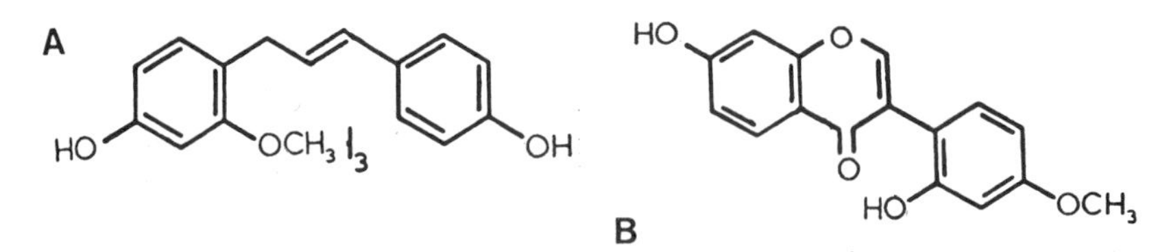


Figure 3. The molecular structure of Xenognosins A and B.

Subsequent characterization of the molecular specificity of Xenognosin A and its analogs (Steffens, et al., 1982; Kamat, et al., 1982) revealed two primary structural features required for activity. These are a meta relationship of the hydroxyl and methoxyl groups and an alkyl branching positioned ortho to the methoxyl substituent. Steffens, et al. (1986) also described the isolation of a haustorial initiation factor from root exudates of Lespedeza. This compound, a triterpene called soyasapogenol B, was less active than the flavanoids Xenognosin A and B and did not account for all of the haustorial inducing activity exhibited by the crude root exudate (Steffens, et al., 1986).

A number of independent studies have examined the ability of various compounds to function as haustorial initiation factors in Striga asiatica and S. hermonthica (MacQueen, 1984; Riopel and Baird, 1987; Chang and Lynn, 1986). There are at least two basic groups which can be distinguished (Table 1). The first are substituted quinones and related phenolics including Xenognosin A; the second are cytokinins and compounds of related structure. In general, the effective concentration range for these molecules is approximately $10^{-5}$ to $10^{-7}$ M.

Table 1. Active Substances in Promoting Primary Haustoria

| Phenolic compounds | Cytokinins |
|---|---|
| Arbutin | |
| Sinapic acid | 6-Benzyl aminopurine |
| Syringic acid | 6-Benzyl aminopurine riboside |
| Vanillin | 6-Benzylmercaptopurine |
| Vanillic acid | 6-N-hexylaminopurine |
| Isovanillin | Dihydrozeatin |
| O-Vanillin | Isopentenyl adenine |
| Vanillin acetate | Kinetin |
| Vanillin azine | Kinetin riboside |
| Umbelliferone | Zeatin |
| Xenognosin A | |
| Gum tragacanth (contains xenognosin A) | |

Haustorial development in response to host root exudate, phenolic compounds, or cytokinins is highly variable. Responses ranging from fully developed haustoria elaborated with multiple hairs to intermediate responses characterized by a slight swelling of the root tip are observed. These variations are in some cases concentration dependent as well as dependent upon the exact structural features of the compound.

Recently, a specific inducer of haustorial development has been isolated from Sorghum root extracts and identified as 2, 6-dimethoxy-p-benzoquinone (2,6-DMBQ) (Chang and Lynn, 1986) (Fig. 4). Analysis of synthetic analogs of 2,6-DMBQ showed that the haustorial inducer in Striga has structural requirements for activity similar to compounds active for Agalinis, notably the required presence of a methoxy functionality. This demonstrable specificity suggests that parasitic plants may have a receptor- mediated recognition system whose isolation and characterization may be a major answer to combat agronomically important parasites like Striga.

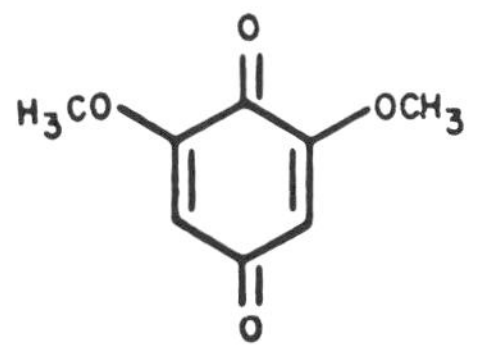

Figure 4. The molecular structure of 2,6-dimethoxy-p-benzoquinone.

Benzoquinones such as 2,6-DMBQ and related compounds are present in most plant species and arise by a number of independent biosynthetic pathways. Dewick and his coworkers (Dewick, 1975 and references therein) have shown that Xenognosin B is a direct biosynthetic precursor to the Leguminosae phytoalexin, medicarpin. Likewise, Xenognosin A has been identified in Pisum sativum as a stress metabolite (Carlson and Dolphin, 1982). These results support an earlier proposal (Atsatt, 1977) that parasitic plants, like herbivorous insects, key on a host's defense chemicals as recognition cues. Intermediates in stress response pathways, such as the lignin synthetic pathway involved in cell wall formation and repair, have been implicated in the induction of other plant pathogenic responses. Phenolic compounds (4-acetyl-2,6-dimethoxyphenol and some related molecules) present in the exudates of wounded or actively metabolizing plant cells have been shown to specifically activate vir gene expression during Agrobacterium tumefaciens infection (Stachel, et al., 1985; 1986). Flavones and isoflavones have similarly been implicated as the inducers of bacterial nodulation gene expression in Rhizobium (Redmond, et al., 1986) and Bradyrhizobium (Kosslak, et al., 1987).

A complex intracellular signalling mechanism by which Striga recognizes and determines its spatial distance from the host root has been proposed (Chang and Lynn, 1986; Smith, et al., 1989). It was suggested that the process of host recognition involves the release of an enzyme with properties similar to the phenol oxidases of pathogenic fungi from the Striga roots into the soil matrix. This enzyme oxidatively cleaves oligosaccharide components present on host root surfaces releasing into the soil matrix an appropriately substituted quinone, such as 2,6-DMBQ. These factors (2, 6-DMBQ or related compounds) diffuse back to the Striga radicle where they induce haustorial development. Through the release of this enzyme Striga presumably screens the environment for host root surfaces, responding only when the product of the enzyme-catalyzed breakdown of that surface reports back to the parasite. This ability to screen the environment is argued to be a more active and effective mechanism of host selection than previously suggested for these parasites. Furthermore, Lynn and his colleagues (Smith, et al., 1989) suggest that the compounds thus far identified which induce haustorial development do so only because they serve as substrate for the exuded enzyme. Although this is an attractive model further studies are clearly necessary to determine its validity. At present it does not take into account the range of compounds active in haustorial induction or all of the described mechanisms by which haustorial initiation is achieved. A fuller experimental characterization of the proposed parasite-derived enzyme and host root surface components will likely contribute to our understanding of these processes.

## CELLULAR ASPECTS OF HAUSTORIAL DEVELOPMENT

We have previously described the cellular aspects of haustorial development in the parasite Agalinis (Riopel and Musselman, 1979; Baird and Riopel, 1983; 1984). More recently we have described a similar series of developmental events in Striga (Riopel and Baird, 1987). The major aspects of this are shown in Fig. 5.

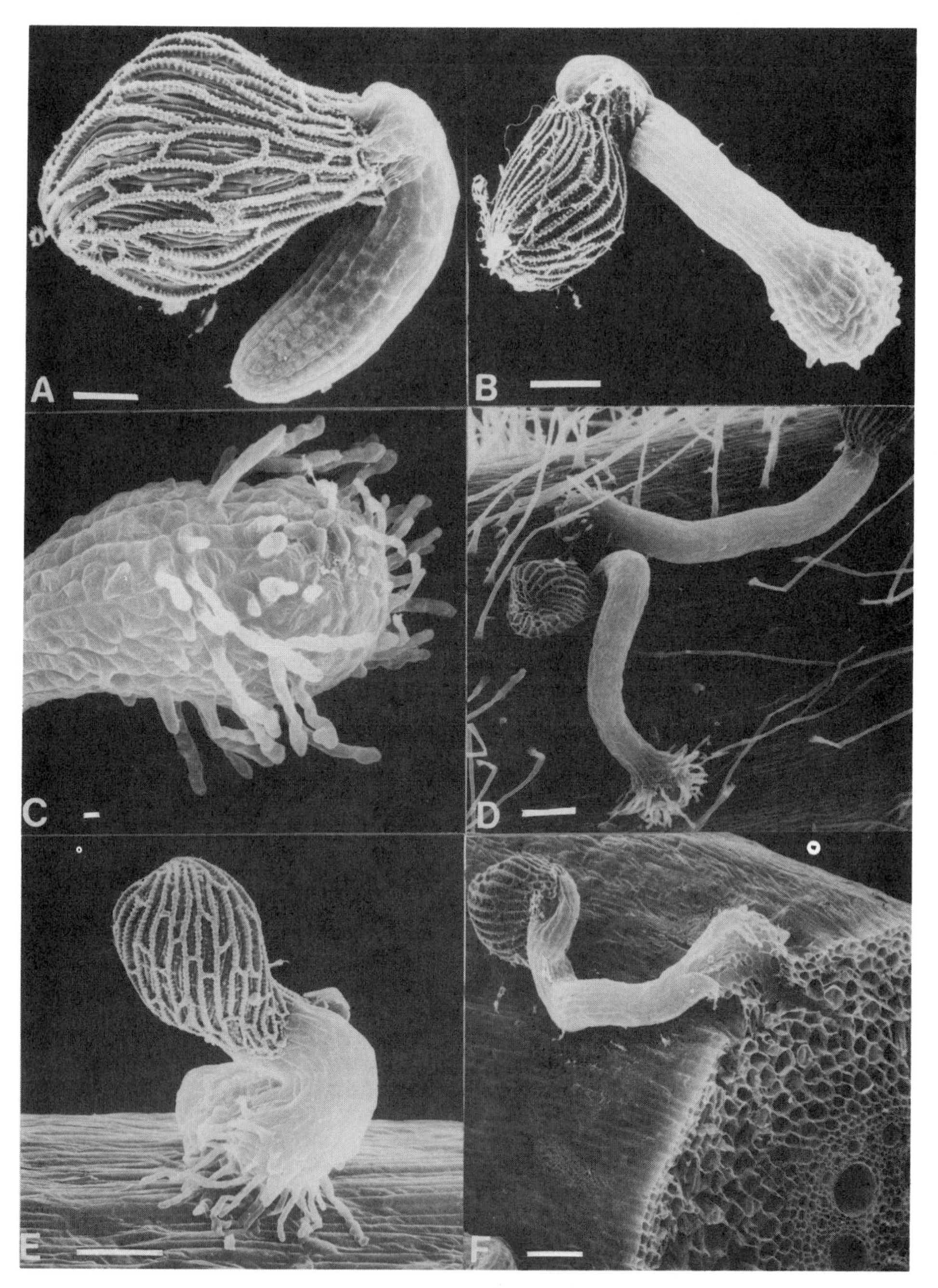


Figure 5. Major developmental stages during the striga life-cycle.

Scanning electron micrographs of germination, haustorial development and attachment stages of Striga asiatica. Panel A. Germination, 1 hr. Strigol, 11 hr., $H_2O$, Bar = 100$\mu$m; Panel B. Haustorium, 16 hr. corn induced. Bar = 100$\mu$m; Panel C. Haustorium, 20 hr. corn induced. Bar = 10$\mu$m; Panel D. Host surface contact, 18 hr. corn. Bar = 100$\mu$m; Panel E. Attachment, 36 hr. corn. Bar = 100$\mu$m; Panel F. Penetration on 5 day-old corn root. Bar = 100$\mu$m.

A primary haustorium develops from the radicle meristem. It is a rapid process with morphological effects seen in 4-6 hours following exposure of the radicle to host exudate or to 2,6-DMBQ. Significant events include localized expansion of cortical cells, the development of epidermal hairs and the appearance of hemicellulose papillae on hair cells. Haustoria show competence to host attach within 24 hours after induction.

We and others have noted that both the age of the radicle and the duration of the exposure to an exogenous signal influence the ability of the parasite to differentiate haustoria. It has been shown that the ability to respond to 2,6-DMBQ or related compounds, while present immediately following germination and radicle expansion is optimal for 36-60 hours post-germination. A drastic reduction in competency is observed within four days post-germination (Chang, 1986; Riopel and Baird, 1987). We have also observed (C. Florea and M.P. Timko, unpublished) that a minimum of 3 - 6 hours of continuous exposure to 2,6-DMBQ is required for formation of mature haustoria. Shorter exposure times (periods of less than 3 hours) or interuption of the exposure results in the cessation of radicle growth and swelling near the root tip, but failure to produce phenotypically mature haustoria.

This response can also be simulated by treatment with various chemicals. While structurally similar to 2,6-DMBQ, a number of compounds (e.g., syringaldehyde, hydroxybenzoic acid) are only capable of stimulating the early response in the radicle characterized by the cessation of radicle elongation and swelling of the root tip due to the initiation of cell division in the inner cortex. These compounds, however, do not provide sufficient information (signal) to induce later events necessary for haustorial maturation (i.e., formation of haustorial hairs).

We interpret these results as indicating that two distinct events occur in haustorial formation. First, there is the termination of radicle elongation and swelling near the root tip. Swelling always precedes further haustorial differentiation and requires only partial stimulation by an inducer. This stage represents an arrest point in development from which further haustorial development cannot occur without full inducer stimulation. The second event is the elaboration of hairs characteristic of the mature haustoria. The inability of some analogs to sustain full haustorial development suggests that initiation events and later events of maturation are developmentally separate. The use of such analogs may provide a useful tool to dissect regulation of these processes or to screen for genetic mutants.

## REQUIREMENT FOR CONTINUED PROTEIN SYNTHESIS IN HAUSTORIAL INDUCTION

To further elucidate the molecular mechanism controlling haustorial formation we examined the requirement for continued protein synthesis in haustorial induction and maturation. One to 2-day germinated seedlings were treated with 2,6-DMBQ for 6 hours to induce haustorial development either in the presence or absence of protein synthesis inhibitors (CHI = cycloheximide; CanA = canavanine A). After 6 hours seedlings in both the control and treatments were washed extensively in sterile water and subsequently placed into either water or a fresh solution containing 2,6-DMBQ. The results of these studies are summarized in Table 2.

As expected, in the absence of 2,6-DMBQ, radicles fail to differentiate haustoria. In contrast, when treated for 6 hours with 2,6-DMBQ we observed quantitative induction of haustorial differentiation. Inhibitor alone was not capable of eliciting a response in the seedlings.

If either CHI or CanA were included in the 6 hour preincubation period before application of 2,6-DMBQ, a dramatic effect on the ability of the radicles to differentiate haustoria was observed. Inhibitor-treated radicles subsequently

Table 2. Effect of Protein Synthesis Inhibitors on Haustorial Development.

ELONGATED RADICLES | SLIGHT SWELLING | LARGE SWELLING | MATURE HAUSTORIA

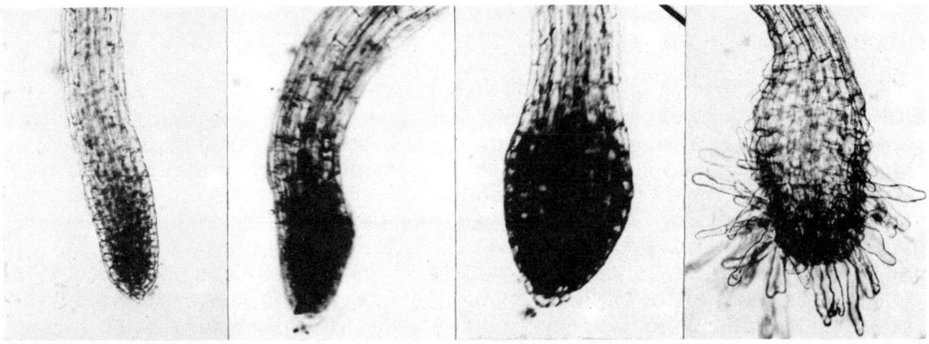

| Six Hour Pre-incubation/ Final Treatment | Elongated Radicles | Slight Swelling | Large Swelling | Mature Haustoria |
|---|---|---|---|---|
| Water / 2,6-DMBQ | 0.0 | 0.0 | 0.0 | 100.0 |
| 2,6-DMBQ / Water | 0.0 | 0.0 | 0.0 | 100.0 |
| $CHI_{(10)}$ / Water | 100.0 | 0.0 | 0.0 | 0.0 |
| $CHI_{(20)}$ / Water | 100.0 | 0.0 | 0.0 | 0.0 |
| CAN $A_{(10)}$ / Water | 100.0 | 0.0 | 0.0 | 0.0 |
| CAN $A_{(20)}$ / Water | 100.0 | 0.0 | 0.0 | 0.0 |
| $CHI_{(10)}$ / 2,6-DMBQ | 58.2 | 2.0 | 7.2 | 32.6 |
| $CHI_{(20)}$ / 2,6-DMBQ | 69.4 | 3.0 | 24.0 | 5.6 |
| CAN $A_{(10)}$ / 2,6-DMBQ | 62.2 | 6.0 | 11.0 | 20.8 |
| CAN $A_{(20)}$ / 2,6-DMBQ | 56.4 | 9.0 | 17.2 | 17.4 |
| $CHI_{(10)}$ + 2,6-DMBQ / water | 74.6 | 13.4 | 12.0 | 0.0 |
| $CHI_{(20)}$ + 2,6-DMBQ / water | 73.8 | 24.6 | 1.6 | 0.0 |
| CAN $A_{(10)}$ + 2,6-DMBQ / water | 36.4 | 21.2 | 34.8 | 7.6 |
| CAN $A_{(20)}$ + 2,6-DMBQ / water | 50.0 | 34.2 | 12.4 | 3.4 |
| $CHI_{(10)}$ + 2,6-DMBQ / 2,6-DMBQ | 47.4 | 2.0 | 44.4 | 6.2 |
| $CHI_{(20)}$ + 2,6-DMBQ / 2,6-DMBQ | 47.8 | 27.8 | 20.0 | 4.4 |
| CAN $A_{(10)}$ + 2,6-DMBQ / 2,6-DMBQ | 28.0 | 1.6 | 22.0 | 48.4 |
| CAN $A_{(20)}$ + 2,6-DMBQ / 2,6-DMBQ | 23.4 | 5.4 | 52.6 | 18.6 |

Values given represent the mean of 10 independent experiments using a total of 500 germinated seedlings and are expressed as percent of total. CHI = cycloheximide; CAN A = canavanine A. Subscripts refer to final concentration of inhibitor in $\mu$M.

exposed to 2,6-DMBQ developed haustoria at levels significantly reduced (approximately one-third) relative to untreated controls. Generally, CHI-treated radicles show greater inhibition than those treated with CanA. This may simply reflect the difference in mode of action of the two inhibitors.

When seedlings were treated with 2,6-DMBQ for 6 hours in the presence of protein synthesis inhibitors, the ability to induce haustoria was either completely lost (CHI treated samples) or substantially diminished (CanA treated samples). A large proportion of the radicles differentiate into intermediate haustoria. Thus, the switch to haustorial expression is achieved, but cannot be sustained through maturation. When additional 2,6-DMBQ is supplied after the preincubation period, large numbers of radicles remain arrested at the intermediate swelling stages. Even with prolonged incubation these radicles did not mature beyond this intermediate point of development.

These studies suggest that haustorial formation is a fragile sequence of events. Any interruption of the normal developmental pattern (either by providing insufficient signal or disrupting protein synthesis) results in most cases in an irreversible arrest at an intermediate stage. We observed that treatment of elongating radicles with protein synthesis inhibitors results in a decrease in the ability to respond to the 2,6-DMBQ. There are a number of interpretations possible for these results including the turnover of one or more proteins involved in either perception of the inducer or necessary in haustorial maturation. Obviously continued protein synthesis is necessary for development beyond radicle swelling. We demonstrate below that a different biosynthetic program is initiated involving new proteins.

To determine the complexity of changes in protein synthesis that accompany haustorial development, protein synthesis in radicles before and following induction with 2,6-DMBQ was analyzed. Uninduced and 2,6-DMBQ-induced radicles were incubated with radioactively labeled amino acids, the radicles and developing haustoria harvested, and the extent of protein synthesis monitored by two-dimensional polyacrylamide gel electrophoresis (Fig. 6). Comparison of the 2D-profiles of 2-day germinated radicles versus radicles of equivalent age but induced to form haustoria by 2,6-DMBQ clearly demonstrate that compositional differences (both qualitative and quantitative) are present. Several major polypeptide species of unknown structure and function accumulated in developing haustoria. Furthermore, changes in protein synthetic activities can be observed as early as 3 hours post 2,6-DMBQ induction.

## HOST ATTACHMENT AND PENETRATION

Some progress has also been made toward understanding the physiological constraints on host attachment and penetration by parasitic plants. In field collections large numbers of normal haustoria do not attach. Host availability is a factor. We also know that the stage of maturation and the presence of hemicellulose papillae on the haustorium affects attachment competency. In Agalinis significant numbers of haustoria attach as early as 12 hours after initiation and reach a maximal level of attachment by 36 to 48 hours. However, haustoria no longer can attach by 60 hours post-induction (Baird and Riopel, 1983). We observed in these studies that without host contact, hairs continue to develop but lack hemicellulose papillae. Thus, even though the haustorium will continue to develop for several weeks, functional attachment is restricted to a very short period.

In culture, haustoria of Agalinis continue to enlarge in the pre-contact stage for up to 2 months. These haustoria consist only of parenchymatous tissue. End point differentiation after host contact involves the differentiation of vascular tissue in the parasite and penetration of the host cortex and vascular system. There is cytochemical evidence for acid phosphatase involvement in cell wall lysis during host invasion by two mistletoe species (Rodriquez and Pannier, 1967; Onofeghara, 1972) and in Comandra umbellata infection (Toth and Kuijt, 1977). Evidence for the involvement of wall degrading enzymes in host penetration comes from the studies by Reddy, et al. (1981) who report pectinases and cellulases in the haustorial tissue of Cassytha. There is little specific information on Striga host penetration. Saunders (1933) reported incomplete enzyme dissolutions of host cell walls and recently Maiti, et al. (1984) described endodermal barriers to haustorial penetration for some Sorghum varieties.

We have also learned that in experiments in which attached haustoria have been dislodged from the host root, they maintain attachment competency longer than haustoria that have never been in host contact. This observation is important since it implies that there are physiological differences in haustoria influenced by host contact. The nature of these influences, whether physical or chemical, is unknown. It also has not been shown whether haustorial attachment

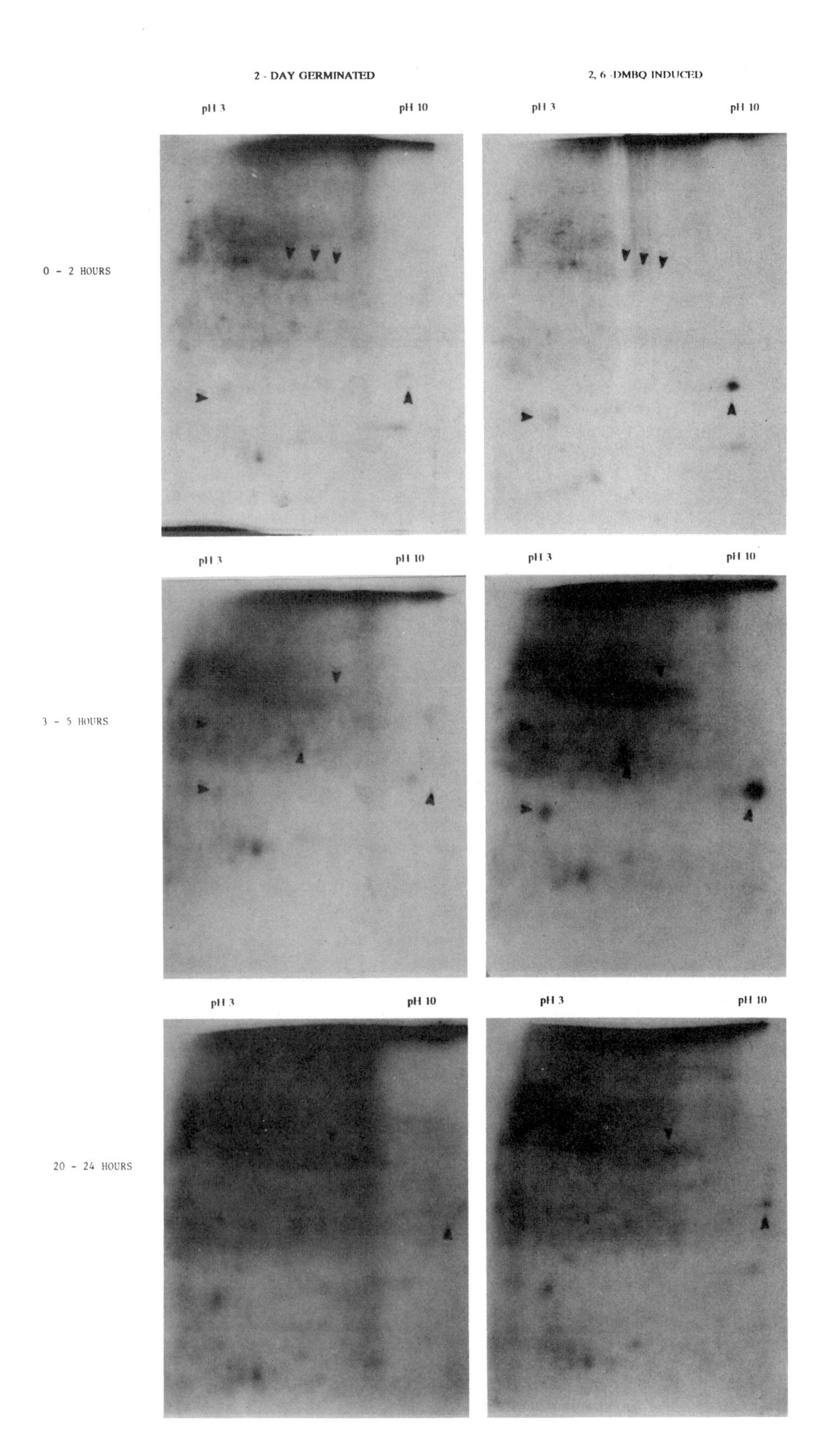
2 - DAY GERMINATED
2, 6 -DMBQ INDUCED
pH 3
pH 10
0 - 2 HOURS
3 - 5 HOURS
20 - 24 HOURS

Figure 6. In vivo protein synthesis in radicles before and after 2,6-DMBQ induction.

Total protein synthesis in radicles before and after treatment with 2,6-DMBQ was measured in the following manner. Two-day germinated seedlings, approximately 50-100 per microtitre well, were incubated in either sterile water or inducer ($1 \times 10^{-5}$ M) for 24 h. At various time points after induction 10 uCI of $^{35}$S-methionine/cysteine (Tran$^{35}$S-label, ICN Radiochemicals, specific activity = 1117 Ci/mmol) was added. Uptake of radiolabeled amino acids was allowed to continue for 2 h incubation after which the seedlings were washed extensively with sterile water and either radicles or radicles with haustoria from 50 seedlings were removed mechanically with a pair of forceps. The samples were solubilized immediately and equivalent amounts of total TCA precipitable counts were loaded (approximately $6 \times 10^4$ cpm). Samples were focused in the first dimension according to the procedure of O'Farrell (1975) and fractionated on SDS-polyacrylamide gels in the second dimension according to Laemmli (1970). Gels were treated for fluorography, dried and exposed to Kodak X-Omat film with an intensifying screen for 1 to 2 wks. Representative autoradiographs from various time points in development are shown.

alone can provide the full signal or, as is more likely, that continued regulation by host chemistry is required. It seems likely that in addition to hemicellulose biosynthesis, a number of other biosynthetic processes are induced by host contact including the formation of new proteins required for haustorial structure and function during intrusive growth.

## SUMMARY

Our understanding of the biology of parasitic plants has moved ahead rapidly in the 1980's (see Weber and Forstreuter, 1987). This is a subject of increasing research interest prompted both by the agronomic importance of these plants as well as the many interesting fundamental questions that remain for study.

We have appreciated for many years that obligate parasites such as Striga and Orobanche require several conditions and host contact for survival. We have only begun to unravel what these conditions are. For Striga, we now know the molecular structure for some of the host signals that promote germination. Striga seeds also germinate in the presence of non-hosts and there have been a variety of germination promoters identified that are common in root exudates. From agar diffusion studies now in progress we know that even with a host species, a very large proportion of Striga seeds germinate outside of their radicle elongation range. We anticipate that similar effects would be observed under field conditions. We feel that discrimination at the level of germination induction is therefore unlikely to be a dominant component in the survival strategy for Striga.

We also know now the identity of chemical signals involved in the initiation of haustorial development. Like the situation discussed above for germination, there is a broad range of chemicals that are capable of inducing haustorial formation. These appear to be mostly phenolics with similarly placed methoxyl groups. It is evident, however, that in the case of Striga haustorial induction is not host discriminate. Striga initiates this organ against about any root we have tried including both host and non-host species. Striga does, however, conduct some very clever chemistry. It maintains rapid elongation of its radicle toward a "potential" host root. The chemical signals now defined, and perhaps others, enable Striga to shut down radicle elongation and initiate haustorial development at a time most favorable for attachment. This process, however, does not confer a high level of specificity to the parasite/host interaction at this level. It would be surprising indeed if the signals identified for Striga have much to do, other than in a general way, with how Orobanche, Epifagus, or Conopholis finds its proper host. Clearly, further detailed analyses of these systems are warranted.

We have also begun to examine the biochemical and molecular genetic changes that occur during germination and haustorial development in Striga. The identification of specific chemical signals capable of precisely controlling development has benefited these studies. Our work described in part here indicates that haustorial formation is a fragile sequence of events. When disrupted, there is a failure to develop a functional parasitic organ. This disruption can occur both at the level of signal perception as well as during the maturation of the haustorium. Continued protein synthesis is required for both of these events and haustorial development involves the activation of a new set of genetic information leading to the formation of new proteins in the parasite. Studies are now in progress in our laboratory to isolate and characterize cDNAs specifically expressed during haustorial formation and the genes which encode them. In combination with our expanding knowledge of the chemistry of parasite/host interactions, these molecular tools will allow us to carefully dissect this complex and interesting plant/pathogen interaction.

## ACKNOWLEDGMENTS

The authors wish to thank Tamela Davis for help in preparing this manuscript and Dr. Susan Wolf for her thoughtful comments. We wish to acknowledge Dr. Wm. V. Baird, Dept. of Horticulture, Clemson University for use of the SEM photographs. This work was support in part by a grant from the Thomas F. and Kate Miller Jeffress Memorial Trust awarded to MPT.

## LITERATURE CITED

Atsatt, P. R., 1977, The insect herbivore as a predictive model in parasite seed biology, Amer. Natur., 111:579.

Baird, W. V., and Riopel, J. L., 1983, Experimental studies of the attachment of the parasitic angiosperm Agalinis purpurea to a host, Protoplasma, 118:206.

Baird, W. V., and Riopel J., 1984, Experimental studies of haustorial initiation and early development, in: Agalinis purpurea, (L.) Raf. (Scrophulariaceae), Am. J. Bot., 71:803.

Bebawi, F.F., Eplee R.E., and Nour R. S., 1984, Effect of age, size, and weight of witchweed seeds on host/parasite relations, Phytopath., 74:1074.

Bkrathalakshmi, B., 1982, Studies on the root parasite - Striga asiatica (L.) Kuntze. Ph.D. Thesis, Bangalore University, Bangalore.

Botha, P. J., 1948, The parasitism of Alectra vogelii Benth. with special to the germination of its seeds, J. South. Afr. Bot., 14:63.

Botha, P. J., 1950, The germination of the seeds of angiospermous root parasites. I. The nature of the changes occurring during pre-exposure of the seed to Alectra vogelii, Benth., J. South Afr. Bot., 16:29.

Brown, R., 1965, The germination of angiospermous parasite seeds. Handburch der pflanzenphysiologie, 15:925.

Brown, R., Johnson, A. W., Robinson, E., and Tyler, G. J., 1952, The Striga germination factor. II. Chromatographic purification of crude concentrates, Biochem. J., 50:596.

Carlson, R.E. and Dolphin, D. H., 1982, Pisum sativum stress metabolites: Two uniannamylphenols and a 2'-methoxychalcone, Phytochem., 21:1733.

Chabrolin, C., 1938, Contribution a l'etude de la germination des graines de l'orobanche de la feve, Ann. Serv. Bot. Agron. Tunis., 14:91.

Chang, M., 1986, Isolation and characterization of semiochemicals involved in host recognition in Striga asiatica, Ph.D. dissertation, Dept. of Chemistry, Univ. of Chicago.

Chang, M. and Lynn D. G., 1986, The haustorium and the chemistry of host recognition in parasitic angiosperms, J. Chem. Ecol., 12:561.

Chang, M., Netzley D. H., Butler L. G., and Lynn D. G., 1986, Chemical regulation of distance: characterization of the first natural host germination stimulant for Striga asiatica, J. Am. Chem. Soc., 108:7858.

Dewick, P. M., 1975, Pterocarpan biosynthesis: 2'-hydroxy-isoflavone and isoflavone precursors of dimethylhomopterocarpin in red clover, J. Am. Chem. Soc. Chem. Commun. pp. 656.

Doggett, H., ed. 1970, "Sorghum", Longmans Green, London pg. 278.

Doggett, H., 1984, in: "Striga Biology and Control", ICSU Press, Paris France.

Erickson, L. C., 1946, The effect of alfalfa seed size and depth of seeding upon the subsequent procurement of stand, J. Am. Soc. Agron., 38:964.
Heinricher, E., 1898, Die grünen Halbschmarotzer. I. Odonities, Euphrasia, und Orthantha, Jahrb. Wiss. Bot., 31:77.
Heinricher, E., 1901, Die grünen Halbschmarotzer. III. Bartschia und Tozzia, nebst Bemerkungen zur Frage nack der assimilalorischen Leistungsfahigbeit der grünen Halbschmarotzer, Idem., 36:665.
Hsiao, A. I., Worsham A. D., and Moreland D. E., 1988a, Effects of temperature and dl-Strigol on seed conditioning and germination of witchweed (Striga asiatica), Ann. Bot. 61:65.
Hsiao, A. I., Worsham A. D., and Moreland D. E., 1988b, Effects of chemicals often regarded as germination stimulants on seed conditioning and germination of witchweed (Striga asiatica), Ann. Bot., 62:17.
Kadry, A. E. R., and Twefic H., 1956, Seed germination of Orobanche crenata, Sv. Bot. Tidskr., 50:270.
Kamat, V. S., Graden D. W., Lynn D., Steffens J., and Riopel, J. L., 1982, A versatile total synthesis of xenognosin, Tetra. Lett. 23:1541.
Koch, L., 1887, Die Entwichlungsgeschichte der Orobanchen mit besonderen Berücksichtigung ihrer Beziehung zu den Kulturpflanzen, C. Winter, ed., Heidelberg.
Kosslak, R. M., Bookland, R., Barkei, J., Paaren, H. E., and Appelbaum, E. R., 1987, Induction of Bradyrhizobium japonicum common nod genes by isoflavones isolated from Glycine max, Proc. Natl. Acad. Sci. USA., 84:7428.
Kusano, S., 1908, Further studies on Aeginetia indica, Beih. Bot. Centralbl., 24:286.
Kuijt, J., 1969, "The Biology of Parasitic Flowering Plants", Berkeley, Univ. Calif. Press. 246 pp.
Lamp, C. A., 1962, Improved turnip crop establishment, Tasm. J. Agric. 33:42.
Laemmli, U. K., 1970, Cleavage of structural proteins during the assembly of the head of bacteriophage $T_4$, Nature, 227:680.
Linke, K. H., and Vogt, W., 1987, A method and its application for observing germination and early development of Striga (Scrophulariaceae) and Orobanche (Orobanchaceae). in: "Parasitic Flowering Plants", Weber, H. Chr. and W. Forstreuter, eds., Marburg, F.R.G. p. 501-509.
Lynn, D. G., Steffens, J., Kamat, V., Graden, D., Shabanowitz, J., and Riopel, J. L., 1981, Isolation and characterization of the first host recognition substance for parasitic engrosperms, J. Am. Chem. Soc. 103:1868.
MacQueen, M., 1984, Haustorial initiating activity of several simple phenolic compounds, in: "Proc. Third Int. Symp. on Parasitic Plants", Aleppo, Syria. p. 118.
Maiti, R. K., Ramaiah, K. V., Bisen, S., and Chidley, V. L., 1984, A comparative study of the haustorial development of Striga asiatica (L.) Kuntze on sorghum cultivars, Ann. Bot., 54:447.
Menetrez, M. L., Fites, R. C., and Wilson, R. F., 1988, Lipid changes during pre-germination and germination of Striga asiatica seeds, J. Am. Oil Chem. Soc., 65:634.
Netzley, D. H., Riopel, J. L., Ejeta, G., and Butler, L. G., 1988, Germination stimulants of witchweed (Striga asiatica) from hydrophobic root exudate of sorghum, (Sorghum bicolor), Weed Science, 36:441.
O'Farrell, P. H., 1975, High resolution two-dimensional electrophoresis of proteins, J. Biol. Chem., 250:4007.
Onofeghara, F., 1972, Histochemical localization of enzymes in Tapinanthus bangwensis: acid phosphatase, Am. J. Bot., 59:549.
Parker, C., 1986, in: "Biology and Control of Orabanche", Workshop Proc. Wageningen, The Netherlands.
Pavlista, A. D., 1981, Why hasn't witchweed spread in the United States? Weeds Today, special isssue, p. 16-19.
Pearson, H. H. W., 1912, On the Rooibloem (Isoma or Witchweed), Agr. Jour. Union S. Afr., 2:1.
Reddy, A. S., Komizaiah, M., and Reddy, S. M., 1981, Production of pectin enzymes by Cassytha filiformis L., Curr. Sci., 50:283.

Redmond, J. W., Batley, M., Djordjevic, M. A., Innes, R. W., Kuempel, P. L., and Rolfe, B. G., 1986, Flavones induce expression of nodulation genes in Rhizobium, Nature, 323:632.
Riopel, J. L., 1979, Experimental studies on induction of haustoria in Agalinis purpurea, Second Symp. Parasitic Weeds., N.C. State Univ., p. 165.
Riopel, J. L., 1983, in: "Vegetative Compatability", R. Moore, ed., Academic Press Inc., New York.
Riopel, J. L., and Musselman, L. J., 1979, Experimental induction of haustoria in Agalinis purpurea (Scrophulariaceae), Am. J. Bot., 66:570.
Riopel, J., and Baird, W. V., 1987, Morphogenesis of the early development of primary haustoria in Striga asiatica. in: "Parasitic Weeds in Agriculture". Vol. 1 Striga., Musselman, L.J., ed., pg. 108-125.
Rodriguez, M., and Pannier, F., 1967, Etude de la distribution de la phosphatase acide dans l'haustorium primaire de Phthirusa pyrifolia (H.B.K.) Eichl. (Loranthaceae), Rev. Gen. Bot., 74:625.
Saunders, A. R., 1933, Union S. Africa Agric., Sci. Bull. No. 128, 56 p.
Sherif, A. M., Hanna, W. W., and Berhane, M., 1987, The problem of Striga Lour. (Scrophulariaceae) in Ethiopia, in: "Parasitic Flowering Plants", Weber, H. Chr. and W. Forsteuter, eds., Marburg, F.R.G. p. 755
Smith, C., Orr, J. D., and Lynn, D. G., 1989, Chemical communication and the control of development, in: "Natural Products Isolation". Cooper, R. and G.H. Wagman., eds., Elsevier Publishers, N.Y.
Stachel, S. E., Messens, E., Van Montagu, M., and Zambryski, P., 1985, Identification of the signal molecules produced by wounded plant cells that activate T-DNA transfer in Agrobacterium tumefaciens, Nature, 318:624.
Stachel, S. E., Nester, E. W., and Zambryski, P. C., 1986, A plant cell factor induced Agrobacterium tumefaciens vir gene expression, Proc. Natl. Acad. Sci. USA, 83:379.
Steffens, J., Lynn, D. G., Kamat, V. S., and Riopel, J. L., 1982, Molecular specificity of haustoria induction in Agalinis purpurca (L.) Raf. (Scrophulariaceae), Ann. Bot., 50:1.
Steffens, J., Lynn, D. G., and Riopel, J. L., 1986, An haustoria inducer for the root parasite Agalinis purpurea, Phytochemistry, 25:2291.
Sunderland, N., 1960, Germination of the seeds of angiospermous root parasites. Brit. Ecol. Soc. Symp., 1:83.
Toth, R., and Kuijt, J., 1977, Cytochemical localization acid phosphatase in endophyte cells of the semiparasitic angiosperm Comandra umbellata (Santalaceae) Can. J. Bot., 55:470.
Vaucher, J. P., 1823, Memoire sur la germination des Orobanches, Memoires due Museum d'Histoire Naturelle, 10:261.
Weber, H. C., and Forstruter, W., (eds), 1987, "Parasitic Flowering Plants", Proc. 4th Int. Symp. Parasitic Fl. Pl., Marburg, F.R.G.
Worsham, A. D., 1987, Germination of witchweed seeds, in: "Parasitic Weeds in Agriculture", Musselman, L., ed., CRC Press p. 46-61.
Worsham, A. D., Moreland, D. E., and Klingman, G. C., 1964, Characterization of the Striga asiatica (witchweed) germination stimulant from Zea mays L., J. Exp. Bot., 15:556.
Yaduraju, N. T., and Hosmani, M. M., 1979, Striga asiatica control in Sorghum, PANS, 25:163.

FACTORS ELICITING THE GERMINATION OF PHOTOBLASTIC KALANCHOE SEEDS

J. A. De Greef*, H. Fredericq, R. Rethy, A. Dedonder, E. De Petter and L. Van Wiemeersch

Laboratory of Plant Physiology, University of Gent, K.L. Ledeganckstraat 35, B-9000 Gent, Belgium. *Dept. Plant Biology University of Antwerpen, UIA-RUCA, Universiteitsplein 1, B-2610 Antwerpen-Wilrijk, Belgium

## INTRODUCTION

In the Series "Monographies de Physiologie Végétale" edited by professor P.E. Pilet of the University of Lausanne, Switzerland, Daniel Côme (1970) wrote in the introductory remarks of his book "Les Obstacles à la Germination" : "... si l'on connaît le plus souvent les causes essentielles de la non-germination d'une semence, et si l'on sait quels traitements appliquer à cette semence pour la rendre apte à germer, les raisons profondes de la non-germination sont les plus souvent inconnues ou seulement très partiellement connues. Les recherches actuelles s'orientent progressivement vers une meilleure compréhension de ces mécanismes. L'expérimentation est difficile du fait que le repos des semences ne correspond pas à un simple arrêt de croissance et que la germination elle-même est encore un phénomène assez mal connu."

At the present, about twenty years later, a large body of experimental work on seed germination has been published, but still we have no profound knowledge of the mechanisms controlling the germination process although we all know that a greater wealth of information has become available with regard to the external and internal cues, - physical as well as chemical, - influencing the physiological conditions of seed germination induction.

Around the same time of the publication of Côme's monograph we started a research project on the photophysiology of seed germination of Kalanchoë blossfeldiana Poellniz cv. Feuerblüte. For the flowering response of this short day plant Fredericq (1965) showed that

1) the degree of the R/FR reversibility in the middle of the night is strongly dependent upon the characteristics of the preceding short day ;
2) the inhibitory action of FR given at the end of day is stronger when the duration or light intensity of the photoperiod is reduced ; the same dependency of terminal FR upon the main white light period was also demonstrated for the vegetative growth of higher (Downs et al., 1957) and lower plants (Fredericq and De Greef, 1968).

Since comparable studies on seed germination were scarce and not extensive at that time (Cumming, 1963 ; Borthwick et al., 1964), we aimed to study the interrelationships between the effects of the main light period and terminal R/FR irradiations also for the germination process.

*Recent Advances in the Development and Germination of Seeds*
Edited by R.B. Taylorson
Plenum Press, New York

Indeed, a better knowledge of such interactions in very different aspects of the plant's life cycle could be helpful to elucidate the mechanism(s) of Pfr mediated physiological responses.

In this paper we give an overview of the experiments of our seed germination project as they proceeded upon the line of our thoughts with time.

DRY SEED WEIGHT versus LIGHT SENSITIVITY (Dedonder et al., 1980)

Kalanchoë seeds are very small ($4.10^{4}$ seeds.$g^{-1}$). In screening experiments concerning the effects of light and GA on germination behaviour, we investigated 250 seed species belonging to 148 genera and 46 families. Among these there were 36% light independent, while 24% of the seeds showed to be partly and 40% absolutely light requiring.

As it is shown in Fig. 1 there is obviously a correlation between light sensitivity and dry seed weight. Very small seeds are most pronounced with respect to their light requirements. This holds also for seeds belonging to the same family. In this way small seeds buried in the soil at greater depth can not germinate and have more chance to survive. Such small seeds have indeed only little food reserves and limited soil penetration capacity.

Analytical and biochemical studies are very difficult in these kinds of seeds. Therefore, most of our germination studies are confined to whole plant physiology.

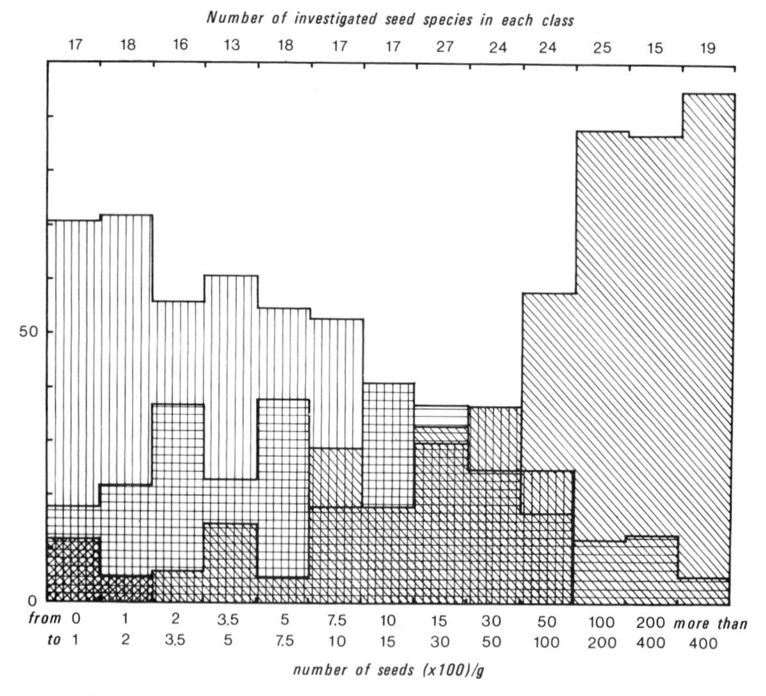


Fig. 1 Correlation between weight and influence of light on the germination of different seed species.
The experiments were performed at 20°C, in darkness or continuous white fluorescent light. A distinction was made between 3 groups vertical lines = light insensitive ; horizontal lines = partly light requiring ; oblique lines = absolutely light requiring.
The numbers of seeds, belonging to the three groups in each weight-class are given as %.

DEPENDENCE OF GERMINATION ON PHOTOPERIOD, LIGHT QUALITY AND TEMPERATURE. (Eldabh et al., 1974 ; Fredericq et al., 1975)

Bunsow and Von Bredow (1958) demonstrated that germination of *Kalanchoë blossfeldiana* seeds is absolutely light-requiring and needs repeated daily inductive light cycles.

Table 1 shows that a minimal number of daily light periods is required. Even 48 hours of continuous WL induced only 23% germination while 18 hours WL given over 6 days as 3 h photoperiods caused 80%. Very sensitive seed batches germinated up to 85% with only 1 min WL daily. In continuous darkness we never met any germinating seed. Exposing the imbibed seeds to longer photoperiods than those minimally required for maximal germination accelerated the germination process. Apparantly, the completion of Kalanchoë germination needs a prolonged period of Pfr action. Phytochrome control became evident when photoperiods potentiating the seeds to full germination, could be negated by terminal FR, while a subsequent R exposure fully restored germination (Eldabh et al., 1974).

Table 1 Minimal photoperiod requirement for germination of Kalanchoë seeds.

| number of photoperiod during 9 subsequent days at 20°C | germination % A 1 min WL.day$^{-1}$ | B 3 h WL.day$^{-1}$ |
|---|---|---|
| 9 | 85 | 80 |
| 5 + CD | 82 | 60 |
| 4 + CD | 48 | 23 |
| 3 + CD | 35 | 2 |
| 2 + CD | 0 | 0 |
| CD | 0 | 0 |

Seeds were imbibed for 18 hours in darkness before taking them in an experimental run for 9 days and temperature was kept at 20°C unless otherwise stated.
WL : during the photoperiods the seeds were exposed to 40 W fluorescent "Phytor" lamps with an intensity of 10.4 μW.cm$^{-2}$ at the peak emission of 660 nm.
Seed batch A was more light sensitive than batch B.

Besides the R/FR photoreversibility it can also be seen in Table 2 that there is an escape from the FR inhibition depending upon the length of the preceding photoperiod, the light intensity during the main photoperiod and the duration of the terminal FR. Thirty seconds FR completely suppressed the beneficial effect of WL photoperiods up to 2 hours. One min FR was fully inhibitory after 4-hour days, but caused no inhibition after 12-hour photoperiods, although the effect of 12-hour photoperiods could be strongly reduced when terminated by 12 hours FR irradiation.

With regard to the light intensity of the main photoperiod (Table 2, C) we notice that 10 min FR strongly inhibits germination after 12 h WL of 0.1 μW.cm$^{-2}$ at 660 nm, but not if the intensity of the photoperiod is 2.8 μW.cm$^{-2}$ at 660 nm or more. Twelve-hour WL photoperiods of 0.1 μW.cm$^{-2}$ at 660 nm result in maximal germination, while 1 h WL photoperiods of 8.5 μW.cm$^{-2}$ at 660 nm could not promote full germination although time x intensity of light exposure exceeded largely that obtained by the 12 h photoperiod.

These results indicate that the final germination response under those experimental conditions is not determined by the Bunsen-Roscoe law, but by the light intensity and the time span of repeated photoperiods. For both VLFR and LFR however, reciprocity holds, as we will see furtheron.

Obviously, the temperature during the photoperiod interacts with the escape from FR-inhibition (Table 2, D). Far-red was also less inhibitory at 25°C than at 20°C with 5 h photoperiods. Germination is optimal at 20°C and 25°C when either 8 h WL photoperiods or continuous WL are applied. When sub-optimal photoperiods are used however, a rise of only 5°C (20°-25°C) during a 1-hour photoperiod while the night temperature was kept at 20°C improved germination. Instead, the same temperature rise during the long night completely suppressed germination. This temperature was more inhibitory in the first than the second half of the night. At a constant temperature of 25°C not a single seed germinated with 1- or 2-hour photoperiods, the unfavorable effect of 25°C during the long night prevailing over the enhancing effect during the very short day.

To sum up at this point, full reversion of Pfr action can be obtained by a short FR exposure only when given immediately after a relatively short photoperiod (up to 4 hours) of high intensity and keeping the seeds at the same temperature all the time. Under these conditions we assume that FR acts on the initial photochemical Pr/Pfr transformation process.

After longer photoperiods of either high or low intensities and at the same physiological temperature (20°C) during the whole experimental run, no or incomplete reversion by the same FR could be obtained. Pfr action has thus proceeded to a state of irreversibility (= escape from FR-inhibition), completely different from the photochemical transformation. We assume therefore that Pfr moved to a further step

$$\text{Pfr} \rightarrow \text{Pfr*} \qquad \text{(at constant T°)}$$

This "reacted" phytochrome is comparable to PfrX, postulated by others (Hartmann 1966, Borthwick et al., 1969).

The formation of reacted phytochrome is dependent upon the number (Table 1), the length, the light intensity (Table 2) of the photoperiods and temperature.

In our system Pfr* formation is very slow, it takes several photoperiodic cycles before its effect can be demonstrated. In photoperiods longer than 4 hours the transformation into Pfr* occurs at the same rate as the FR inhibition is decreased. The final germination response is not determined by the total light energy, but by the light intensity and the continuity of a minimum number of photoperiods.

Based on its physiological expression we suppose that reacted phytochrome can only be formed after a minimal length of the photoperiod and a minimum duration of Pfr present in light or darkness. The length depends upon the sensitivity of the seeds and a threshold value of light intensity during the photoperiods. Since even 1 min photoperiods are inductive in very sensitive seeds, we conclude that Pfr* is formed in light as well as in darkness. The higher the light intensity during the photoperiod, the faster the onset of reacted phytochrome formation and its completion. After prolonged periods of darkness (several days), following short or long photoperiods, Pfr* disappears again.

The longer the exposure to terminal FR, the more inhibitory (Table 2,B). Similar results were obtained for germination of other seeds (Hendricks et al., 1959 ; Rollin, 1963 ; Mohr and Appuhn, 1963).

By the same token, the considerable enhancement of germination, when the temperature is increased from 20 to 25°C during a suboptimal photoperiod, is due to biochemical events subsequent to the photochemical

Table 2 R/FR reversibility and escape from FR inhibition (in % germination)

A. PHOTOREVERSIBILITY

| WL photoperiod | control | 5 min FR | 5 min FR + 5 min R |
|---|---|---|---|
| 10 min | 66 | 0 | 70 |
| 8 h | 93 | 24 | 90 |
| 10 h | 96 | 77 | 90 |

B. EFFECT OF TERMINAL FR vs LENGTH OF PHOTOPERIOD

| WL photoperiod | control | FR (duration) | | |
|---|---|---|---|---|
| 10 min | 66 | 25 (5 s) | 3 (10 s) | |
| 30 min | 77 | 27 (5 s) | 3 (10 s) | 0 (30 s) |
| 1 h | 85 | 70 (5 s) | 44 (10 s) | 0 (30 s) |
| 2 h | 89 | 88 (5 s) | 72 (10 s) | 0 (30 s) |
| 4 h | 90 | 92 (10 s) | 32 (30 s) | 0 (60 s) |
| 8 h | 94 | 82 (10 min) | 80 (20 min) | |
| 12 h | 91 | 88 (10 min) | 24 (12 h) | |

C. EFFECT OF 10 MIN FR AFTER 12 HOURS WL PHOTOPERIODS OF VARIOUS LIGHT INTENSITIES.

| $\mu W.cm^{-2}.nm^{-1}$ (at 660 nm) | control | 10 min FR |
|---|---|---|
| 0.1 | 85 | 23 |
| 0.25 | 87 | 45 |
| 1.2 | 90 | 68 |
| 2.8 | 91 | 92 |
| 8.5 | 92 | 87 |

D. INTERACTION OF PHOTOPERIOD, TERMINAL FR AND TEMPERATURE

| daily treatment | | | |
|---|---|---|---|
| 1 h WL (25°C) | + 23 h CD (20°C) | | 83 |
| 1 h WL (20°C) | + 23 h CD (25°C) | | 0 |
| 2 h WL (20°C) | + 22 h CD (20°C) | | 85 |
| 2 h WL (20°C) | + 22 h CD (25°C) | | 0 |
| 2 h WL (20°C) | + 10 h CD (25°C) | + 12 h CD (20°C) | 9 |
| 2 h WL (20°C) | + 10 h CD (20°C) | + 12 h CD (25°C) | 68 |
| 2 h WL (25°C) | + 22 h CD (20°C) | | 93 |
| 4 h WL (20°C) | + 20 h CD (20°C) | | 96 |
| 4 h WL (20°C) + 10 min FR | + 20 h CD (20°C) | | 2 |
| 4 h WL (25°C) | + 20 h CD (20°C) | | 97 |
| 4 h WL (25°C) + 10 min FR | + 20 h CD (20°C) | | 8 |
| 5 h WL (20°C) | + 19 h CD (20°C) | | 96 |
| 5 h WL (20°C) + 10 min FR | + 19 h CD (20°C) | | 6 |
| 5 h WL (25°C) | + 19 h CD (20°C) | | 97 |
| 5 h WL (25°C) + 10 min FR | + 19 h CD (20°C) | | 44 |

R : 9.4 $\mu W.cm^{-2}$ at 660 nm ; FR : 3.3 $\mu W.cm^{-2}$ at 730 nm (see details in : Fredericq and De Greef, 1968 ; De Greef and Fredericq, 1969) phytochrome transformation. The complete inhibition, caused by the same temperature rise during very long nights can be due to enhanced Pfr destruction and/or to the induction of secondary dormancy.

The same effects were established by Taylorson and Hendricks (1969) for Amaranthus retroflexus L. seeds, whose germination is not absolutely light-requiring, but still phytochrome-mediated : the inactivation of Pfr proceeded 4-times faster at 25°C than at 20°C after a saturating R exposure.

SECONDARY DORMANCY (Rethy et al., 1983)

After overnight imbibition (14 hours) on water the seeds were either immediately treated with photoperiods of different lengths or after a dark interval of increasing duration. In Table 3 the results are summarized.

Table 3 Induction of secondary dormancy by dark incubation.

| daily RL treatment | % germination in function of dark interval (in days) | | | |
|---|---|---|---|---|
| | 0 d | 5 d | 10 d | 20 d |
| 5 min | 78 | 11 | 0 | 0 |
| 2 h | 93 | 85 | 27 | 0 |
| 4 h | 95 | 97 | 91 | 1 |
| 8 h | 95 | 96 | 95 | 17 |
| 16 h | 92 | 93 | 96 | 61 |
| 24 h | 95 | 96 | 96 | 91 |
| 2 h cy R | 86 | 28 | 0 | 0 |
| 4 h cy R | 84 | 85 | 22 | 0 |
| 8 h cy R | 96 | 95 | 89 | 0 |
| 16 h cy R | 94 | 96 | 91 | 50 |
| 24 h cy R | 94 | 94 | 91 | 84 |

Seeds were imbibed in water for 14 hours overnight and then exposed immediately or after a dark interval indicated to daily R light exposures during 9 days at a constant temperature of 20°C. The cyclic R (see cy R) light consists of 1 min R + 29 min darkness, repeated over the time indicated and terminated by another 1 min R irradiation.

The water soaked seeds become easily dormant which is most obvious when suboptimal photoperiods are used.

Secondary dormancy can also be broken, when the continuous R irradiation is substituted by cyclic R (1 min R + 29 min darkness), but it is less efficient than continuous R of the same duration.

Since seeds of other light requiring species such as those of Lactuca sativa cv. Grand Rapids, which show some, albeit a low dark germination in water, are strongly promoted by $GA_3$ (Kahn et al., 1957 ; Ikuma and Thimann, 1963a ; Evenari, 1965 ; Chawan and Sen, 1970), we studied the effect of gibberellic acid on the phytochrome-mediated seed germination of Kalanchoë.

Indeed, it has been suggested that $GA_3$ could substitute for light (Köhler, 1966 ; Kahn, 1968). However, light does not appear to induce germination of lettuce seeds through an increased production of gibberellins (Ikuma and Thimann, 1960 ; Scheibe and Lang, 1965 ; Hsiao and Vidaver, 1973 ; Lewak and Kahn, 1977 ; Berrie and Taylor, 1981), but this problem is not yet solved unequivocally (Carpita and Nabors, 1981).

When Kalanchoë seeds were incubated in $GA_3$ solutions at concentrations ranging from 10 $mg.liter^{-1}$ to 1000 $mg.liter^{-1}$ no germination occurred in continuous darkness, neither at 15°C, nor at 20°C, nor at 25°C.

From our results and those of Bunsow and von Bredow (1958) we may conclude that the seeds of Kalanchoë blossfeldiana are absolutely light requiring even in the presence of $GA_3$. Also, in the photoblastic seeds of Begonia evansiana the administration of $GA_3$ brought about no germination in complete darkness (Nagao et al., 1959).

SYNERGISM BETWEEN $GA_3$ AND VERY LOW Pfr LEVELS (Fredericq et al., 1983 ; De Greef and Fredericq, 1983 ; Dedonder et al., 1983)

With very light sensitive seed lots only 2 min WL given daily induces optimal germination on water and $GA_3$ (Fig. 2). In the absence of $GA_3$ 30 s FR irradiations fully reverse the effect of the preceding 2 min WL. In the presence of $GA_3$ however, there is no longer R/FR reversibility since all seeds germinate. Moreover, there is a strong synergism between GA and FR, while each of both treatments is completely ineffective for germination.

In combination with $GA_3$ even one single FR treatment induces up to 60% germination after longer incubation times preceding FR irradiation. When FR is given during the first 40 hours after sowing, it is ineffective (= lag-phase).

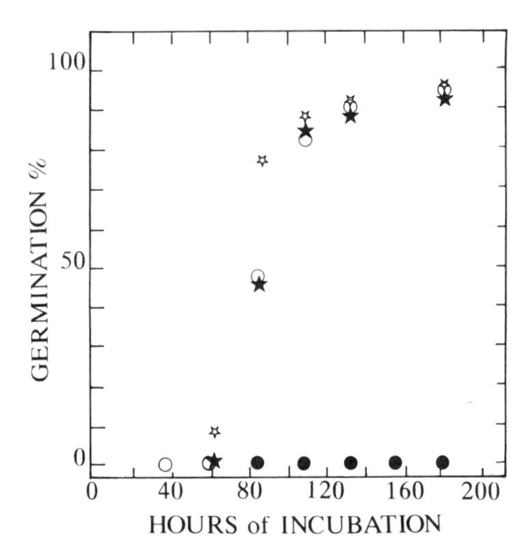


Fig. 2 Germination rate of Kalanchoë seeds incubated either on water (circles) or on $2.10^{-3}$ M $GA_3$ (stars) and irradiated daily with 2 min WL (open symbols) or with 2 min WL + 30 s FR (solid symbols).

The length of the lag-phase cannot be explained by a long preparatory action of $GA_3$ because transfer of punctured seeds to $GA_3$ only 6 hours before the first FR exposure is given after the lag-phase, hardly affects germination. The lag-phase must consequently be considered as a period needed for reaching a specific developmental state of the embryo as a condition for occurrence of subsequent germination after irradiation.

When the first of two FR irradiations, separated by 24 h, is given during the lag phase, the irradiations act synergisticaly as compared to the sum of the effects of both irradiations given separately. This synergism between both FR irradiations reaches a maximum between 40 and 50 h after sowing (time of first irradiation) and disappears quickly afterwards.

Action spectra taken in the presence and absence of GA strongly indicate phytochrome involvement. The light sensitivity of the seeds is enhanced drastically in the presence of GA.

We may conclude from these results that $GA_3$ does not substitute for high Pfr levels, but it appears to increase the physiological activity of Pfr to the extent that a very low threshold level of Pfr is sufficient for full germination in the presence of $GA_3$, assuming that the efficiency of Pr - Pfr phototransformations is not altered by the hormone.
$GA_3$ may exert its amplification effect either by increasing the number and/or the efficiency of Pfr receptor sites, or by stimulating the chain of events leading to germination after the formation of reacted Pfr. The first alternative is also suggested in studies of the mechanism of phytochrome dependent seed germination of Lactuca sativa and Betula papyrifera by pre-chilling (Vanderwoude and Toole, 1980 ; Bevington and Hoyle, 1981).

MEMORY EFFECTS IN SECONDARY DORMANT SEEDS (Fredericq et al., 1978, 1980 ; Rethy et al., 1976, 1980)

As already shown, seeds kept in darkness on water during several days become secondarily dormant at 20°C since they do not respond any longer to daily 5 min R exposures.

This dormancy can be broken either by prolonged R irradiation (Table 3) or by short R and FR when the dormant seeds are transferred to $GA_3$, but a larger number of daily FR exposures is needed than for non secondarily dormant seeds on $GA_3$. Red is much more effective than FR in dormant seeds transferred to $GA_3$ and there is complete R/FR reversibility.

In secondarily dormant seeds two short R or FR irradiations, given respectively 1 and 2 days after transfer to $GA_3$, are almost ineffective, but they become highly promotive if followed by a third light exposure (being ineffective in itself). The synergism between the first two and the third light exposure still occurs if the last one is given after a very long dark period (several weeks). The first two irradiations induce events which are maintained over long periods of time in darkness. It is not very likely that this memory effect is due to the persistent presence of Pfr although preservation of a very low Pfr amount, physiologically active by a very low fluence response, can not be excluded.

The memory effect on the germination level is strongly temperature dependent (Table 4).

Table 4 Temperature dependency of the memory effect.

| Treatment | | | Germination % |
|---|---|---|---|
| 2 min R | 15°C | 2 min R | |
| D8 + D9 | D10 | - | 5 |
| " | " | D15 | 98 (1) |
| " | " | D20 | 93 (6) |
| " | " | D25 | 84 (10) |
| " | " | D30 | 74 (24) |
| | 25°C | | |
| D8 + D9 | D10 | - | 9 |
| " | " | D15 | 72 (0) |
| " | " | D20 | 64 (2) |
| " | " | D25 | 28 (1) |
| " | " | D30 | 12 (1) |

Kalanchoë seeds are made secondarily dormant in darkness on water at 20°C for 6 days and then transferred to $2.10^{-3}$ M $GA_3$ on day 7. Two R irradiations of 2 min are given on days 8 and 9 (D8 + D9), followed by a transfer to 15°C or 25°C on day 10 (D10). After the third R irradiation of 2 min, given on the day indicated, seeds are brought back to 20°C. Between brackets germination % when the third R exposure is given only.

At 15°C there is a gradual decrease of secondary dormancy on $GA_3$, in contrast with the slow, gradual increase of secondary dormancy on $GA_3$ at 20°C, as found earlier, and with the results obtained at 25°C. Nevertheless, the difference between the germination induced by all three R irradiations and the sum of the germination % caused by the first two and the third irradiation given separately (the memory effect), is much better maintained at 15°C than at 25°C.

The occurrence of an analogous memory effect was also shown in non secondarily dormant seeds, sown directly on $GA_3$, with both R and FR, but the effect was smaller.

In absence of $GA_3$ there is no long lasting physiological memory effect.

In accordance with these light effects on GA-treated seeds we found an expression of the memory effect on energy metabolism in secondarily dormant seeds (Fig. 3).

The increase in ATP content, induced by the two first R irradiations is partly maintained over a long dark period. The rise in ATP content after the third irradiation is higher than when this irradiation is not preceded by the first 2 R pulses. As seen before, the third light exposure makes germination visible on the macromorphological scale.

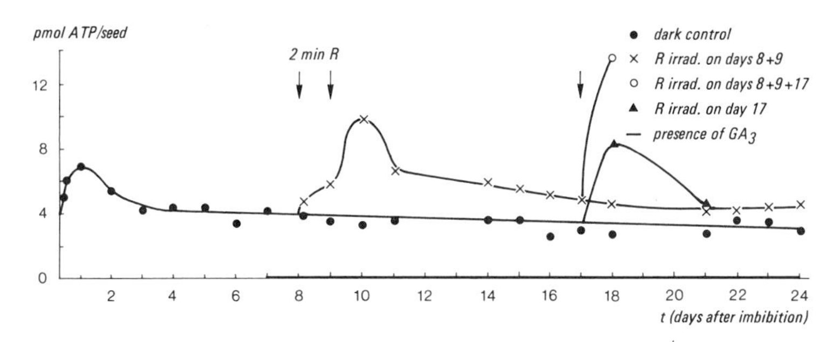


Fig. 3 Effect of an irradiation program, inducing a memory effect on the ATP level of seeds made secondarily dormant by a 6 days dark period in water, followed by transfer to $GA_3$ (20°C).

These effects on ATP levels are also temperature-dependent, but they are more significant. Indeed, whereas the memory effect on the germination level has completely disappeared after an intervening dark period of 3 weeks at 25°C, it is still present at the ATP-level.

Since these ATP changes are very significant and occur relatively fast after the first two R exposures without any visible physiological response, they may be related to the early events of the light-mediated germination effect, namely the transduction chain following signal perception and receptor activation (see further). There is indeed growing evidence that the pathway of signal transduction in plants involves activation of an ATP-ase, proton-pumping, a protein kinase and the NADH oxidase of the plasma membrane. All these biochemical activities have a great demand of energy supply. In this regard, research might be directed towards such intermediates present between and immediately after the first two inductive R irradiations, and after the third irradiation displaying the germination response.

On the other hand, we would like to investigate the mechanism(s) by which the light-stimulated energy supply is brought about and whether or not these reactions of energy metabolism are associated with the reactions of dormancy breakage. These ideas can be substantiated by analysis of the contribution of the two cytoplasmic respiratory pathways (HMP and EMP-TCA), on the one hand, and by analysis of the contribution of the mitochondrial respiratory pathways (CN-sensitive and CN-resistant), to germination by light, on the other.

ABA-GA ANTAGONISM IN PHYTOCHROME MEDIATED SEED GERMINATION (Fredericq et al., 1980 ; Rethy et al., 1976 ; Dedonder et al., 1983).

Increasing concentrations of ABA are applied in the absence or presence of $GA_3$ ($2.10^{-3}$ M) to phytochrome treated Kalanchoë seeds (Table 5).

In the absence of GA, ABA concentrations higher than $5.10^{-7}$ M gradually inhibit the germination induced by three R irradiations on days 1, 2 and 3 (during and after the lag phase).

In the presence of GA higher ABA concentrations are required for analogous effects with the same irradiation program. There is an obvious GA-ABA antagonism.

Table 5 ABA - GA antagonism in R/FR treated Kalanchoë seeds expressed in % germination.

| Irradiation program | | Hormone treatment | | | | | | | |
|---|---|---|---|---|---|---|---|---|---|
| | | $H_2O$ | $GA_3$ | ABA ($10^{-7}$M) | | | | | |
| | 5 min | | $2.10^{-3}$M | 5 | 10 | 50 | 100 | 500 | 1000 |
| D1+2+3 | R | 92 | 100 | 87 | 77 | 18 | 8 | 7 | 0 |
| | | | | in the presence of $2.10^{-3}$ M $GA_3$ | | | | | |
| D1+2+3 | R | 92 | 100 | 99 | 100 | 83 | 59 | 12 | 0 |
| D1+2+3 | FR | 0 | 86 | 94 | 92 | 55 | 26 | 2 | 0 |
| D2+3+4 | FR | 0 | 98 | 98 | 97 | 73 | 53 | 12 | 0 |
| D4+5+6 | FR | 0 | 100 | 98 | 99 | 79 | 66 | 10 | 6 |
| D7 | R | 0 | 80 | - | 76 | 71 | 61 | - | - |
| D7 | FR | 0 | 87 | - | 71 | 63 | 33 | - | - |

Comparing the effect of 3 R or FR irradiations given the 3 initial days (partly during the lag phase), ABA is more inhibitory when FR exposures are administered. So, during the lag phase, when the amount of absorbed $GA_3$ by the embryo is still limiting, a partial compensation of the inhibitory effect of ABA by a high Pfr level seems to occur. Indeed, if 3 FR irradiations are given after the lag phase (when more GA has penetrated in the seed), ABA inhibition in combination with exogenous GA is the same as with 3 R irradiations administered partly during the lag phase.

If however a single irradiation is given after the lag phase (on day 7, D7), at a time that Pfr becomes more limiting, ABA inhibits more the FR effect than that of R.

In conclusion Pfr interacts with $GA_3$ and ABA, since the ABA-GA antagonism can be modified by the Pfr-level.

In the context that we have discussed above in relation to energy metabolism, it would be worthwile to put the hypothesis of Berrie (1984) to test. According to this author seeds save $O_2$ to oxidize germination inhibitors such as ABA by increasing the $CN^-$-resistant pathway, thus oxygen consumption associated with electron flow would be reduced and made available for other oxygen consuming processes.

LFR AND VLFR IN PHYTOCHROME-TREATED AND GA-SENSITIZED SEEDS (Rethy et al., 1976, 1986, 1987 ; De Petter et al., 1985a,b, 1987, 1988, 1989)

When two low fluence R or FR pulses, separated by 24 h, are given to seeds incubated on a range of $GA_3$ concentrations, the effect of R is higher than that of FR in the 0.1 to 0.01 mM concentration range. These results suggest the possibility of a LFR, additional to the VLFR, which is induced by two R but not by two FR irradiations at suboptimal $GA_3$ concentrations.

In the presence of GA's Kalanchoë blossfeldiana seeds show two photoresponses, VLFR and LFR.

We could prove the existence of these photomorphogenic phenomena experimentally and the data could be handled mathematically. In contrast to our earlier findings that the Bunsen-Roscoe law is not applicable when inductive photoperiods are used, reciprocity holds for both VLFR and LFR.

The fraction of the seed population exhibiting a VLFR and/or a LFR is determined by the experimental conditions, e.g. $KNO_3$ or $GA_3$ concentration, moment and/or number of irradiations. The remaining seed fraction, which is insensitive to short irradiation(s), can be induced with prolonged irradiation resulting in a so-called high fluence response, HFR (De Petter et al., 1988).

The involvement of phytochrome as photoreceptor for the LFR has been extensively illustrated by R/FR reversibility. LFR action spectra for the induction of several seed species, e.g. Arabidopsis thaliana seeds (Cone et al., 1985 ; Cone and Kendrick, 1985 ; Shropshire et al., 1961) and Lactuca sativa seeds (Blaauw-Jansen and Blaauw, 1976 ; Borthwick et al., 1954 ; Small et al., 1979a,b), show similarities to the absorption spectrum of purified phytochrome (Pr) (Kelly and Lagarias, 1985).

LFR and VLFR are analogous with the two photoresponses described for Lactuca seed germination by Blaauw-Jansen and Blaauw (1975), Small et al. (1979a) and Vanderwoude (1983). Induction of a VLFR is obtained for Lactuca, Rumex and Arabidopsis seeds as a consequence of some specific shock treatment, e.g. osmotic shock, ethanol treatment, various thermal treatments, i.e. prechilling or exposure to high temperature (Blaauw-Jansen and Blaauw, 1975 ; Blaauw-Jansen, 1983 ; Cone et al., 1985 ; Kendrick and Cone, 1985 ; Small et al., 1979a,b ; Takaki et al., 1985 ; Vanderwoude, 1983, 1985).

Induction of a VLFR by auxins has been demonstrated for the R-stimulated growth of etiolated subapical Avena coleoptile sections and the R-inhibited growth of etiolated Zea mesocotyl sections (Shinkle and Briggs, 1983, 1984, 1985 ; Shinkle, 1986).

VLFR and LFR requirements of the various systems (seed germination, growth responses, anthocyanin synthesis, Chl accumulation) differ by a factor of about 10,000.

For Kalanchoë blossfeldiana seeds the VLFR is only observed in the presence of gibberellic acid. This is the first time that induction of a VLFR by GA's is demonstrated. Induction of a VLFR by GA's causes the observed synergism between GA's and very low Pfr amounts. The LFR is observed after several daily irradiations when the seeds are incubated either in water or in $KNO_3$. In the presence of GA's 2 daily R-irradiations are required for induction of the LFR. The analogy is even more striking if we consider that under specific $GA_3$ treatments, Kalanchoë blossfeldiana seeds show a biphasic fluence-response relation consisting of a VLF as well as a LF component. Indeed, the fluence-response curves for the effect of two R pulses separated by 24 hours on the germination of Kalanchoë blossfeldiana

Poelln. cv. Vesuv seeds incubated on $GA_3$ are biphasic for suboptimal concentrations (Fig. 4). In the sub-optimal $GA_3$ concentration range (0.01-0.1 mM) increasingly biphasic relations between log R fluences and germination are obtained. The first response component, promoted by VLF in the $10^{-9}$ to $10^{-6}$ $mol.m^{-2}$ range, is called VLFR. The second response component occurs at LF in the $3.10^{-5}$ to $3.10^{-2}$ $mol.m^{-2}$ range and is called LFR. At 1 mM $GA_3$, nearly the entire seed population responds in the VLF range.

With 2 FR pulses, for different $GA_3$ concentrations monophasic (increasing) relations between log FR fluences and germination are obtained. However, in comparison with R, considerably more light is required to obtain the VLFR (no LFR is induced by FR fluences as high as $2.10^{-1}$ $mol.m^{-2}$). Dark germination is very poor (<5%).

Far-red induces and saturates the VLFR since the maximal germination is the same as that of the R induced VLFR. At the optimum GA concentration ($10^{-3}$ M) the entire seed population responds in the VLF range. The LFR induced by broad-band R (total fluence of $7.2.10^{-4}$ $mol.m^{-2}$), is reversed by FR to the level of maximum VLFR induced by saturating FR alone. The maximal germination response as a consequence of VLF plus LF component (see second maximum in the curves of Fig. 4) increases with the $GA_3$ concentration. The concentration of $GA_3$, - the factor sensitizing the seeds by induction of VLFR, - also affects the LFR.

Seeds sown in water and given 2 R or FR irradiations on D7 and D8 after sowing did not germinate.

The response in the low fluence (LF) range corresponds with a classical R/FR reversible phytochrome mediated reaction. The VLFR is a GA-sensitized and very low Pfr requiring reaction, promoted both by R and FR. Indeed, the sensitivity to Pfr is increased about 20,000-fold, so that even FR fluences become saturating.

As germination is a quantal (all-or-none) response, monophasic dose-response relations are sigmoid in a response versus logarithm of promotor (light or GA) plot. Hence, the sub-optimal (monotonically increasing) segments of the dose-response curves can be analysed by means of probit analysis in order to calculate the seed population parameters, e.g. the zero response, the response range (maximal response minus zero response), the dose for half-maximal response and the slope of the curve.

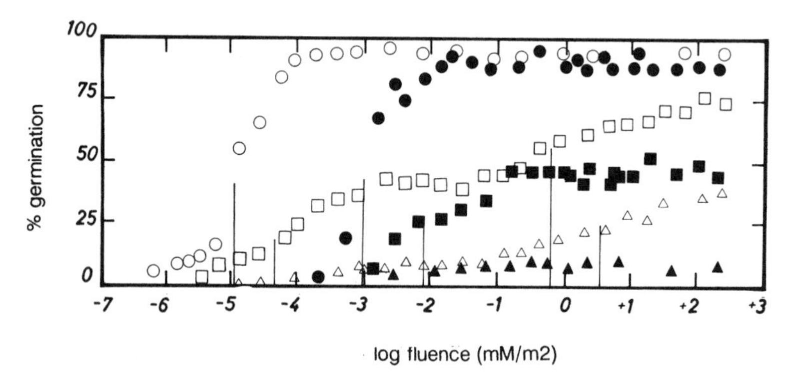


Fig. 4 Fluence-response curves for germination of Kalanchoë seeds for two narrow-band R pulses (open symbols) or two narrow-band FR pulses (solid symbols) on D7 and D8, each consisting of half the total fluence indicated, on different $GA_3$ concentrations : 1 mM (circles), 0.07 mM (squares), 0.02 mM (triangles).
All germination percentages are based on six batches of about 100 seeds.
Vertical lines indicate the median sensitivity of the system to the stimulus (= fluence for half-maximal response).
mM/m2 = millimol photons per square meter.

Irrespective of the $GA_3$ or $KNO_3$ concentrations there is a linear relation between the logarithm of the photon fluence and the probit of the germination response as illustrated by the respective LDP lines (Logarithm of Dose versus Probit regression line). Moreover, the slopes for the VLFR as well as for the LFR are very similar. The slope of the probit regression line reflects the apparent efficiency rate of the photosystem in relation to the light stimulus (Borthwick et al., 1954 ; Shropshire et al., 1961 ; Frankland, 1976 ; Bartley and Frankland, 1984). The similarity of the slopes indicates the involvement of the same photoreceptor, phytochrome. This is corroborated by action spectroscopy (Rethy et al., 1980 ; De Petter et al., 1989).

From two current models, accounting for the occurrence of VLFR and LFR (Blaauw-Jansen, 1983 ; Vanderwoude, 1985), the model of Blaauw-Jansen, based on phytochrome destruction, seems not compatible with our present results. Indeed, the different characteristics of VLFR and LFR with regard to the number of irradiations required indicate that they are two physiologically distinct responses. VLFR and LFR may be related to the respective formation of two active phytochrome-dimer-receptor complexes, PrPfr.X and PfrPfr.X, as proposed by Vanderwoude (1983, 1985).
This model has been sustained to a large extent by Brockmann et al. (1987). $GA_3$ sensitization of germination behaviour to VLF irradiations is consistent with the dimer model.

Although $GA_3$ and $KNO_3$ modulate VLFR and LFR respectively, there is no direct influence on the phytochrome phototransformations. For a given Pfr concentration, independently of its origin (as preserved in dry seeds or induced by light in a VLF or LF reaction) the response is proportional to the GA or $KNO_3$ concentration.

These facts strongly suggest a similar and direct action of $GA_3$ and $KNO_3$ on the transduction chain of the phytochrome signal leading to germination.

## CONFIGURATIONAL INTERPRETATION OF RECEPTOR REGULATION IN SEED GERMINATION (De Greef, 1989).

Seed germination can be deeply influenced by various environmental conditions acting as stimulating or inhibiting factors such as storage conditions, after-ripening, light (short-day SD, and long-day LD, light regimes ; continuous and intermittent irradiations ; spectral quality), osmoregulation, alternating temperatures, scarification, GA's, nitrate, ethanol, kinetin, coumarin, ... (Ottenwalder, 1914 ; Gassner, 1915a,b ; Gardner, 1921 ; Axentjev, 1930 ; Isikawa, 1954 ; Black and Wareing, 1955 ; Bunsow and von Bredow, 1958 ; Nagao et al., 1959 ; Fujii, 1962 ; Isikawa and Yokohama, 1962 ; Cumming, 1963 ; Ikuma and Thimann, 1963b ; Borthwick et al., 1964 ; Yokohama, 1965 ; Steiner, 1968 ; Hartman, 1970 ; Karssen, 1970 ; Wulff and Medina, 1971 ; Eldabh et al., 1974 ; Blaauw-Jansen and Blaauw, 1975 ; Small et al., 1979a,b ; Blaauw-Jansen, 1983 ; Vanderwoude, 1983, 1985 ; Cone et al., 1985 ; Kendrick and Cone, 1985 ; Takaki et al., 1985 ; Ensminger and Ikuma, 1987).

The physiological nature and the regulatory molecular implications of such interactions are not clearly understood at this time (Bewley and Black, 1982 ; Frankland and Taylorson, 1983).

In light-dependent seed germination of the two related species *Arabidopsis thaliana* and *Sisymbrium officinale* R light has a dual effect : light effect I induces a chain of events leading to GA biosynthesis, light effect II seems to enhance the sensitivity of the seeds to GA's. In the second effect the co-action of R and exogenous $GA_{4+7}$ is clearly additive (Hilhorst and Karssen, 1988). In several photoblastic seeds Pfr action can be mimicked by exogenous GA (Taylorson, 1982). In other seeds, e.g. *Kalanchoë blossfeldiana*, light cannot be replaced by GA treatment, although germination is obviously influenced by phytochrome-GA interactions, as we have discussed earlier extensively.

How can we understand the mode of action of these modulator molecules in plant photomorphogenesis in general, and in light promoted seed germination, in particular ?

Signals from outside such as light are absorbed by appropriate elicitor molecules (as phytochrome e.g.) that regulate from within the cellular system a particular physiological response depending upon the developmental program of the plant's life cycle.

As seen before it is obvious that the stimulatory or inhibitory action of light through phytochrome can be replaced by hormones in light sensitive seeds. In most cases exogenous hormones are used to elicit the physiological response, while phytochrome mediated changes of endogenous hormone levels correlated with germination behaviour are poorly documented.

In spite of these facts there is abundant evidence that phytochrome and phytohormones interact during the germination process of photoblastic seeds.

What kind of molecular complexes convey these elicitors and how do these intermediaries act upon the cellular functions. On the other hand, it is generally known that Pfr has manifold effects in photomorphogenesis what might suggest a multiple action of the active form of phytochrome.

In recent years it has often been shown that there is more than one pool of functional phytochrome (see review Cordonnier, 1989 ; Tokuhisa and Quail, 1989). So, it is intriguing to speculate that these multiple phytochrome populations might each be resposible for a single function.

Another explanation that seems feasible to us, is the existence of very specific receptor proteins bound to a stratum of various kinds and sensitive to phytochrome and/or plant growth regulators (Fig. 5). This stratum can consist of different compartments of the cell belonging to both the genome and the cytoplast. This stratum can be part of the regulator-operator-operon model (compatible with the actual promotor model), regulating transcription and also translation by feed back systems of primary metabolic events. The receptor proteins might also be involved in the control of mRNA turnover.

At the cytoplasmic level the receptor proteins acting in concert with other effectors as regulatory elements might have strata at membrane and MTL (micro trabecular lattice) levels. Despite their apparently different cellular nature they can unify their differential action when the regulatory elements have control of membrane molecular conformation, fluidity, permeability, symmetry, ...

At the MTL level receptor proteins can be located at strategic sites (such as MTOC's microtubuli organization centers) controlling intracellular positioning of compartments, on the one hand, and polyribosome sequential reaction chains, on the other. Presuming the existence of such specific receptor proteins then they are very comparable with the concept we have about the properties of enzymes and carrier molecules in membranes. In the late fifties Sterling B. Hendricks suggested already the existence of a master enzyme (proteins) upon which phytochrome exerts its molecular action to controlling photomorphogenic processes in higher plant systems.

As shown in Fig. 5,B-D we suggest that the receptor protein has 1) a binding site that can be joined to the active center which is coupled to the transduction chain ; 2) a regulatory site that can change the conformation of the binding site and thus disconnecting the binding site from the active center. Elicitor molecules themselves can also have the wrong configuration to fit the binding site (e.g. the inactive form of phytochrome, Pr).

In this way, cooperative enhancement in higher plant developmental processes, e.g. seed germination, can proceed from receptor activation by phytochrome and plant growth regulating substances through allosteric interaction.

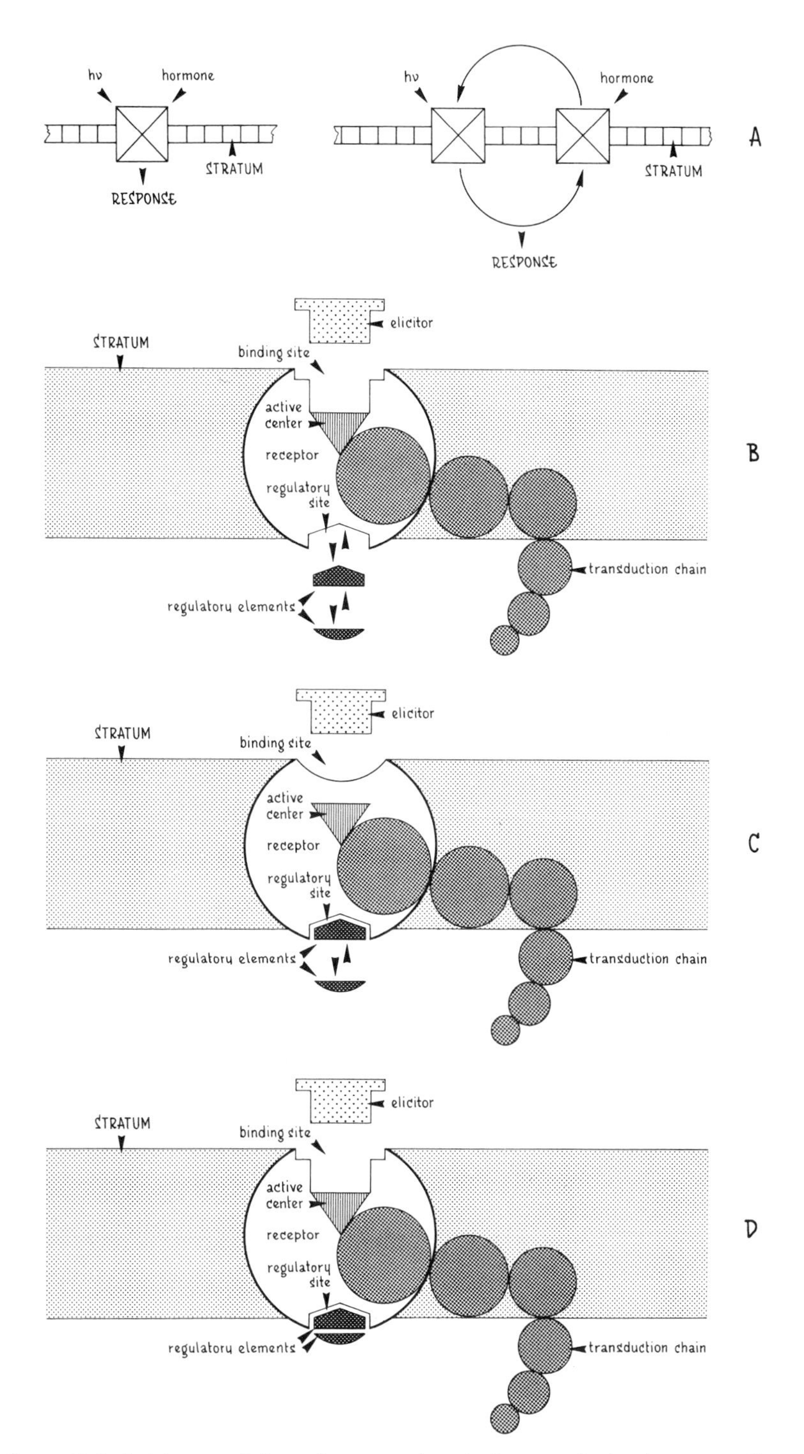


Fig. 5 Modulation models of synergism between phytochrome and growth regulating substances in higher plant systems.

ACKNOWLEDGEMENT

This research project was supported by the Belgian National Science Foundation (F.K.F.O. grants Nos. 2.9009.75, 2.9009.81, 2.0083.83).

As one of the senior authors of this review, and responsible of the whole team, I am very grateful to Dr. Roger Rethy for his alert criticism with regard to the factual data of this research project going for so many years in full agreement and good understanding of the so intricate problems involved in seed germination.

Special thanks go to Mrs. R. Pype who typed the manuscript with patience and accuracy, and to Mr. A. Audenaerde and Mr. A. Cuylits for appropriate illustrations of the published material in the subsequent steps of the research project.

REFERENCES

Axentjev, B. V., 1930, Über die Rolle der Schalen von Samen und Früchten, die bei der Keimung auf Licht reagieren. Bot. Zentralbl., 46:119.

Bartley, M. R. and Frankland, B., 1984, Phytochrome intermediates and action spectra for light perception by dry seeds, Plant Physiol., 74:601.

Berrie, A. M. M., 1984, Germination and dormancy, in: "Advanced Plant Physiology", M.B. Wilkins ed., Arnold Heinemann, London.

Berrie, A. and Taylor, G., 1981, The use of population parameters in the analysis of germination of lettuce seed, Physiol. Plant., 51:229.

Bevington, J. and Hoyle, M, 1981, Phytochrome action during prechilling induced germination of Betula papyrifera March, Plant Physiol., 67:705.

Bewley, J. D. and Black, M., 1982, "Physiology and Biochemistry of Seeds", vol. 2, Springer-Verlag, New York.

Blaauw-Jansen, G., 1983, Thoughts on the possible role of phytochrome destruction in phytochrome-controlled responses, Plant Cell and Environment, 6:173.

Blaauw-Jansen, G. and Blaauw, O. H., 1975, A shift of the response threshold to red irradiation in dormant lettuce seeds, Acta Bot. Neerl., 24:199.

Blaauw-Jansen, G. and Blaauw, O. H., 1976, Action spectra for phytochrome-mediated germination of lettuce seeds (Lactuca sativa L.). Acta Bot. Neerl., 25:149.

Black, M. and Wareing, P. F., 1955, Growth studies in woody species. VII. Photoperiodic control of germination in Betula pubescens Ehr., Physiol. Plant., 8:300.

Borthwick, H. A., Hendricks, S. B., Toole, E. H. and Toole, V. K., 1954, Action of light on lettuce-seed germination, Bot. Gaz., 115:205.

Borthwick, H. A., Toole, E. H. and Toole, V. K., 1964, Phytochrome control of Paulownia seed germination. Israel J. Bot., 13:122.

Borthwick, H. A., Hendricks, S. B., Taylorson, M. J., Toole, V. K., 1969, The high-energy light action controlling plant responses and development. Proc. Natl. Acad. Sci. U.S., 64:479.

Brockmann, J., Rieble, S., Kazarinova-Fukshansky, N., Seyfried, M. and Schäfer, E., 1987, Phytochrome behaves as a dimer in vivo. Plant Cell Environm., 10:105.

Bunsow, R. and Von Bredow, K., 1958, Wirkung von Licht und Gibberellin auf die Samenkeimung der Jurztagpflanze Kalanchoë blossfeldiana. Biol. Zentralbl., 77:132.

Carpita, N. and Nabors, M., 1981, Growth physics and water relations of red-light-induced germination in lettuce seeds. V. Promotion of elongation in the embryonic axes by gibberellins and phytochrome. Planta, 152:131.

Chawan, D. D. and Sen, D. N., 1970, Role of some growth regulating substances on seed germination and seedling growth of Asteracantha longifolia Nees, Biochem. Physiol. Pflanz,. 161:417.

Côme, D., 1970, Les obstacles à la germination. N°6, in "Monographies de physiologie végétale" P. E. Pilet, ed. Masson et Cie, Publishers, Paris (France).
Cone, J. W. and Kendrick, R. E., 1985, Fluence-response curves and action spectra for promotion and inhibition of seed germination in wildtype and long-hypocotyl mutants of Arabidopsis thaliana. Planta, 163:43.
Cone, J. W., Jaspers, P. A. and Kendrick, R. E., 1985, Biphasic fluence-response curves for light induced germination of Arabidopsis thaliana seeds, Plant Cell Environment, 8:605.
Cordonnier, M.-M., 1989, Monoclonal antibodies : Molecular probes for the study of phytochrome, Photochem. Photobiol, 49:821.
Cumming, B. G., 1963, The dependence of germination on photoperiod, light quality and temperature in Chenopodium spp, Can. J. Bot., 41:1211.
Dedonder, A., Rethy, R., De Petter, E., Fredericq, H. and De Greef, J., 1980, Preliminary screening experiments on the effects of light and $GA_3$ on the germination of different seed species, in : "Photoreceptors and Plant Development", Proceedings of the Ann. Eur. Symp. Plant Photomorphogenesis, J. De Greef, ed., Antwerpen University Press, Antwerpen.
Dedonder, A., Rethy, R., Fredericq, H. and De Greef, J. A., 1983, Interaction between Pfr and growth substances in the germination of light-requiring Kalanchoë seeds. Physiol. Plant., 59:488.
De Greef, J., 1989, unpublished results.
De Greef, J. and Fredericq, H., 1969, Photomorphogenic and chlorophyll studies in the bryophyte Marchantia polymorpha. II. Photobiological responses to terminal irradiations with different red/far-red ratios, Physiol. Plant., 22:462.
De Greef, J. and Fredericq, F., 1983, Photomorphogenesis and Hormones, in: "Encyclopedia of Plant Physiology", New Series, vol. 16, W. Shorpshire Jr. and H. Mohr, eds., Springer-Verlag, Berlin Heidelberg.
De Petter, E., Van Wiemeersch, L., Rethy, R., Dedonder, A., Fredericq, H., De Greef, J. A., Steyaert, H. and Stevens, H., 1985a, Probit analysis of low and very-low fluence-responses of phytochrome-controlled Kalanchoë flossfeldiana seed germination, Photochem. Photobiol., 42:697.
De Petter, E., Van Wiemeersch, L., Fredericq, H. and De Greef, J. A., 1985b, Importance of probit analysis of fluence-response data for phytochrome-controlled germination of Kalanchoë, Biol. Jb. Dodonaea, 53:177.
De Petter, E., 1987, Fotobiologische studie van de zaadkieming van Kalanchoë blossfeldiana Poelln. Ph.D. Thesis, State University of Ghent, Belgium.
De Petter, E., Van Wiemeersch, L., Dedonder, A., Rethy, R., Fredericq, H. and De Greef, J. A., 1988, Mathematical approach to effects of repeated treatments in the study of very low fluence, low fluence and high fluence germination responses, Physiol. Plant., 72:36.
De Petter, E., Van Wiemeersch, L., Rethy, R., Dedonder, A., Fredericq, H and De Greef, J., 1989, Fluence-response curves and action spectra for the very low fluence and the low fluence response for the induction of Kalanchoë seed germination, Plant Physiol. in press.
Downs, R.J., Hendricks, S. B. and Borthwick, H. A., 1957, Photoreversible control of elongation of pinto beans and other plants under normal conditions of growth, Bot. Gaz., 118:199.
Eldabh, R., Fredericq, H., Maton, J. and De Greef, J., 1974, Photophysiology of Kalanchoë seed germination. I. Interrelationship between photoperiod and terminal far-red light, Physiol. Plant., 30:197.
Ensminger, P. A. and Ikuma, H., 1987, Photoinduced seed germination of Oenothera biennis L. I. General characteristics. Plant Physiol., 85:879.
Evenari, M., 1965, Light and seed dormancy, Encycl. Plant Physiol., 15(2):804.
Frankland, B., 1976, Phytochrome control of seed germination in relation to the light environment, in: "Light and Plant Development", H. Smith ed., Butterworths, London.

Frankland, B. and Taylorson, R., 1983, Light control of seed germination. in: "Encyclopedia of Plant Physiology", vol. 16A, W. Shropshire and H. Mohr eds., Springer-Verlag, New York.

Fredericq, H., 1965, Action of red and far-red light at the end of the short day and in the middle of the night on flowering induction in Kalanchoë blossfeldiana, Biol. Jb. Dodonaea, 33:66.

Fredericq, H. and De Greef, J., 1968, Photomorphogenic and chlorophyll studies in the bryophyte Marchantia polymorpha. I. Effect of red, far-red irradiations in short and long term experiments. Physiol. Plant., 21:346.

Fredericq, H., Eldabh, R., De Greef, J. and Maton, J., 1975, Photophysiology of Kalanchoë Seed Germination. II. Effects of short- and long-term irradiations with different red/far-red ratios and of gibberellic acid. Physiol. Plant., 34:238.

Fredericq, H., Rethy, R., De Greef, J. and Cappelle, M., 1978, Occurrence and characteristics of a memory effect in non-dormant Kalanchoë seeds, sown on gibberellic acid, Arch. Intern. Physiol. Biochim., 86:942.

Fredericq, H., Rethy, R., Dedonder, A., De Greef, J. and De Petter, E., 1980, Photocontrol of Kalanchoë blossfeldiana seed germination, In: "Photoreceptors and Plant Development", Proceedings of the Ann. Eur. Symp. Plant Photomorphogenesis, J. De Greef, ed., Antwerpen University Press, Antwerpen.

Fredericq, H., Rethy, R., Van Onckelen, H. and De Greef, J., 1983, Synergism between gibberellic acid and low Pfr levels inducing germination of Kalanchoë seeds, Physiol. Plant., 57:402.

Fujii, T., 1962, Studies on photoperiodic responses involved in the germination of Eragrostis seeds. Bot. Mag., 75:56.

Gardner, W. A., 1921, Effect of light on germination of light-sensitive seeds. Bot. Gaz, 71:249.

Gassner, G., 1915, Über die keimungsauslösende Wirkung der Stickstoffsalze auf lichtempfindliche Samen, Jahrb. Wiss. Bot., 55:259.

Hartmann, K., 1966, A general hypothesis to interpret "high energy" phenomena of photomorphogenesis on the gasis of phytochrome. Photochem. Photobiol., 5:349.

Hartmann, W., 1970, Untersuchungen zur Lichtabhängigen Samenkeimung von Oenothera biennis L. Biodhem. Physiol. Pflanzen, 161:368.

Hendricks, S. B., Toole, E. H., Toole, V. K. and Borthwick, H. A., 1959, Photocontrol of plant development by the simultaneous excitation of two interconvertible pigments. III. Control of seed germination and axis elongation, Bot. Gaz., 121:1.

Hilhorst, H. W. M. and Karssen, C. M., 1988, Dual effect of light on the gibberellin- and nitrate-stimulated seed germination of Sisymbrium officinale and Arabidopsis thaliana, Plant Physiol., 86:591.

Hsiao, A. and Vidaver, W., 1973, Induced requirements for gibberellic acid and red light in grand rapids lettuce seeds, Plant Physiol., 51(suppl.):36.

Ikuma, H. and Thimann, K. V., 1960, Action of gibberellic acid on lettuce seed germination, Plant Physiol., 35:557.

Ikuma, H. and Thimann, K. V., 1963a, Action of kinetin on photosensitive germination of lettuce seed as compared with that of gibberellic acid, Plant Cell Physiol., 4:113.

Ikuma, H. and Thimann, K. V., 1963b, The role of the seed-coats in germination of photosensitive lettuce seeds. Plant Cell Physiol., 4:169.

Isikawa, S., 1954, Light sensitivity against germination. I. 'Photoperiodism' of seeds. Bot. Mag., 67:51.

Isikawa, S. and Yokohama, Y., 1962, Effect of 'intermittent irradiations' on the germination of Epilobium and Hypericum seeds. Bot. Mag., 75:127.

Kahn, A., 1968, Inhibition of gibberellic acid-induced germination by abscissic acid and reversal by cytokinins. Plant Physiol., 43:1463.

Kahn, A. A., Goss, J. A. and Smith, D. E., 1957, Effect of gibberellin on germination of lettuce seeds, Science, 125:645.

Karssen, C. M., 1970, The light promoted germination of the seeds of Chenopodium album L., IV. Pfr requirement during different stages of the germination process. Acta Bot. Neerl., 19:297.

Kelly, J. M. and Lagarias, J. C., 1985, Photochemistry of 124-kilodalton Avena phytochrome under constant illumination in vitro, Biochemistry, 24:6003.

Kendrick, R. E. and Cone, J. W., 1985, Biphasic fluence response curves for induction of seed germination, Plant Physiol., 79:299.

Köhler, D., 1966, Veränderungen des Gibberellinsgehaltes von Salatsamen nach Belichtung, Planta, 70:42.

Lewak, S. and Khan, A., 1977, Mode of action of gibberellic acid and light on lettuce seed germination, Plant Physiol., 60:575.

Mohr, H. and Apphun, U., 1963, Die Keimung von Lactacu-Achänen under dem Einfluss des Phytochromsystems und der Hochenergiereaktion der Photomorphogenese, Planta, 60:274.

Nagao, M., Esashi, Y., Tanaka, T., Kumagai, T. and Fukumoto, S., 1959, Effects of photoperiod and gibberellin on the germination of seeds of Begonia evansiana Andr., Ibid., 1:39.

Ottenwalder, A., 1914, Lichtintensität und Substrat bei der Lichtkeimung. Z. Bot., 6:785.

Rethy, R., Fredericq, H., De Greef, J., Van Onckelen, H. and Maton, J., 1976, Long-lasting light effects in secundary dormant seeds of Kalanchoë blossfeldiana (cv. Feuerblüte), treated with gibberellic acid ($GA_3$), Arch. Intern. Physiol. Biochim., 84:1102.

Rethy, R., De Petter, E., Dedonder, A., Fredericq, H. and De Greef, J., 1980, Effect of gibberellic acid on light-sensitivity of Kalanchoë seeds, in: 8th. Intern. Congress on Photobiology, Strasbourg, France, book of abstracts p.87.

Rethy, R., Dedonder, A., Fredericq, H. and De Greef, J. A., 1983, Factors affecting the induction and release of secondary dormancy in Kalanchoë seeds, Plant Cell Environm., 6:731.

Rethy, R., Dedonder, A., De Petter, E., Van Wiemeersch, L., Fredericq, H., De Greef, J.A., Steyaert, H. and Stevens, H., 1987, Biphasic fluence-response curves for phytochrome-mediated Kalanchoë seed germination, Plant Physiol., 83:126.

Rethy, R., Dedonder, A., Van Wiemeersch, L., De Petter, E., Fredericq, H. and De Greef, J.A., 1986, Factors affecting the very low and low fluence versus germination responses of Kalanchoë seeds, in: Proc. XVI Yamada Conference, Okazaki, Japan, p. 120.

Rollin, P., 1963, Observations sur la différence de nature de deux photo-réactions contrôlant la germination des akènes de Lactuca sativa var. "Reine de Mai", C.R. Acad. Sci. Paris, 257:3642.

Scheibe, J. and Lang, A., 1965, Lettuce seed germination : evidence for a reversible light-induced increase in growth potential and for phytochrome mediation of the low temperature effect, Plant Physiol., 40:485.

Shinkle, J. A., 1986, Photobiology of phytochrome-mediated growth responses in sections of stem tissue from etiolated oats and corn, Plant Physiol., 81:533.

Shinkle, J. R. and Briggs, W. R., 1983, Auxin increases coleoptile section sensitivity to red light, In: "Annual Report of the Director, Department of Plant Biology 82-83", W.R. Briggs, ed., Carnegie Institution, Stanford.

Shinkle, J. R. and Briggs, W. R., 1984, Indole-3-acetic acid sensitization of phytochrome-controlled growth of coleoptile sections. Proc. Natl. Acad. Sci. USA, 81:3742.

Shinkle, J .R. and Briggs, W. R., 1985, Physiological mechanisms of the auxin-induced increase in light sensitivity of phytochrome-mediated growth responses in Avena coleoptile sections, Plant Physiol., 79:349.

Shropshire, W. Jr., Klein, W. H. and Elstad, V. B., 1961, Action spectra of photomorphogenic induction and photoinactivation of germination in Arabidopsis thaliana, Plant Cell Physiol., 2:63.

Small, J. G., Spruit, C. J. P., Blaauw-Jansen, G. and Blaauw, O. H., 1979a, Action spectra for light-induced germination in dormant lettuce seeds. I. Red region, Planta 144:125.

Small, J. G., Spruit, C. J. P., Blaauw-Jansen, G. and Blaauw, O. H., 1979b, Action spectra for light-induced germination in dormant lettuce seeds. II. Blue region, Planta, 144:133.

Steiner, E., 1968, Dormant seed environment in relation to natural selection in Oenothera. Bull. Torr. Bot. Club, 95:140.

Takaki, M., Heeringa; G. H., Cone, J. W. and Kendrick, R. E., 1985, Analysis of the effect of light and temperature on the fluence response curves for germination of Rumex obtusifolius. Plant Physiol., 77:731.

Taylorson, R. B., 1982, Interaction of phytochrome and other factors in seed germination. in: "The Physiology and Biochemistry of Seed Development, Dormancy and Germination", A. A. Khan, ed., Elsevier Biomedical Press, Amsterdam, pp. 323.

Taylorson, R. B. and Hendricks, S., 1969, Action of Phytochrome during prechilling of Amaranthus retroflexus L. seeds, Plant Physiol., 44:821.

Tokuhisa, J. G. and Quail, P. H., 1989, Phytochrome in green-tissue : partial purification and characterization of the 118-kilodalton phytochrome species from light grown Avena sativa L., Photochem. Photobiol., 50:143.

Vanderwoude, W. J., 1983, Mechanisms of photothermal interactions in phytochrome control of seed germination, in: "Strategies of Plant Reproduction", W. Meudt, ed., Beltsville Symp. Agric. Res. vol. 6, Beltsville.

Vanderwoude, W. J., 1985, A dimeric mechanism for the action of phytochrome : evidence from photothermal interactions in lettuce seed germination, Photochem. Photobiol., 42:655.

Vanderwoude, W. J. and Toole, E. H., 1980, Studies of the mechanism of enhancement of phytochrome dependent lettuce seed germination by prechilling, Plant Physiol., 66:220.

Wulff, R. and Medina, E., 1971, Germination of seeds in Hyptis suaveolens Poit. Plant Cell Physiol., 12:567.

Yokohama, Y., 1965, Analytical studies on the variation of light dependence in light-germinating seeds. Bot. Mag., 78:452.

# FACTORS INFLUENCING THE EFFICACY OF DORMANCY-BREAKING CHEMICALS[1]

Marc Alan Cohn

Department of Plant Pathology and Crop Physiology
Louisiana Agricultural Experiment Station
Louisiana State University Agricultural Center
Baton Rouge, Louisiana 70803, USA

## INTRODUCTION

Seed dormancy is advantageous for the survival of wild species but presents serious obstacles for the evaluation of seed performance, stand establishment, and efficient control of weeds necessary for cost-efficient agricultural practices. Significant time and effort have been invested in attempts to derive simple and uniform chemical dormancy-breaking treatments with only limited success.

While various substances are known to break seed dormancy, a characteristic response to applied chemicals is the germination of only a part of the treated seed population. Such a partial response makes it difficult to recommend chemical application as an effective method to alleviate dormancy and complicates studies concerning the biochemistry of the dormancy-breaking process. In mechanistic work, seeds must be sampled for analysis prior to visible germination when it is impossible to identify which will germinate and which will remain dormant. With a heterogeneously responding population, e.g. 60% germinating and 40% dormant, the situation is analogous to homogenizing 60 bean plants and 40 corn plants together in a blender and then trying to determine which constituents were originally present in each species from examination of the homogenate.

While one of the goals of the LSU Developmental Arrest and Activation Unit is to determine the mechanisms of seed dormancy and its alleviation under agricultural conditions, our first objective was to find a system with a relatively uniform response to applied chemicals. Red rice, a wild *Oryza* species that possesses physiological dormancy and is present in Southern USA rice production areas, has been employed as a model in these investigations. In the course of studying seed response as a function of solution pH, new dormancy-breaking chemical treatments have been identified. These experiments led to an evaluation of the chemical properties that endow a substance with dormancy-breaking activity. In

---

[1]Approved for publication by the Director of the Louisiana Agricultural Experiment Station as manuscript number 89-38-3287.

this report, the evolution of this research program, its current status, and its future prospects will be reviewed.

## EXPERIMENTAL

Details of methods employed have been published (Cohn et al., 1983; Cohn and Hughes, 1986). Briefly, a uniform biotype of red rice, based upon one dimensional gel electrophoresis of individual seeds and whole plant characteristics that we have named the 'Sonnier' biotype, has been utilized for these studies. Fresh seeds were harvested from production plots at Crowley, LA through the cooperation of the Louisiana Rice Research Station. All aspects of growth, harvest, and storage were personally monitored by this laboratory's personnel, and seeds were hand-harvested by shattering at or slightly past physiological maturity. Plants were grown under flooded conditions. In the experiments summarized here, only individually hand-dehulled seeds were employed. Dormancy-breaking chemicals were applied as a pulse under pH-buffered conditions followed by incubation on water for at least 7 days at 30C. Lots of 20 seeds were placed in 50 ml Erlenmeyer flasks with 2 layers of Whatman No. 1 filter paper and 2 ml of solution. Flasks were capped with rubber septa to minimize evaporation. Germination was considered to have occurred upon radicle protrusion, but in most experiments significant seedling development had occurred by day 7. When optimizing treatments, seed viability was always evaluated. All treatments reported here yield > 90% viability.

## pH-DEPENDENT ACTIVITY OF DORMANCY-BREAKING CHEMICALS

Initial experiments with common inorganic dormancy-breaking chemicals yielded variable germination percentages. After a lengthy process of eliminating potential variables, it was evident that the need for pH

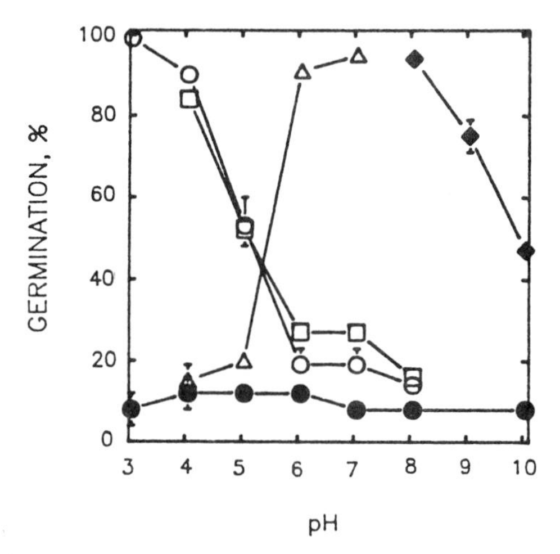


Figure 1. Effect of incubation medium pH during contact with 10 mM $NaNO_2$ (○)(4h contact); 0.5 mM $NaN_3$ (□)(8 h contact); 10 mM $NH_2OH$-HCl (△)(3d contact); 1 mM KCN (◆)(1d contact) at 30C. Buffer controls (●).

control of the incubation medium had been generally overlooked. Thereafter, it was unambiguously shown that seed response to nitrite, azide, cyanide, and hydroxylamine was pH-dependent (Cohn et al., 1983; Cohn and Hughes, 1986). In each case, activity was observed at pH values which favored the uncharged form of each chemical (Figure 1). Because of the varied dissociation constants of these substances, it was clear that the pH vs. activity relationship was attributable to the equilibrium of the charged vs. uncharged forms of the chemicals rather than a general influence of acidic pH (scarification), which would weaken the pericarp. Consistent with this point, minimal germination was obtained with nitrate, which is completely dissociated in solution, at any pH value between 3 and 9. While ammonium chloride was initially reported as being inactive (Cohn et al., 1983), recent work (Footitt and Cohn, unpublished) demonstrated dormancy-breaking activity for ammonia providing reasonable buffer capacity was employed under basic conditions.

Based upon results obtained by 1986, it seemed that the weak acid character of many previously identified, but otherwise chemically dissimilar, dormancy-breaking compounds could have contributed to their physiological activity. As a consequence, any weak acid should possess dormancy-breaking activity if it could be applied at a pH value where the undissociated form was present in solution. To test this idea, linear monocarboxylic acids of 1 to 6 carbon chainlength were tested at their pK values for dormancy-breaking activity. These weak acids elicited concentration- and pH-dependent dormancy-breaking activity consistent with previous data, which indicated that the neutral form was the active chemical species (Cohn et al., 1987).

BYPASSING pH-DEPENDENCY WITH $NO_2$ APPLIED TO UNIMBIBED SEEDS

Can the apparent lack of permeability to dissociable substances be circumvented? One possible approach to this question was to expose seeds to a dormancy-breaking substance in a gaseous form such as nitrogen dioxide (Cohn and Castle, 1984). While the application of nitrogen

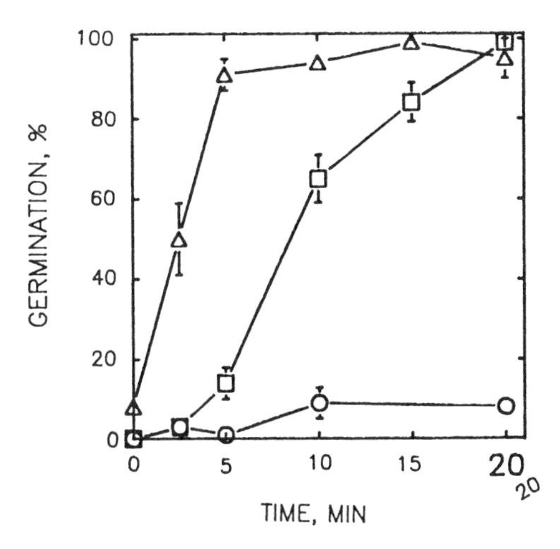


Figure 2. Effect of contact time with $NO_2$. Gas exposure was prior to imbibition: (○) 2 ml/L, (□) 4 ml/L, (△) 8 ml/L.

dioxide gas to imbibing seeds is merely a roundabout way of exposing seeds to nitrite and is pH-dependent, further work focused upon the activity of the gas when applied to unimbibed seeds. Nitrogen dioxide at 4 to 8 ml/L broke dormancy of dehulled, unimbibed seeds with exposure times to the gas of 20 minutes or less (Figure 2). Seed response was independent of the incubation medium pH following gas treatment but varied with initial seed moisture content (9 to 12% water content). This protocol provided a means by which a dormancy-breaking chemical could be introduced into seeds in the absence of a liquid solvent and circumvented the pH-dependent barrier to dissociable compounds. Other than red rice, there are no other reports indicating that other species will respond in a similar fashion to nitrogen dioxide. However, a similar strategy has been employed to break dormancy of wild oats with ammonia gas (Cairns and De Villiers, 1986a, 1986b).

## STRUCTURE-ACTIVITY STUDIES: THE ROLE OF LIPOPHILICITY

One puzzling feature of these experiments was the wide range of concentrations required to elicit the same germination response by different weak acids and bases. When considering that substances with other functional groups (i.e. aldehydes, esters, lactones, alcohols, nitriles, and ketones) also break seed dormancy (French and Leather, 1979; French et al., 1986), the effective concentration range is even broader. How can this be accounted for, and why do chemicals with such diverse functional groups all break dormancy? Recent results of a structure-activity study (Cohn et al., 1989) indicated that the efficacy of dormancy-breaking compounds (Table 1) is an inverse function of their lipophilicity (Log $K_{o/w}$) as measured by octanol/water partition coefficients. Substances with the highest Log $K_{o/w}$ require the lowest concentration to stimulate 50% germination (Figure 3). However, sufficient lipophilicity alone does not guarantee a physiological response since alkanes do not break dormancy (Abeles, 1986; Cohn et al., 1989; Taylorson, 1979). Furthermore, if chemicals possess small enough molecular dimensions (i.e. cyanide, azide, nitrite, formic acid, and methyl formate), activity can be related to their size rather than lipophilicity (Cohn et al., 1989).

Data analysis suggested that the nature of the functional group of active compounds also modified the efficacy of the dormancy-breaking treatments in some manner in addition to alteration of lipophilicity. Why this should be so remains mysterious as is the reason why an incredibly diverse array of chemicals elicit germination. However, the recent proposal that there are two distinct, primary dormant states [based upon differential chemical response during afterripening of wild oats] (Adkins and Simpson, 1988) may simply be a function of relative permeability of applied chemicals.

## SPECULATION AND FUTURE PROSPECTS

In hindsight, it is distressing that it has taken more than a decade to discover these principles for dormancy-breaking chemicals, which have existed in the drug design literature for close to thirty years (reviewed by Ariens, 1971). For the first time, the broad concentration ranges required for the activity of various dormancy-breaking compounds can be rationalized in terms of a structure-activity framework. Essential to the derivation of this relationship has been: (1) the development of a routine bioassay which could reliably accommodate the chemicals of interest; (2) performance of several hundred time-consuming, repetitive experiments required to pin-point the concentrations eliciting 50% germination; and (3) monitoring of seed viability after chemical treatments to differentiate between lack of physiological response *versus* seed death. This is particularly important for the alcohols, which are well known for their preservative properties. Poor correlations between membrane/buffer partition coefficients and the relative dormancy-breaking activity of alcohols for *Amaranthus retroflexus* (Taylorson, 1989) may be improved by further attention to these points.

Implicit in the current state of the art is the possibility that chemical structure principles which have been helpful in the design of

Table 1. Chemicals tested for dormancy-breaking activity.

| Code | Chemical | Code | Chemical |
|---|---|---|---|
| A | salicylic acid | N | butyrolactone |
| B | caproic acid | O | propanol |
| C | benzoic acid | P | acetaldehyde |
| D | valeric acid | Q | ethyl acetate |
| E | isovaleric acid | R | butanol |
| F | propionic acid | S | glycolic acid |
| G | butyric acid | T | dimethadione(DMO) |
| H | isobutyric acid | U | lactic acid |
| I | trimethylacetic acid | V | succinic acid |
| J | pentanol | W | ethanol |
| K | acetic acid | X | isopropanol |
| L | methyl propionate | Y | methanol |
| M | propionaldehyde | | |

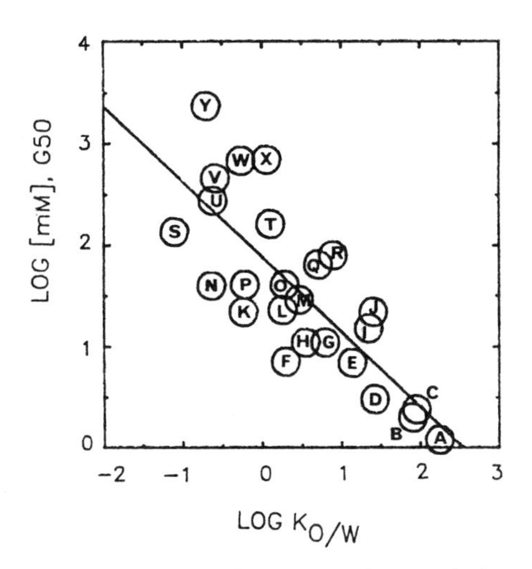


Figure 3. Correlation between chemical lipophilicity and concentrations required for 50% germination ($y = -0.73x + 1.86$; $r = -0.79$; $P < 0.001$).

pharmaceuticals and herbicides may guide the synthesis and identification of substances of extraordinary dormancy-breaking activity which will find substantial agricultural application. It is amusing and somewhat painful to reflect on our current battle to obtain funds for continuation of this research: Industry has indicated that we are too far removed from a product and, thus, will provide only limited support. On the other hand, federal granting agencies have indicated that the work is too applied and that industry should be doing this work.

What, if anything, do these results suggest concerning the mechanism by which chemicals break dormancy? Most active substances are weak acids or could be easily converted to a weak acid enzymatically. One can speculate that acid-loading of the tissue could trigger cellular activation similar to that observed in other systems. This prospect is currently under investigation in my laboratory. The underlying assumption of this hypothesis is that some common chemical feature is required to elicit dormancy-breaking activity. If a common chemical structural feature (i.e. a dissociable proton) is not essential, one could suggest that all of the active chemicals impose a shock to the system which would induce a response analogous to heat-shock. Alternatively, each active substance may directly interact with membranes as has been proposed by Taylorson (1988). Hopefully, further experimentation will endow one of these ideas with some semblance of reality.

Interest in the biochemical regulation of seed processes has increased in recent years, particularly with the availability of the new tools of molecular biology (Goldberg et al., 1989). Certainly, this approach must be applied to attempt an understanding of seed dormancy. However, in seeds such as red rice and others which are water permeable yet dormant, how can one account for the suspension of normal biochemical activity if many enzymes are present constituitively? While gene expression will certainly be involved in the germination process, the regulation of biochemical pathways at the metabolic level should not be overlooked.

## CONCLUSIONS

Utilizing red rice (*Oryza* sp.) as a model system, parameters that influenced the effectiveness of dormancy-breaking chemicals have been identified. In addition to chemical concentration and seed exposure time, the dormancy-breaking response to weak acids or bases was highly dependent upon incubation medium pH values. Conditions that favored the presence of the undissociated form of each substance elicited greater than 90% germination. This apparent permeability barrier to ionized substances was circumvented by application of nitrogen dioxide gas to dormant, unimbibed seeds.

A chemical structure-physiological activity study showed an inverse correlation between seed response and lipophilicity for substances possessing at least one functional group: mono- and dicarboxylic acids, aldehydes, esters, alcohols, and hydroxyacids broke dormancy while alkanes were inactive. Another modulating functional group effect was also apparent but could not be characterized in detail. The dormancy-breaking activity of substances with very small molecular dimensions such as methyl formate, formic acid, nitrite, azide, and cyanide was better correlated with molecular size rather than lipophilicity.

ACKNOWLEDGEMENTS

This work has been supported, in part, by the United States Department of Agriculture Competitive Grants, Southern Regional, and Special Grants Programs; the Louisiana Education Quality Support Fund, and the American Seed Research Foundation. The technical assistance of J.A. Hughes, L.A. Chiles, J. Ranken, and K.L. Jones (Boullion) has been gratefully appreciated.

LITERATURE CITED

Abeles, F. B., 1986, Role of ethylene in Lactuca sativa cv 'Grand Rapids' seed germination, Plant Physiol., 81:780.

Adkins, S. W., and Simpson, G. M., 1988, The physiological basis of seed dormancy in Avena fatua. IX. Characterization of two dormancy states, Physiol. Plant., 73:15.

Ariens, E. J., 1971, A general introduction to the field of drug design, in: "Drug Design, Vol. 1," E.J. Ariens, ed., Academic Press, New York.

Cairns, A. L. P., and De Villiers, O. T., 1986a, Breaking dormancy of Avena fatua seed by treatment with ammonia, Weed Res., 26:191.

Cairns, A. L. P., and De Villiers, O. T., 1986b, Physiological basis of dormancy-breaking in wild oats (Avena fatua L.) seed by ammonia, Weed Res., 26:365.

Cohn, M. A., Butera, D. L., and Hughes, J. A., 1983, Seed dormancy in red rice. III. Response to nitrite, nitrate, and ammonium ions. Plant Physiol., 73:381.

Cohn, M. A., and Castle, L., 1984, Dormancy in red rice. IV. Response of unimbibed and imbibing seeds to nitrogen dioxide, Physiol. Plant., 60:552.

Cohn, M. A., Jones, K. L., Chiles, L. A., and Church, D. F., 1989, Seed dormancy in red rice. VII. Structure-activity studies of germination stimulants, Plant Physiol., 89:879.

Cohn, M. A., Chiles, L. A., Hughes, J. A., and Boullion, K. J., 1987, Seed dormancy in red rice. VI. Monocarboxylic acids: a new class of pH-dependent germination stimulants, Plant Physiol., 84:716.

Cohn, M. A., and Hughes, J. A., 1986, Seed dormancy in red rice. V. Response to azide, cyanide, and hydroxylamine, Plant Physiol., 80: 531.

French, R. C., Kujawski, P. T., and Leather, G. R., 1986, Effect of various flavor-related compounds on germination of curly dock seed (Rumex crispus) and curly dock rust (Uromyces rumicis), Weed Sci., 34:398.

French, R. C., and Leather, G. R., 1979, Screening of nonanal and related flavor compounds on the germination of 18 species of weed seed, J. Agric. Food Chem., 27:828.

Goldberg, R. B., Barker, S. J., and Perez-Grau, L., 1989, Regulation of gene expression during plant embryogenesis, Cell, 56:149.

Taylorson, R. B., 1979, Response of weed seeds to ethylene and related hydrocarbons, Weed Sci., 27:7.

Taylorson, R. B., 1988, Anaesthetic enhancement of Echinochloa crus-galli (L.) Beauv. seed germination: possible membrane involvement, J. Exp. Bot., 39:50.

Taylorson, R. B., 1989, Responses of redroot pigweed (Amaranthus retroflexus) and witchgrass (Panicum capillare) seeds to anesthetics, Weed Sci., 37:93.

# PHYSIOLOGICAL MECHANISMS INVOLVED IN SEED PRIMING

Cees M. Karssen[1], Anthony Haigh [1,2], Peter van der Toorn[3] and Rolf Weges [1,4]

[1] Department of Plant Physiology
Agricultural University
Wageningen, The Netherlands

[2] Faculty of Horticulture
University of Western Sydney, Hawkesbury
Richmond, N.S.W., Australia

[3] Seed Technology Department
Nunhems Seeds
Haelen, The Netherlands

[4] Seed Technology Department
Incotec
Enkhuizen, The Netherlands

## INTRODUCTION

The high degree of mechanisation in modern plant cultivation systems demands fast, uniform and complete germination. However, this agricultural demand is the extreme opposite of the evolutionary adaptations which wild species have made to ensure success in their natural environment. Gradients of dormancy in the annually produced seeds of most wild species guarantee the spread of germination over periods up to many years. This wild background is often still noticeable in crop seeds that, therefore, germinate slowly, irregularly and to low percentages.

Seed quality can be improved in many ways. Breeding and selection are the most fundamental approaches, but these methods are expensive and time consuming. Treatments of plants during seed production have only incidentally improved seed quality. Therefore, seed priming is to date the most promising method to improve the quality of seeds. The method was described before, but it was brought to general attention by Heydecker et al (1973). Osmotic priming consists of the incubation of seeds for a specific period of time at a certain temperature in an osmoticum of -1.0 to -1.5 MPa usually salt or polyethyleneglycol (PEG) (molecular weight 6000) dissolved in water. Priming is generally followed by redesiccation of the seeds to allow storage and handling.

## MECHANISMS OF PRIMING

An understanding of the water relations of germination and cell expansion is necessary before mechanisms of seed priming can be examined. Germination may be regarded as a specialized growth phenomenon. Expansive growth of plant cells results from cell wall yielding and water uptake. Expansion is initiated with a yielding of the cell wall and continues under the effects of turgor. Wall relaxation reduces the cell water potential by dissipation of turgor and gives rise to water influx, which in turn increases cell volume (Crosgrove, 1986). For seeds with structures enclosing the embryo additional processes will be necessary to remove the restraint imposed on embryo cell expansion.

Under optimal conditions of supply the water uptake by seeds is triphasic. Phase I, or imbibition, is largely a consequence of matric forces. Phase II is the lag phase of water uptake, the water potential ($\psi$) of the seed is in balance with the environment. During this phase major metabolic events prepare the seed for radicle emergence. Only germinating seeds enter phase III, which is concurrent with radicle elongation. The duration of each phase depends on certain inherent properties of the seed and the conditions of water supply. Phase II is crucial for the timing of germination. In some species, such as the garden pea, phase II hardly exists, radicles protrude directly following imbibition. In most other species certain preparations for the renewed water uptake have to proceed during phase II. In the case of dormant seeds it requires special environmental conditions to remove blockages to germination. In all species phase II is temperature sensitive. Evidently, the control of water relations is essential to the control of germination.

Osmotic priming prevents the start of phase III and thus permits prolonged reaction times for processes during phase II. The benefits of priming are seen after reimbibition. They can be diverse. (1) In general, priming may cause a shorter lag phase because part of the preparation to phase III needs no repetition. (2) In a population of seeds the start of radicle protrusion is often better synchronized. (3) The rate of the growth process may be increased and, (4) more seeds in a population may germinate. In the latter case priming has been instrumental in the breaking of dormancy that requires specific temperature conditions just like those that occur in the field during the seasons with low or high temperatures. (5) A few studies have reported an improvement of seed vigour in low quality seeds due to priming.

The general mechanism of priming seems to be that irreversible products are formed that are tolerant of desiccation and stimulatory to germination. Attempts to identify the physiological mechanisms involved in seed priming have been mainly restricted to a few species. In this paper we will review recent developments in studies on priming of celery, tomato, and lettuce seeds in our laboratories. The perspectives for future research will be discussed.

## CELERY

A celery "seed", being morphologically a schizocarpic fruit, consists largely of endosperm surrounded by a thin testa and relatively thick pericarp. The small linear embryo is located at the micropylar end of the endosperm and measures about one third of the length of the endosperm. The embryo grows to at least twice its size before visible germination occurs. The need for embryo growth in the mature seed may be the main reason that germination of celery and other Umbelliferae is slow and irregular. Osmotic priming more than doubles germination rate, increases the uniformity of germination and raises the upper-temperature limit for germination (see e.g. Brocklehurst and Dearman, 1983).

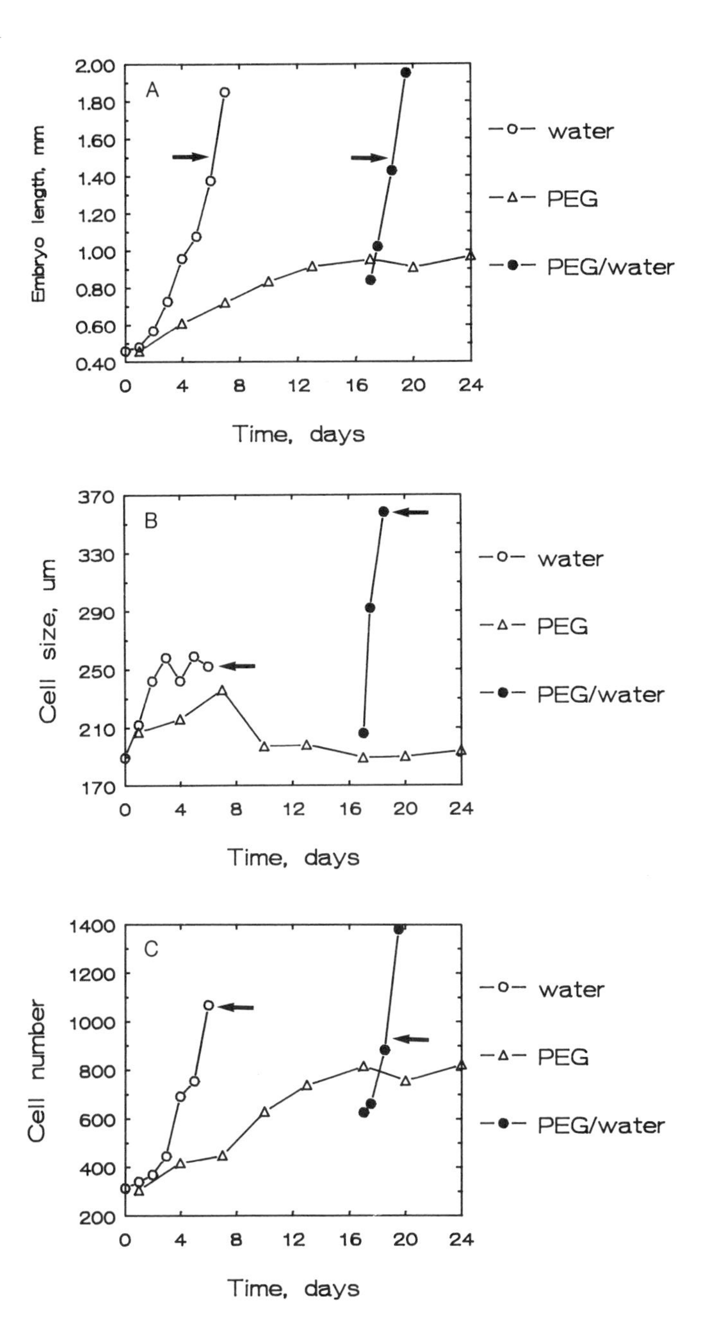


Fig. 1. Embryo growth (A), changes in mean embryo cell size (B) and increases in total embryo cell number (C) in celery seeds cv. Monarch during germination in water at 15 °C in 12 h light, 12 h dark of non-primed seeds (○) or primed seeds (●) and during priming in -1.2 MPa PEG (△). Primed seeds were surface dried before transfer to water. Embryo length was measured in longitudinally cut seeds. The number of cells was calculated from counts in a standard surface area of the hypocotyl just below the apical meristem. The mean cell size was calculated from these counts. Arrows indicate moment of germination (from Van der Toorn, 1989).

In water 50% of celery seeds cv. Monarch germinated after 6.1 day at 15 °C in light (Van der Toorn, 1989). Prior to germination the embryo length increased from 0.45 to 1.5 mm (Fig 1A). In -1.2 MPa PEG embryos grew at a much slower rate than in water, moreover the growth process ceased at an embryo length of 0.95 mm. When, after 17 days in PEG, seeds were rinsed and transferred to water, embryos extended from 0.9 to 1.5 mm in 1.7 day. For non-primed seeds this growth took 2.7 days. Evidently, priming not only permitted embryos to grow to twice their original size, as was also reported for carrot seeds (Austin et al. 1969), but it also improved their growth rate during subsequent incubation in water.

To further analyze embryo growth the number and mean size of embryo cells was determined by microscopic observation of longitudinal sections of a standard area of the hypocotyl just below the apical meristem. In water the mean cell size increased during the first 3 days from 190 to 250 $\mu m^2$ and stabilized thereafter (Fig 1B). The original number of 300 cells increased from the second day to reach 1100 cells at the moment of protrusion (Fig 1C). Incubation in PEG reduced the cell division rate to about a quarter of that in water. After 16 days the number of cells stabilized at about 800. Directly after transfer to water the cell division rate increased to that observed in non-primed seeds. (In the transfer experiments the number of cells after 17 days was smaller than in the continious PEG treatment probably due to a slight decrease in the osmotic potential of the PEG solution).

The size of cells formed during PEG incubation varied. Initially larger cells were formed but in the latter half of the incubation cells formed were of the same size as observed at the start. Directly after transfer to water at day 17 cell size increased dramatically. This reduced the number of cells that were required to reach the critical embryo length of 1.5 mm. Thus, during osmotic priming newly formed cells hardly expanded but preparations were made for very quick cell expansion after transfer to water.

For expansion embryos need space and nutrients. Celery embryos grow at the expense of endosperm tissue. Van der Toorn (1989) showed that the hydrolytic activity continues in endosperm cells during priming. Endosperm cell walls probably consist largely of galactomannans. During priming the activity of endo-β-mannanase, one of the galactomannan hydrolyzing enzymes, was hardly reduced below that in germinating non-primed seeds. Therefore, progress in endosperm hydrolysis might be another benefit of primed over non-primed seeds that might explain quicker germination of the former seeds.

It is not known yet why embryos need to achieve a length of 1.5 mm before starting protrusion of the testa and pericarp layers opposing the radicle tip. In non-primed seeds after 4 days of incubation in water changes were observed in the 2 to 3 cell layers at the micropylar end of the endosperm which indicated that hydrolysis and breakdown had occurred. We suppose that these endosperm weakening processes occur independently of the general endosperm breakdown required for earlier embryo growth.

## TOMATO

Tomato seeds are flattened and ovoid in shape, up to 4 mm in length, with a curved linear embryo embedded in non-starchy endosperm. In such seeds the tissues enclosing the embryo may influence germination by mechanically restraining the expansion of the embryo. Recent studies of germinating tomato seeds have revealed much about the mechanisms involved in the control of germination and have thus indicated possible mechanisms involved in the priming of tomato seeds.

As with capsicum seeds (Watkins and Cantliffe, 1983), the germination of tomato seeds was associated with a weakening of the resistance offered by the endosperm opposing the radicle tip. In seeds of a gibberellin-deficient mutant in the background of cv. Moneymaker weakening did not occur and germination was prevented (Groot and Karssen, 1987).

A study of the changes in the water relations of germinating tomato seeds cv. UC 82B has provided confirmatory evidence for the role of the endosperm (Haigh and Barlow, 1987). During imbibition of tomato seeds water uptake continued until water potential equilibrium was established between the seed and the imbibitional solution. The embryo was found to be capable of expansive growth prior to its emergence, but the endosperm tissue enclosing the embryo restricted further hydration until such weakening occurred so as to permit radicle emergence. Priming of tomato seeds may lead to more rapid germination by modifying these mechanisms.

Tomato seeds mainly benefit by priming in a strong reduction of the germination lag-time, particularly at moderate temperatures around 15 °C (Haigh, 1988). The rapid germination of primed and redried tomato seeds was found to result from changes to all three major steps involved in germination: imbibition, radicle cell-wall loosening and endosperm weakening. Water uptake by primed seeds was more rapid than that by non-primed seeds and may have resulted from the slightly lower osmotic potential of primed seeds during imbibition or from improved hydraulic conductivity. Some solute accumulation occurred during priming as was suggested by a number of authors (see Bradford, 1986, for references).

During priming cell-wall loosening occurred in the radicle. It was initiated 30h after the beginning of the treatment. As this occurred at the same time as in non-primed seeds in water this process must not be inhibited by the low water potential. The embryos from primed tomato seeds were capable of expansion at the earliest time at which they could be excised

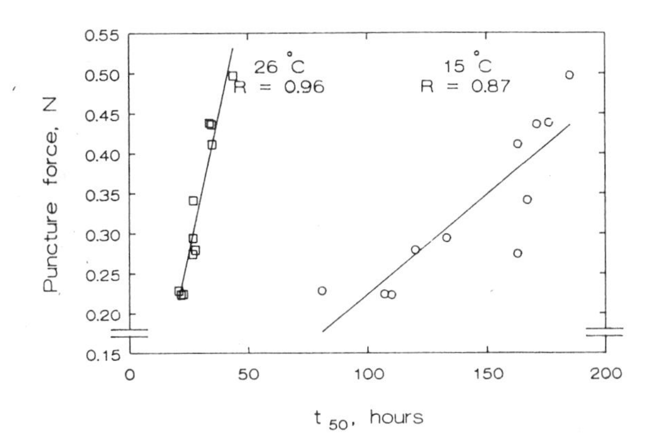


Fig. 2. Correlation between the time needed for 50% germination and the puncture force required to break through the layers that enclosed the tip of the radicle of primed tomato seeds cv. Moneymaker. Seeds were primed in -1.2 MPa PEG at 26 °C during 1 to 15 days and dried before germination occurred in distilled water at 15 °C (O) or 26 °C (□). Endosperm resistance was determined according to Groot and Karssen (1987). (T. Heimgartner and C.M. Karssen, original data).

from the seeds. This indicated that the cell wall yielding which had occurred during priming was not reversed by drying. Thus the time normally necessary to initiate cell wall yielding during germination of non-primed seeds had been completely eliminated by priming. However, as the timing of radicle emergence was controlled by endosperm weakening, the changed embryo cell wall properties may have not affected the timing of germination, but rather the rate of expansion subsequent to radicle emergence. The more rapid expansion of embryos from primed seeds was attributed to changes in radicle cell wall extensibility during priming (Haigh, 1988).

During priming the mechanical resistance of the enclosing tissues decreased to the same extent, but at a slower rate, as during incubation in water of non-primed seeds. As a consequence, at the start of imbibition the resistance in primed and dried seeds was much lower than in non-primed seeds (Haigh, 1988). In primed seeds of the cultivar Moneymaker the reduction of the time to 50% germination was a function of the reduction of the mechanical resistance (Fig. 2).

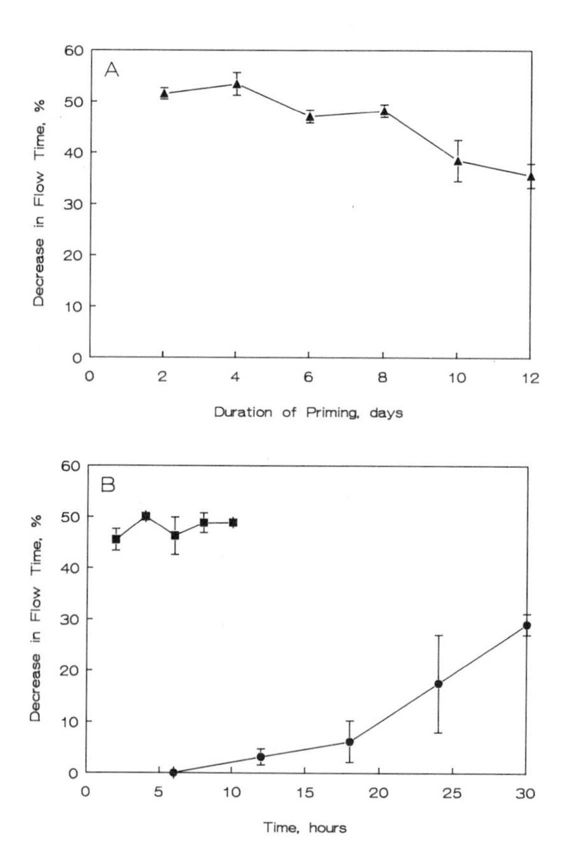


Fig. 3. Endo-β-mannanase activity in tomato seeds cv. Moneymaker during (A) priming in -1.4 MPa $KNO_3$ at 25 °C (▲)and (B) germination in water at 25 °C for non-primed (●) and 6 day primed and dried seeds (■). Enzyme activity was assayed viscometrically according to Groot et al. (1988). (D.C. Keulen and A.M. Haigh, original data).

The decrease in the mechanical restraint of the endosperm layers enclosing the radicle tip prior to radicle emergence is under the control of gibberellins produced by the embryo (Groot and Karssen, 1987). The endosperm cell walls were found to be rich in galactomannan and endosperm weakening was associated with the induction of endo-β-mannanase activity and an increase in the activity of mannohydrolase (Groot et al., 1988).

Endo-β-mannanase activity was studied in relation to priming. Priming seeds and germinating primed seeds showed uniformily higher activity than observed in germinating non-primed seeds (Fig. 3). The endosperm of tomato seeds was found to consist of two distinct cell types found in separate locations within the seed. At the micropylar end of the seed the endosperm cells had thin walls, whereas those in the rest of the seed had thickened walls. All cells, except those of the root cap, contained protein bodies. During priming protein body breakdown was more extensive in the micropylar region endosperm cells than was observed prior to germination in non-primed seeds (Haigh, 1988). These changes during priming appear to associate with the endosperm weakening and may be part of the enhancement of germination caused by priming. Similar morphological changes were observed in Moneymaker seeds (I. Zingen-Sell, personal communication).

However, as in both non-primed and primed seeds the mechanical resistance never fully disappeared, an additional step would appear to be necessary for radicle emergence to occur. This second step would immediately precede radicle emergence and would probably resemble a cell separation process similar to abscission rather than be an extensive breakdown of wall material. As it is impossible to predict the exact time of radicle emergence in individual seeds, this second step may be very difficult to identify.

These studies have shown that tomato seeds prime because the endosperm does not weaken sufficiently to permit expansion of the radicle. The mechanism by which some endosperm weakening was permitted but the final weakening for radicle emergence was prevented was not identifiable. A main area for further investigation is what mechanisms are involved in the removal of the last barrier for radicle protrusion.

## LETTUCE

Germination of lettuce achenes ("seeds") is often restricted to rather low maximum temperatures of around 20 to 25 °C. This phenomenon is often termed thermoinhibition or thermodormancy but essentially it is an example of a phenomenon common to most wild species: dormant seeds only germinate at a restricted range of temperatures, dormancy breaking widens the range and dormancy induction narrows it (Karssen, 1982). In some species like lettuce the maximum temperature (Tmax) for germination varies, in other species like muskmelon (Bradford et al., 1988) the minimum temperature.

Priming of lettuce seeds is thus basicly breaking of dormancy by rather low temperatures. In experiments with seeds of cv. Musette, which had an extremely low T50 (the temperature for 50% germination) of 15 °C (Table 1), the Tmax increased to over 34 °C when pretreatment occurred at 2 °C. Unfortunately, such a treatment also caused seed vernalization leading to bolting (Weges, 1987). Therefore, moderate temperatures around 15 °C are more suitable.

If pre-incubation occurred in -0.5 MPa PEG instead of water the treatment could be extended and consequently resulted a higher Tmax. Moreover, the priming effect was more stable to drying. The reduction of the improved Tmax was reversible by a second incubation at moderate temperatures (Weges,

1987). Therefore, this effect of drying differs from the damage that is caused when seeds that have started extension growth are dehydrated. Comparative studies of different seed lots of different cultivars indicated a positive correlation between the T50 of non-primed seeds and the optimal temperature for priming.

Table 1. The temperature for 50% germination of lettuce seeds cv. Musette. Seeds were pre-incubated at different conditions in water or PEG without or with drying to a moisture content of 4.1% (From Weges, 1987).

| pre-incubation conditions | $T_{50}$, °C not dried | dried |
|---|---|---|
| none | 15 | - |
| water, 20 h 15 °C | 23.5 | 16.0 |
| water, 40 h 10 °C | 26.5 | 21.0 |
| water, 5 d 2 °C | 34 | 25.0 |
| PEG -0.5 MPa, 40 h 15 °C | 25.5 | 21.5 |
| PEG -0.5 MPa, 72 h 10 | 27.5 | 23.5 |
| PEG -0.5 MPa, 17 d 2 °C | 34 | 28.0 |

Psychrometric measurements of osmotic potentials ( $\psi\pi$ ) showed that the increase of T50 was not associated with osmotic adjustment of cells (Weges, 1987). Bradford et al. (1988) point to a comparable situation in osmotically stressed seedling roots where osmotic adjustment to restore turgor was very slow compared to the rapid restoration of the growth rate. The authors concluded that maintenance of cell growth despite decrease in turgor implicates changes in cell wall properties.

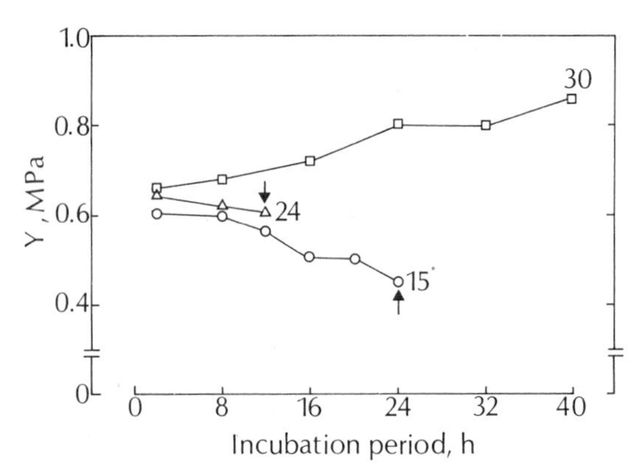


Fig. 4. The yield threshold Y of lettuce seeds cv. Capitan at 24 °C following pre-incubation in water at 15 ° (○), 24 ° (△), or 30 °C (□). The arrow indicates the moment of germination (from Weges, 1987).

In an analysis of the water relations of lettuce seeds cv. Capitan, Weges (1987) preincubated seeds in water at 15, 24 or 30 °C. At 15 °C T50 was increased and at 30 °C it decreased, 24 °C hardly changed T50. The rise and fall of T50 was negatively correlated to $\psi$50 (the external osmotic potential that enables 50% germination). During pretreatment at 15 °C the resistance to osmotic stress (tested at 24 °C) increased, at 30 ° it decreased.

As we concluded before, germination is a specialized growth phenomenon. Therefore, inhibition of germination by osmotic stress may be regarded as a steady-state growth rate of zero. According to Lockhart (1965), as developed by Schopfer and Plachy (1985), at zero growth $\psi 50 = \psi\pi + Y$, where Y is the minimum turgor at which growth occurs (the yield threshold). Since $\psi\pi$ was not influenced by any pre-incubation temperature, the changes in $\psi$50 indicated changes in cell wall extensibility. Y decreased during pre-incubation at 15 °C but increased at 30 °C (Fig. 4). Thus it seems that lower temperatures specifically increases T50, i.e. breaks dormancy, because only at those temperatures cell wall extensibility can be decreased. Higher temperatures have an opposite effect.

For seeds with tissues that enclose the radicle, Y will be composed of a component due to the radicle cell walls (Yr) and one due to the enclosing endosperm, testa and pericarp (Ye). Bradford et al. (1988) concluded that priming of lettuce seeds cv. Empire at 20 °C lowered Yr at 34 °C, whereas Ye was not effected. Data obtained by Weges (1987) do not permit accurate determinations of the two components of Y, but evidently, pre-incubation at 15 °C increased the resistance to osmotic stress of both intact seeds and isolated embryos of cv. Musette, suggesting a decrease of both Yr and Ye. It has to be realised that Empire seeds were clearly non-dormant, 100% germination occurred at 34 °C, whereas Musette seeds were dormant, the T50 was 20 °C. Comparisons between cultivars showed that the deeper the dormancy the larger the contribution of the enclosing structures to the restraint of radicle protrusion. Therefore, it is most likely that priming increases cell wall extensibility in both seed parts of lettuce. Endosperm cells opposing the radicle were highly vacuolated prior to radicle protusion and showed clearly mobilized storage materials. Cells at the lateral and cotyledonary end of the endosperm were not changed (Psaras et al., 1981). Enzymatic studies of endosperm loosening have only described changes associated with visible germination (Bewley et al., 1983). Priming related changes are not known yet.

## LONGEVITY AND REPAIR

In the range of conditions involved in commercial seed storage and longterm storage for the conservation of genetic resources the logarithm of any measure of longevity is a negative linear function of the logarithm of percentage moisture content (Roberts and Ellis, 1989). In lettuce seeds it extends from 2.6 to 15% moisture content. Above the critical moisture content, which varies between species, the trend is reversed and, providing oxygen is present, seed longevity increases with increase in moisture content until full hydration is reached. Therefore, storage at very low moisture contents may be as good as in fully hydrated state.

The benefits of hydrated storage were also noticed with aged seeds. Partial hydration of aged lettuce seeds by exposure to moist air (up to 34% moisture content) or to an osmotic priming solution (up to 44% moisture content) reduced the proportion of seeds which would otherwise germinate abnormally, increased subsequent root lengths and reduced the frequency of visible chromosome damage at first mitoses (Rao et al., 1987). Liou (1987) reached similar results by priming aged cabbage seeds in -1.5 MPa PEG.

Thus, in addition to the benifits discussed so far, priming also gives repair processes an opportunity to restore storage damage.

Priming is not always beneficial to seeds. Recently it has been reported that primed lettuce seeds show reduced storability at moisture contents between 5.0 and 16% and temperatures between 22° and 55 °C. (Weges, 1987). Bradford et al. (1988) found that primed tomato seeds retained their high quality for years when stored below 10 °C, but longevity was reduced by storage at high temperatures even at low moisture contents. In contrast, Thanos et al. (1989) reported that priming of pepper seeds in 0.4 M mannitol at 25 °C improved the resistance of seeds to storage at 25 °C at 96% moisture. Further study is required to explain these conflicting results. Better knowledge of the processes involved in ageing and repair will greatly benefit our understanding of priming effects.

## CONCLUSIONS

The significant benefits arising from seed priming have been readily explored and are being applied in the cultivation systems of many crops. However, the study of the physiological mechanisms of priming is left with many open questions, particularly at the molecular level. Nevertheless, some general conclusions are starting to appear.

It is evident that incubation in osmotic solutions only prevents the processes directly involved in radicle protusion. All other reactions seem to proceed albeit often at a reduced rate as is shown for instance, for respiration in lettuce (Weges, 1987), cell division in celery (Van der Toorn, 1989), and protein and DNA synthesis in leek seeds (Bray et al., 1989). The reduced rates ask for the prolonged priming times that are standard in most protocols.

As a result of priming, seeds may contain increased levels of certain products. When the product is an enzyme, reactions in primed seeds start quicker after imbibition and at a much higher rate (Fig. 3).

It is an open question whether priming stimulates "more of the same" or induces the synthesis of specific proteins, i.e. controls at the level of gene expression. Some evidence appears in favour of the latter conclusion (Bray et al., 1989). Future research has to show whether the increased molecular activity is indeed essential for the effect of priming.

Comparison of the three species that were described in this paper show both similarities and differences. In all three species priming prepared a quick expansion of radicle cells. In celery embryos new cells also were formed. In all species the embryo expansion depended to a certain extent on changes in the endosperm. The endosperm of celery had to be hydrolyzed and its products had to be absorbed by the embryo to permit growth and to make space. Endosperm breakdown was not prevented by priming conditions. Radicle protusion in celery seeds depended both on a critical length of the embryo (1.5 mm in cv. Monarch) and on hydrolytic changes in the few cell layers of endosperm at the micropylar end. Also in tomato and lettuce seeds radicle protrusion was preceeded by specific hydrolytic changes that only occurred in the endosperm cells that oppose the radicle tip. In tomato seeds endosperm weakening could be measured. The process also occurred during priming at a slightly reduced rate. In lettuce, endosperm resistance seems to differ between dormant and non-dormant seeds.

In general, priming in all three species seems to permit a quick and uniform germination by stimulating cell wall extensibility in the radicle and specific weakening of essential endosperm cell walls.

Priming induced repair processes are most probably of a different character than those involved in stimulating fast and synchronous germination.

## REFERENCES


Austin, R.D., Longden, P.C., and Hutchinson, J., 1969, Some effects of 'hardening' carrot seed, Annals of Botany, 33:883.

Bradford, K.J., 1986, Manipulation of seed water relations via osmotic priming to improve germination under stress conditions, HortScience, 21:1105.

Bradford, K.J. Argerich, C.A., Dahai, P., Somasco, O., Tarquis, A., and Welbaum, G.E., 1988, Seed enhancement and seed vigor, in: "Proc. Int. Conf. on Stand Establishment for Horticultural Crops," M.D. Orzolek, ed., Lancaster, Pennsylvania.

Bray, C.M., Davison, P.A., Ashraf, M., and Taylor, R.M., 1989, Biochemical changes during osmopriming of leek seeds, Annals of Botany, 63:185.

Brocklehurst, P.A., and Dearman, J., 1983, Interactions between seed priming treatments and nine seed lots of carrot, celery and onion. I. Laboratory germination, Ann.Applied Biol., 102:577.

Cosgrove, D.J, 1986, Biophysical control of plant cell growth, Annual Rev.Plant Physiol., 37:377.

Groot, S.P.C., and Karssen, C.M., 1987, Gibberellins regulate seed germination in tomato by endosperm weakening: a study with gibberellin-deficient tomato seeds, Planta, 171:525.

Groot, S.P.C., Kieliszewska-Rokicka, B., Vermeer, E., and Karssen, C.M., 1988, Gibberellin induced hydrolysis of endosperm cell walls in gibberellin-deficient tomato seeds prior to radicle protusion, Planta, 174:500.

Haigh, A.M., 1988, "Why do Tomato Seeds Prime? Physiological Investigations into the Control of Tomato Seed Germination and Priming", Ph.D. Dissertation, Macquire University, North Ryde, Australia.

Haigh, A.M., and Barlow, E.W.R., 1987, Water relations of tomato seed germination, Aust.J.Plant Physiol., 14:485.

Heydecker, W., Higgins, J., and Gulliver, R.L., 1973, Accelerated germination by osmotic seed treatment, Nature, 246:42.

Karssen, C.M., 1982, Seasonal patterns of dormancy in weed seeds, in: "The Physiology and Biochemistry of Seed Development, Dormancy and Germination", A.A. Khan, ed., Elsevier, Amsterdam.

Liou, T.S., 1987, "Studies on Germination and Vigour of Cabbage Seeds," Ph.D. Dissertation, Agricultural University, Wageningen, The Netherlands.

Lockhart, J.A., 1965, Analysis of irreversible plant cell elongation, J.Theor.Biol., 8:264.

Psaras, G., Georhiou, K., and Mitrakos, K., 1981, Red-light-induced endosperm preparation for radicle protusion of lettuce embryos, Bot.Gaz., 142:13.

Rao, N.K., Roberts, E.H., and Ellis, R.H., 1987, The influence of pre- and poststorage hydration treatments on chromosomal aberrations, seedling abnormalities, and viability of lettuce seeds, Annals of Botany, 60:97.

Roberts, E.H., and Ellis, R.H., 1989. Water and seed survival, Annals of Botany, 63:39.

Schopfer, P. and Plachy, C., 1985, Control of seed germination by abscisic acid. III. Effect on embryo growth potential (minimum turgor pressure) and growth coefficient (cell wall extensibility) in Brassica napus L., Plant Physiol. 77:676.

Thanos, C.A., Georghiou, K., and Passam, H.C., 1989, Osmoconditioning and ageing of pepper seeds during storage, Annals of Botany, 63:65.

Van der Toorn, P., 1989, "Embryo growth in Mature Celery Seeds," Ph.D.

Dissertation, Agricultural University, Wageningen, The Netherlands.

Watkins, J.T., and Cantliffe, D.J., 1983, Mechanical resistance of the seed coat and endosperm during germination of *Capsicum annuum* at low temperature, *Plant Physiol.*, 72:146.

Weges, R., 1987, "Physiological Analysis of Methods to Relieve Dormancy of Lettuce Seeds," Ph.D. Dissertation, Agricultural University, Wageningen, The Netherlands.

# GERMINATION RESEARCH TOWARDS THE NINETIES:

# A SUMMARY AND PROGNOSIS

Alfred M. Mayer

Botany Department
The Hebrew University of Jerusalem
Jerusalem, 91904, Israel

## INTRODUCTION

This is the third meeting dealing with seeds, germination and seed development. The first was held nine years ago in Jerusalem and the second four years ago in Wageningen. One of the things which strikes me is that although there is a core of people here who attended the first meeting, there are a lot of new faces, of younger scientists, and that is as it should be. We have come here not just to present data but to discuss problems and see how things are developing. The presence of new faces is proof that the subject is dynamic, is developing and that we are succeeding in rejuvenating our approaches. When I was asked to give this summing up lecture, I was in somewhat of a dilemma. When I summed up the Wageningen meeting, it was not recorded, so nobody can prove today that all my predictions were wrong. Unfortunately, this summing up lecture will be put on paper, and its predictions can therefore be faulted at some future time, as indeed I am sure they will be. Instead of discussing the many papers, I am going to use some general headings, which group together certain subjects and see what they have taught us. The first such heading is molecular biology.

## MOLECULAR BIOLOGY

It is the use of molecular biology which, perhaps more than any other methodology, has brought the most striking advances, if only because four years ago its use with regard to seeds was only in its infancy. The reports we have heard in this area have shown us that it is possible to analyse certain well defined processes in or related to seed development or germination. Using the techniques of transgenic yeasts, plants, or protoplasts it is possible to study how genomic information is expressed, modified or moderated by such various factors as protein binding to DNA, by phosphorylation reactions, by ABA and other factors or processes. We have learned something about the way proteins produced during seed development are targeted to specific sites and how such

*Recent Advances in the Development and Germination of Seeds*
Edited by R.B. Taylorson
Plenum Press, New York

targeting is regulated. All these regulatory mechanisms are extremely complex but are becoming amenable to direct study. We have seen that the techniques allow us to investigate biosynthetic pathways, and using mutants, to work out the biogenesis of, for example, ABA. Such studies have also shown that certain previously unsuspected relationships may exist between various pathways. Such apparent linkages raise new questions about the interrelations between such pathways. We have heard that it is possible to pull out specific proteins characteristic of certain developmental stages. Although there is not yet an example of an identified, germination specific protein, one which is characteristic of a post-germination stage has been discovered. Because it was a rather special protein, the researchers were able to identify this protein and assign it a location in the cell. One may expect further advances in this area, particularly on germination specific proteins, including the identification of at least some of them.

The use of molecular biology has also taught us that it is not only the expression of DNA during seed development and germination that should concern us. Equally, the deterioration of DNA during seed storage and its repair during the initial stages of germination imbibition may be critical for its subsequent behaviour.

Lastly we have seen that ABA is not the general stop signal as we thought at first. ABA is indeed a red stop-light with a filter, i.e. it allows certain things to continue. But perhaps, even more important is the fact that ABA apparently initiates a whole lot of new processes by controlling the expression of the genome.

One aspect of protein synthesis that has not been considered as yet in the germinating seed is protein synthesis in sub-cellular organelles. An important step early in germination is the replication, or multiplication of mitochondria. The time seems ripe to study this problem, applying the techniques of molecular biology and finding out whether this protein synthesis behaves in the same way as it does in other cases.

## WATER

The next heading I want to consider is water. We have heard a great deal about the effect of water on seeds, beginning with the importance of water content as a factor in storage properties of seeds and in affecting their deterioration, about water damage to seeds during imbibition and about the desiccation tolerance of seeds. Water acts in two ways. One effect is due to the chemical activity of the water molecule and the second can be ascribed to the water potential, which is related to the chemical activity of water. But in addition, the rate at which water enters a seed or leaves it is clearly also an important factor in the behaviour of seeds. The forces exerted during entry or exit of water can be very considerable. In addition we have learned about another, perhaps previously unsuspected effect. This is the effect of water content of the soil, particularly water-logging. Besides creating anoxia, this can directly change the microflora, which in turn can effect seed behaviour by competition for gases, nutrients etc. Such effects may also occur in the laboratory if care is not taken to keep the seeds aseptic and sterilise them before use. In fact, it is not always easy to maintain sterile conditions, because of the presence of bacteria and fungal spores below the seed coat. It may be important to pay more attention to this point than has been done in the past.

## MEMBRANES

Next, I would like to consider the role of membranes. In some ways progress in this important field has been considerable. First of all, it is clear that we have to revise some of our views on the structure of membranes in seeds. The transition of membranes from the gel to the liquid form is changed by their water content, clearly relating membrane behaviour to water. Furthermore, the stability of membranes during drying is modified by the presence of sugars, some of which can stabilise the membranes and protect them against desiccation. An additional important aspect of membranes in seeds and germination is the probable location of light receptors on the membranes thereby making the nature of membranes an important aspect of the light response in seeds. Membranes have also been shown to respond to dormancy breaking substances, such as anesthetics. Although we have heard a great deal about the direct effect of temperature on membranes *per se*, we have heard little on the role of membranes in the temperature response of germination. The area of metabolism of membranes has not been discussed in this meeting, yet there is a lot to be learned about it.

## LONGEVITY

Under the heading of longevity, the chief aspect to be considered is its prediction, which is of enormous practical importance. We have seen that certain models and equations now permit us to predict the longevity of a number of species, under known conditions, with reasonable accuracy. At the same time, unexpectedly, the melt temperature of lipids *in vivo* of certain seeds appears to correlate extremely well with storage properties of the seeds. Probably, this indicates something about the location and form of the lipids *in situ*, which relate to such factors as membrane integrity and protection against water. We have also seen that certain recalcitrant seeds, or rather their embryos, can be dried down using a flash drying technique and their longevity prolonged. All these are very promising developments. Nevertheless, a word of caution seems appropriate. The total number of species we, as seed biologists, are investigating in any detail is perhaps forty, yet the total number of plant species with seeds runs into the tens of thousands. We should be careful about generalisations and be clear that what appears to apply to the few species studied need not be true of all species. Some modesty in our claims would be indicated. We must broaden our investigations in order to test our models and predictions.

## PRIMING

I must confess that the heading "seed priming" has caused me a great deal of confusion. Basically, we are dealing with processes which start the seed to germinate under conditions which prevent germination. The "start in life" can effect the subsequent behaviour of the seed, sometimes promoting its keeping quality, sometimes reducing it. The events occurring during "priming" or "osmoconditioning" are apparently very diverse. They could involve the unequal distribution, or redistribution of water between different parts of the seed, as apparently occurs during the ethylene evoked precocious germination of some seeds. Sometimes repair processes occur, e.g. membrane or DNA repair. In other cases the resistance of the seed coat is specifically reduced opposite the point of emergence of the radicle and in yet other cases, embryo or extension growth may begin. What happens depends on the species studied. Perhaps we can conclude that the diversity in the subsequent response is a function of the initial process occurring during priming. If repair processes occur, the seed will survive better because it will take longer to deteriorate again. If, however, priming causes irreversible changes on the way to germination then such a primed seed will be far more vulnerable than it was before priming. The priming of seeds is clearly of enormous potential importance in agriculture and horticulture. It seems important therefore, to clear up some of the outstanding questions in this field of our endeavour.

## DORMANCY

The heading of dormancy breaking is inevitable in any seed meeting. The phenomenon of dormancy is an essential part of seed behaviour. Dormancy breaking is achieved by a wide variety of ways, physical and chemical. The study of structure-activity relationships of chemical dormancy breaking substances can provide some insight with regard to their site of action and perhaps give clues to the mechanism of dormancy breaking. However, we should recall that we had high hopes of achieving similar insights by studying structure-activity relations of germination inhibitors and at least in the latter we have been disappointed. Nevertheless, such studies are needed and careful attention to detail may provide some answers. Dormancy is not a single defined phenomena. Dormancy can be imposed in different ways, by endogenous or exogenous factors, and therefore there cannot be a single simple way of explaining dormancy in biochemical, morphological or anatomical terms.

## METABOLIC REGULATION

Regulation of seed germination and seed metabolism is clearly an important general heading, which has many facets. We have seen that some new regulatory systems must be considered. The possible role of $Ca^{++}$, the function of fructose-2,6-biphosphate, of cyanide and of ethanol are some which have been discussed. The indications are that at least some of these might be extremely important in regulating some parts of seed metabolism and might perhaps be important in dormancy phenomena. Many points about the formation and metabolism of the major phosphate storage form, phytin, still remain obscure and we have heard about advances in the way phytin is formed and transported. At the same time energy metabolism still remains a focus for attention. It is now clear, and need not really be discussed further, that oxidative phosphorylation is the primary source of energy for the germinating seed and that functional mitochondria can be prepared even from dry seeds, albeit by an aqueous isolation technique. At the same time it may be necessary to revise the view that carbohydrates are the first source of carbon for energy metabolism. It is possible that fatty acids, after beta-oxidation might provide the carbon for the tricarboxylic acid cycle. More observations are obviously needed and on more species, e.g. on non-lipid containing seeds or those in which lipids are not the main storage material.

## GROWTH REGULATORS

An obvious heading in a summary lecture is that of hormones or growth regulators. The main emphasis throughout the meeting has been on ABA. In the first seed meeting, nine years ago, we were discussing whether ABA had any role in seed dormancy and most of the things described in this meeting were not even postulated. In this meeting it has been discussed as a "signal" in the transition from seed development to seed germination; as being involved in some way with dormancy; as perhaps determining tolerance to water loss; and its effects on nuclear division. In 1980, the discussion centered chiefly on gibberellic acid. This time we have heard very little about GA. We have heard reports about ethylene and its role in dormancy and its relation to vigour. It is important to remember that not so long ago the involvement of ethylene in germination and its formation by seeds, as opposed to the effect of exogenous ethylene, was still debated. Surprisingly little has been said about kinetin or cytokinins except as a substance involved in regulating the formation of IAA or ethylene. I think it is worth recalling quite old findings which indicated that cytokinins induce light sensitivity. Perhaps this should be looked at again. Another substance not mentioned at all was IAA. The general concept has been that IAA has no role in germination. Yet, there is plenty of evidence that seeds are loaded with bound IAA, that this bound IAA can be metabolised and can release free IAA. If, in addition, we remember that part of germination is the result of cell extension I think there is every reason to look again at the role of IAA in germination. It is possible that, although it has a clear function, it is never limiting, but even proof of this would be worthwhile.

## LIGHT

The effects of light on germination are another obvious heading in this discussion. We have been presented with a number of models, dealing with various aspects of the light effect. These include a highly sophisticated model to explain the LFR and the VLFR reactions of seeds based on phytochrome interactions, as well as models to explain the interaction of light, GA and nitrate. We also have been introduced to refined analyses of the light inhibition reactions, based on the rate of cycling between two forms of phytochrome $P_{FR}$ and $P_R$. The models and analyses which were presented are an important advance because they can be tested experimentally, and hopefully next time we meet we will hear reports on these models. One point was missing. There has not even been a suggestion about the nature of the phytochrome receptor or acceptor molecule or its location within the cell. The fact that this has eluded all those working in this area, and not only in seed germination, suggests that new approaches and new ways of thinking are needed in order to achieve a breakthrough in this very complex and intriguing reaction.

## PARASITIC PLANTS

Parasitic seed germination is a problem in its own right. Two aspects are especially remarkable in such seeds. One is that for seedling establishment two distinct developmental processes must take place. One is the actual germination step. The second is the differentiation of the radicle into an "haustorium" which attaches to the host root as in *Striga* and probably *Orobanche*. The second aspect is the quite remarkable sensitivity of each of these steps to precise, highly specific chemical signals. These two aspects make the germination of parasitic seeds particularly fascinating. In addition, of course, the losses to agriculture caused by these noxious, destructive weeds in many areas of the world makes their study particularly important.

## SOMATIC EMBRYOS

My last heading relates to an intruder in our midst: somatic embryos. Somatic embryos are an alternative way of propagating plants and their behaviour resembles, but is not identical with, that of seeds. As seed biologists we should on the one hand welcome the study of a system so similar to our own. At the same time we should also beware of being carried away by the similarities. Somatic embryos are not seeds. They have no seed coat, which is a characteristic trait of seeds. In addition, the very limited study of their biochemistry indicates that they differ from seeds; for example in the composition of their storage materials. Other aspects of their biochemistry have not really been investigated, but will probably reveal additional features which differentiate them from seeds. I certainly take the somewhat biased view that they will never replace seeds as the ideal way of propagating a species and ensuring its survival in time and space. Attempts to coat such embryos and make them otherwise more similar to seeds appear to be a little self-defeating.

## THE FUTURE

Can we hazard some guesses about progress in the next decade and point to some areas which seem to be particularly important for study? The transition from seed formation to seed germination is an example of differentiation which is of exceptional interest and particularly well suited for study. I expect very considerable progress in coming years. It should be possible to identify those parts of the genome which are switched off and those which are switched on during the transition and to identify with greater certainty which are really important steps in the transition and which are merely the result. This may provide us with means to regulate certain aspects of germination. It may perhaps help us to manipulate recalcitrant seeds, which appear to lack some specific mechanism present in "normal" seeds. We seem to be well on the way to understanding what allows a "normal" seed to tolerate drying, at least at the level of the underlying chemistry and the molecular phyisco-chemical events involved. Some of the basic features appear to be associated with membranes which interact with other cellular components such as sugars, although other factors are no doubt involved.

More progress can be expected in clarifying the difference between dormant and non-dormant seeds. There is little doubt that the techniques of molecular biology will provide clear answers to questions about the basic differences between the two kinds of seeds, e.g., what parts of the genome are involved in the difference. Whether we will also find the answers to the question of why these differences exist and how are they brought about is less clear. For example, we must eventually ask why in a specific situation one lot of seeds becomes dormant and their embryos respond to ABA, while another lot is non-dormant and its embryos do not respond, or rather require concentrations of ABA at least two orders of magnitude higher in order to respond. It is extremely likely that the molecular biologists among us will be able to identify certain regions of DNA associated with such differences, but less obvious that they will be able to tell us how factors outside the nucleus cause these differences. Whether our methodology and our techniques are sufficiently refined remains to be seen, but there certainly is a challenge to be met.

Much will be discovered about the basic biochemical processes going on during seed germination. Current views on some metabolic pathways will have to be reconsidered and revised. I believe one aspect is still eluding us - how are the different chemical pathways coordinated, first of all in any given part of a seed, and secondly, between different parts of the seed? Advances in the general understanding of the action of plant hormones is going to help us, but I would suspect that not only hormones are involved in the coordination between seed parts. For example, the fact that during seed "priming" different processes take place in different species indicates a diversity in the response of seeds. We are still only scratching the surface with regard to diversity among species in their germination behaviour. Attempts to generalise from the relatively few observations we have are doomed to failure, because too many details are lacking. I think it might be rewarding to select a finite number of species, which are clearly divergent in their anatomy, morphology, overall size and shape and in their germination behaviour and study them in depth from certain pre-determined aspects. After sufficient data become available comparisons can then be made in a meaningful manner.

An area in which progress is still sketchy is that of the interaction of the imbibed seed with its environment. Much progress is being made in the study of the light response of seeds and phytochrome action. The seed seems to be a particularly suitable object for such studies. The putative "X" which we all insert in our schemes of phytochrome action is still unknown. Indeed we do not know whether there is one such "X" or several. The seeds response to temperature during its germination remains basically unknown and the effect of gases such as $CO_2$ or $O_2$ are not even investigated today except with regard to the response of seeds to anoxic conditions.

An area of research virtually at a standstill is that of the action of germination inhibitors. There seems to be little doubt that many of these are quite specific in their action, are widespread in occurrence and have a regulatory function in germination. Why this should be so is intuitively understood, but mechanisms are totally unknown. Wild species of plants contain a large array of secondary metabolites, largely removed by domestication of crop plants, which most of us study. These metabolites, at least in part, determine the fate of the species in face of the vicissitudes of attack by bacteria, fungi, insects and other predators. This aspect of seed physiology merits much more attention, particularly as agriculture in the future will be less likely to use exogenous chemicals to protect plants. We clearly have a dilemma, whether to opt for tasty edible products, which lack internal protectants or for products with protectants which makes them less attractive for us to eat or utilise. Sooner or later we must resolve this dilemma.

Can we design a better seed, better protected against attack, edible, with the preferred chemical composition from the nutritional point of view and with desirable germination characteristics? At present I strongly suspect that the answer is no. But perhaps by looking at more species, we will find some which meet at least some of our requirements. We have been quite conservative in the choice of plants which we use. If we find more species and understand them better, we can try to modify them by genetic engineering.

We will be able to get around some of the problems of seed germination by the use of other techniques, particularly by using somatic embryogenesis to reproduce plants. The technique is developing rapidly, but it is doubtful whether it can replace the seed except in certain very specific and unusual cases.

How does the mother plant and its nutrition affect the seed and its germination behaviour? A fair amount of information on this subject is available. The basic facts about the existence of the phenomena are no longer in doubt. However, this subject has not really been researched in depth and no mechanisms have been proposed, let alone tested. This is clearly closely linked with the questions about seed development: how are nutrients and regulatory substances transferred from the mother plant to the seed?, what are the barriers?, where do they occur?, and so on. Systematic studies in this area could be very rewarding. It may help us to manipulate seed behaviour without resorting to genetic engineering. The ecology of seed germination is also somewhat neglected and requires more sophisticated quantitative approaches, although the description of the ecological response of more species is also valuable and can provide us with new systems to work with.

Much progress has clearly been made since our last meeting four years ago and I feel quite optimistic for the future. If we, in Jerusalem, ever get around to writing a fifth edition of our book on germination, I am sure we will have to scrap a very considerable part of the existing one because we have obtained new insights. Equally, I am sure we will be able to again point to large areas in which progress has been slight. Research must continue along different lines in parallel - molecular biology, seed ecology, conventional biochemistry and physiology and, let us not forget: the simple accumulation of new data on species not previously studied.

INDEX